Brush with Death

CHRISTIAN WARREN

Brush with Death

A SOCIAL HISTORY OF LEAD POISONING

The Johns Hopkins University Press
BALTIMORE AND LONDON

© 2000 The Johns Hopkins University Press
All rights reserved. Published 2000
Printed in the United States of America
on acid-free paper
9 8 7 6 5 4 3 2 1

The Johns Hopkins University Press
2715 North Charles Street
Baltimore, Maryland 21218-4363
www.press.jhu.edu

Library of Congress Cataloging-in-Publication
Data will be found at the end of this book.
A catalog record for this book is available from
the British Library.

ISBN 0-8018-6289-2

You will see . . . that the Opinion of this mischievous Effect from Lead is at least above Sixty Years old; and you will observe with Concern how long a useful Truth may be known and exist, before it is generally receiv'd and practis'd on.

BENJAMIN FRANKLIN
TO BENJAMIN VAUGHAN,
31 JULY 1786

And shall we not signally fail in our guarantee of a reasonable surety for the health and happiness of all those within our borders, if this cry of suffering, now audible after so many years of dumb endurance, remains unheeded and unanswered?

GORDON THAYER,
"THE LEAD MENACE,"
1913

The problem is so well defined, so neatly packaged with both causes and cures known, that if we don't eliminate this social crime, our society deserves all the disasters that have been forecast for it.

RENÉ DUBOS,
NEW YORK CITY,
26 MARCH 1969

Contents

Figures and Tables

TABLES

Acknowledgments

Writing acknowledgments is similar in a way to receiving treatment for lead absorption: a single course of therapy is supposed to abate a burden that has accumulated over many years. The patient feels better, some of the burden is lifted; but the bulk remains forever bound up in the bone. The burdens one acquires in writing a book are of course nontoxic; most, in fact, are thoroughly benign. Still, they are no easier to discharge. So I welcome the therapy of formally enumerating some of the debt I have accumulated since I took up the subject of lead poisoning over ten years ago.

This book, and the doctoral dissertation from which it sprang, would not have been possible without the financial support I received from the Irving and Rose Crown Fellowship at Brandeis University. Two Dibner Summer Research Grants in the History of Science paid for research into pediatric and environmental lead poisoning, and a Mellon Resident Research Fellowship at the American Philosophical Society enabled me to spend a month in Philadelphia, the home of America's lead-paint industry. An Andrew W. Mellon postdoctoral fellowship at Emory University made it possible—pleasant, in fact, given the supportive environment—to revise the manuscript for publication.

I have benefited from the patience, helpfulness, and endurance of many librarians and archivists. I am pleased to thank the librarians at the Nelson Poynter Library of the University of South Florida and the research staff at Brandeis University. The staff of the government documents sections of the University of South Florida library and the Boston Public Library were patient and willing tutors. The staff at the National Archives politely withstood my repeated forays into familiar ground as well as my searches of ever-more tangential record groups. Thanks also to Boston Children's Hospital archivist Joan D. Krizack, Richard Wolf at Harvard's Countway

Library in Boston, Beth Caroll-Horrocks at the American Philosophical Society in Philadelphia, and Billie Broaddus at the Cincinnati Medical Heritage Center.

A number of health professionals aided and abetted me. At Children's Hospital in Boston, John Graef was very helpful in the early stage of my thinking about pediatric lead poisoning, and Dr. John Kozakevich gave me access to the Neurology Department's earliest records. Rich Rabin of the Massachusetts Department of Labor and Industries discussed his work in occupational lead poisoning and offered constructive advice. Herbert Needleman and Julie Riess of the University of Pittsburgh offered both encouragement and bibliographic assistance. Margaret Wood shared her memories and the memoirs of her father, Randolph Byers. I am grateful to Howard Frumkin of the Emory University School of Public Health for thought-provoking discussions and for exemplifying what occupational and environmental medicine can aspire to.

At the early stages of my research, many friends and colleagues provided useful insights and theoretical and historiographical caveats, carefully read parts of the manuscript, or simply offered their companionship and conviviality. Pierre Cennerelli, Jeffrey Kahana, Rich Rath, Mark Rennella, and Mark Taylor read portions of my dissertation or listened and argued patiently as I tried out new ideas. Martha Gardner read several chapters and responded with encouragement and clear editorial advice. Chris Sterba's contributions are difficult to summarize. He read drafts of almost every chapter, filling the margins with incisive comments, editorial suggestions, and hearty endorsements.

At Emory University, Randall Packard welcomed me into the bustling and stimulating milieu of the Center for the Study of Health, Culture and Society. Thanks go to Margot Finn, Judith Miller, Jeffrey Reznick, Robert Silliman, and Matt Payne for making me feel at home in the history department. I learned a lot from the students in my course in the social history of American medicine. Their rapt attention and their insightful questions and essays confirmed the goal of this book: that the history of medicine may be employed to address a wide array of historical issues in a compelling manner.

As a teacher now myself, it is especially gratifying to thank my teachers. Stating simply that Ray Arsenault directed my master's thesis on occupational lead poisoning fails to account for the many intellectual, temperamental, and inspirational gifts he bestowed. At Brandeis, Jacqueline Jones immediately saw the potential of what must have seemed at the time a somewhat quirky and nebulous topic. She encouraged me to begin with the stories of the workers, parents, and children whose direct experiences with lead poisoning made the topic such compelling social history. The project matured and developed its own story, yet I have tried to keep those individuals' voices audible. Morton Keller encouraged me to adopt a deliberate but flexible framework of analysis . . . and carefully edited every chapter anyway. As the project neared completion and the underlying framework became visible, Professor Keller's enthusiasm doubled. Allan Brandt frequently crossed the boundary that labeled him the dissertation's "outside reader": his generous gifts of time, energy, and ideas profoundly shaped my thinking about this project.

Many scholars and specialists assisted in revising the manuscript. Yair Goldstein and John Low-Beer of the New York City Law Department shared many ideas about the history of regulation and litigation. Alan Loeb and William Kovarik gave me salient comments, sharp questions, and very interesting readings pertaining to leaded gasoline. Herbert Needleman offered sound advice on the chapters concerning pediatric plumbism in the postwar period. Christine Rosen and Christopher Sellers read over much of my material on Progressive Era occupational medicine and paint legislation, suggesting new questions I should ask of my documents and data, and better ways to state the answers I had already found. Janet Golden and Philip J. Pauly were very generous in their formal comments on conference papers. David Rosner read the second draft very carefully, responding with a healthy balance of probing questions and endorsements. Jack Davis has been a good friend and colleague since I started graduate school. His most direct contribution to the book may seem like a little thing, but if not for his title, I would still be sending lists of pun-laden proposals to my editor!

Jackie Wehmueller endured dozens of tortured title suggestions—and much else—with good humor and patience. Of course, she is a splendid editor. I sent her a dissertation I brazenly believed was a book manuscript. Her handwritten comments, filling many a margin with her delicate red-penciled orthography, excised pounds of unsightly "diss.-ease" and turned many an awkward phrase into forceful prose. She has been more than an editor, though. Throughout the past months she has been a friendly advisor to the author. Many other people have been very helpful, especially Anne Fullerton, who pored over the revised manuscript with a remarkable combination of acuity and perspicacity, finding errors of omission, commission, documentation, and argument, and suggesting structural improvements often requiring movement of passages across several chapters.

Though the debts I owe to my immediate and extended family are of a different sort, anyone who has lived with someone in the throes of writing and revising a dissertation knows how great those debts are. Thanks to Gunnar and Nell Nilson, Gail and Scott Oler, and Dawn and Luken Potts for putting the graduate student and his family up on repeated trips to visit archives. My parents, Joan and Lyman Warren, combined medical careers with keen and wide interests in cultural and social issues, a combination which I hope this study reflects. Lyman Warren read the manuscript with a keen eye for errors of fact and interpretation. A father's pride did not always mute his critic's voice.

In uncountable and unaccountable ways, Janis Nilson Warren has sustained me through this project. She creates a space in the world, and a place in the heart, where I am always happy to return. I dedicate this study to our children, Jeremy, Johanna, and Justin, for whom lead poisoning will always carry a very different meaning than for most children. I am sure they will never forget this project; some of their best artwork is drawn on the backs of rough drafts.

I wanted this study to be, in Alice Hamilton's words, "fair, but not too fair." The advice of the many friends and colleagues whom I thank here enhanced both its fairness and accuracy. The errors that remain are no doubt due in part to low-grade lead exposure during my childhood.

Brush with Death

What's Lead
in the Bone . . .

The United States is lead poisoned.

Such a declaration in a book exploring the social history of lead poisoning may unsettle readers. Some will fortify themselves against the onslaught of a dozen tired metaphors. Others, perhaps having read ecological diatribe passed off as scholarly analysis, may dismiss the statement as the hyperbole of the muckraker. But in declaring that the United States is lead poisoned I am not speaking metaphorically about a diseased body politic, although the nation's stumbling and often stuporous responses to this toxic element do suggest aspects of lead poisoning. Nor do I mean that the people of the United States are lead poisoned, although it can be reasonably argued that millions of Americans are and that tens of millions who grew up in the years of leaded gasoline, arsenic-laced apples, lead pipes, and chipped paint continue to pay the price for that exposure—in kidney diseases, hypertension, and diminished mental capacity.[1]

Instead, this statement rests on the application of modern clinical diagnostic standards to the nation's infrastructure. Those standards define lead poisoning in living creatures by the quantity of the metal present in the body—ions of lead in tissues, blood, and bone—rather than by any symptoms arising from that lead.[2] The lead in the nation's buildings, its water pipes, the dust and dirt along its highways, and the air of its cities certainly constitutes a major absorption. Applying the same standards to the inanimate as to the quick suggests that the United States—or more specifically, the physical environment and especially the built environment where its people live—is itself lead poisoned.

The standards behind this "diagnosis" are relatively new. Lead researchers fifty years ago would have found them ridiculous. In 1950, doctors had no interest in treating children whose blood-lead levels were as much as three times higher than those that today prompt aggressive "de-leading." In fact, average blood-lead levels for urban children then were close to those pediatricians now routinely treat with powerful drugs, levels that cause parents to fear that lead exposure may have impaired their children's mental health.[3] To a significant extent, the changes that brought today's definition of lead poisoning, or plumbism, were due to improved diagnosis and treatment, and these improvements are important subjects in this study.[4] Of far greater interest to the social historian, however, is the cultural shift in attitudes toward health, safety, and risk that accompanied—indeed, that mandated—the new definitions. This shift transformed medicine and health care and gave purpose and power to modern environmentalism. It fundamentally altered jurisprudence. Its effects on social behavior are equally profound: what were once bad habits—smoking in public, drinking when pregnant, and other personal and private choices—are now cause for litigation or even criminal prosecution. Our blossoming risk aversion has changed the way we eat, the way we drive, the way we raise our children—almost every aspect of modern life in the United States.[5] A study of lead poisoning offers a significant opportunity to explore this cultural change.

The lead industry often described lead as "the useful metal." In the last hundred years, many industries have exploited its useful qualities, driving consumption to unprecedented levels and making it a nearly ubiquitous toxin in the physical environment. At any point in the century, lead exposure was at the center of concern in a number of issues, including foods and drugs, occupational safety and employment rights, air and water pollution, housing and civil rights, transportation, health care, and education. In whatever problems lead was implicated, ensuing debates sought to balance the perceived needs of producers and consumers with the potential health consequences—in short, to negotiate the definition of acceptable risk. Hence, the history of lead poisoning is useful for getting at broad changes in the culture because efforts to deal with it have made its presence in the regulatory arena nearly as widespread as in the physical environment.

Figure 1-1 was intended to make comprehensible the pathways lead follows in the environment and into humans. But rather than clarifying matters, the image suggests the nearly bewildering complexity and the dynamic interrelatedness of the problems of regulating lead exposure. Although this study addresses only a handful of these regulatory issues, it does not shy away from complexity. On the contrary, it seeks to highlight interdependence, contingency, and the transitory nature of "solutions" in

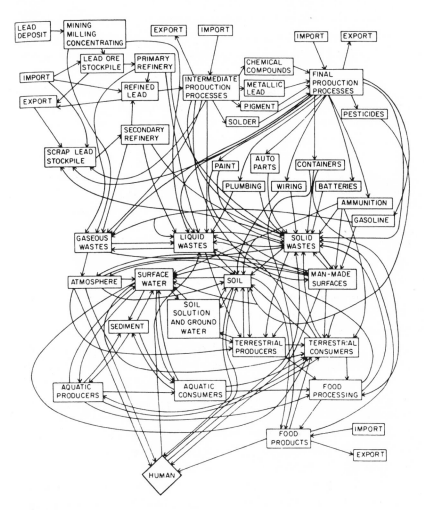

FIGURE I-1
Sources and Environmental Pathways of Anthropogenic Environmental Lead
National Research Council, Committee on Lead in the Human Environment, *Lead in the Human Environment* (Washington: National Academy Press, 1980), 35.

the face of an ever-shifting foundation of acceptable risk. An overview of this study's analytical framework may prevent these complexities from obscuring the book's subject.

Although lead is essentially one toxin, the sources of exposure are as varied as Figure I-1 suggests. This study distinguishes three primary modes of exposure: occupational, pediatric, and environmental. Each has its own dynamic regulatory calculus but draws intermittently from and feeds the regulation of the other two. The mode with the longest history is occupational, stemming from work in lead-using industries. Before the 1940s, almost all lead-poisoning research focused on occupational plumbism. Standards established by lead industry doctors for diagnosing and treating adult workers formed the basis for understanding, diagnosing, and treating children, as well as for setting standards for lead pollution in the general environment.

The risks specific to children constitute a second mode of lead exposure. Certain physiological traits associated with rapid physical and mental development make young children especially susceptible to lead. Furthermore, the cultural, medical, and regulatory issues raised by pediatric lead poisoning are distinct in many ways from those related to occupational lead exposure. Still, until recently, childhood plumbism was thought to share two important traits with the latter: it occurred within a prescribed location, and the exposure was usually limited to one clear source.

The third mode is often called environmental exposure, though a better term might be *universal*, since it is the uniform distribution, not the specific source, that distinguishes it from occupational and pediatric lead exposure. Here, children and lead workers face the same exposure as the rest of the population. The sources include lead pipes; food tainted by leaded tin-can solder, insecticides, and contaminated soil; cosmetics and drugs; and air pollution. Truly "environmental" exposures, such as those from leaded gasoline or smelters, compose a subset of "universal" exposures.

Distinguishing between the three modes of exposure is important because they have prompted such very different regulatory responses. Crucial factors in determining action at any given time include the severity of exposure; the degree of certainty as to its source and the manner in which culpability can be assigned; the size of the affected population (as recognized by the epidemiology of the time) and that population's sociopolitical power; and, always, the costs of reducing the risk. For example, great strides in occupational hygiene were made in the first quarter of the twentieth century—despite workers' anemic sociopolitical power—because clearly culpable plant owners could drastically reduce severe exposures from obvious sources, with little expense (in fact, profitably in many cases). No such clarity remains in today's leaden regulatory agenda.

As the chapters of this book suggest, the regulatory regimes that rose around each of these avenues of exposure interacted dialectically. Often it appears that the issues did not address one another. But studying occupational, pediatric, and environmental exposure side by side reveals an ongoing, if halting, three-way conversation. Occupational and pediatric lead poisoning can be studied in isolation, but referring each to the other and both to larger political, scientific, and social issues in which they were embroiled is far more informative. For example, one cannot explain the absence of research into and awareness of pediatric lead poisoning at the beginning of the century without understanding the dynamics of occupational research at the time. Today, the subjects of occupational and pediatric lead poisoning are inextricably mixed. Any debate about addressing the lead hazard in children's homes quickly runs into the costs of protecting the health of the workers who will remove the dusty poison. And the past two decades' most pressing occupational issue in lead factories involves protecting the unborn children of lead workers. Workers, children, and workers' children—these subjects must be integrated.

One cumulative effect of these conversations has been the gradual enlargement of the circle of concern. Increased interest—on the part of researchers, government agents, or the affected community itself—has prompted increased case-finding efforts, enlarging arithmetically the number of potential lead-poisoning victims identified. In addition, new appreciation of lead's effects at lower levels several times forced new definitions of lead poisoning, geometrically enlarging the populations considered "at risk." Neither of these processes occurred in a cultural vacuum. Rather, social and political necessities variously promoted or inhibited case finding. And despite their claims to the contrary, researchers and regulators have continuously weighed potential costs when defining thresholds of concern.

A second long-term trend that emerges in this study follows from the regulatory requirements for standards and thresholds. At the beginning of the century, lead poisoning was defined by clinical symptoms, not quantitative measures. Improvements in the ability to measure small quantities of lead in body fluids or tissues radically altered lead-poisoning epidemiology. But these quantitative definitions have hardened to the point where patients' symptoms might be ignored if their blood-lead levels do not "measure up." This increased reliance on biometrics has affected public policy as well. As the thresholds dropped far below levels associated with clinical symptoms of lead poisoning, opponents of further regulation began to criticize the increasingly abstract reasoning undergirding the new definitions. Today's high standards are not a reaction to grave new risks posed by lead. In fact, lead exposures are now lower than at any time in the past hundred years. Instead, the new definitions reflect the culture's transformed attitude toward the built environment and a new, but deeply felt, cultural aversion to preventable risks.

There has been no history of lead poisoning that incorporates the occupational, pediatric, and environmental issues involved.[6] The history of occupational plumbism has received a fair amount of attention, largely because of its central place in the furor stirred up by Progressive Era reformers. Patricia Vawter Klein's 1989 dissertation examined occupational lead poisoning in women in lead industries in order to compare so-called protective (exclusionary) labor policies in the Progressive Era and the 1960s. Christopher Sellers's study of the development of occupational medicine in the early twentieth century deals less with lead poisoning as experienced by its victims than with the disease as a subject studied by doctors. It contains critical insights and a wealth of information about the scientists and policymakers who "created a body of knowledge and standards of practice for an expert audience in the factories."[7]

In the late 1960s, commentators often referred to pediatric lead poisoning as "the silent epidemic." Childhood plumbism persists but is now thoroughly documented, rigorously and bountifully researched by medical science. The popular press reports it perennially with breathless sensationalism. Given the ongoing ballyhoo and the glut of state and federal legislation, lead poisoning, which federal health officials call the "most prevalent environmental threat to children in the United States," might be more fittingly titled "the shrieking pandemic."[8]

Historians have paid little attention to how pediatric lead poisoning came to cause such political and public furor. Most of the scholarship falls roughly into three overlapping categories: social policy analysis, "medical office" history, and environmental advocacy.[9] Richard Rabin, an occupational health specialist for the Massachusetts Department of Labor and Industries, has written a useful summary of childhood lead poisoning from the beginning of the twentieth century that emphasizes how doctors and public health advocates ignored warnings about lead-paint poisoning. A study by Barbara Berney, of the Boston University School of Health, finds three rounds of regulatory interest in childhood lead-paint poisoning since the 1950s. Historian Elizabeth Fee's study of the Baltimore Health Department's lead-poisoning prevention campaigns locates the social, economic, and medical changes that produced that city's aggressive and highly visible programs from the 1930s to 1970s.[10] My research has benefited from the insights and the bibliographies of all these works.

Leaded gasoline is another saturnine topic that has had considerable attention from historians, political scientists, and legal scholars.[11] The best historical overview of the origins of tetraethyl lead remains David Rosner and Gerald Markowitz's study focusing on the 1925 conference to determine whether leaded gasoline posed a health hazard. William Kovarik's detailed examination of the press coverage of the tetraethyl lead scandal adds a sharp and deeply researched analysis of the reasons the nation came to depend on such a toxic fuel additive.[12] My study of tetraethyl lead seeks to show its centrality in what became an environmental and toxicological epidemic.

It is impossible to understand the course of an epidemic without analyzing the culture in which it occurs. Conversely, studying epidemics reveals a great deal about culture.[13] Lead poisoning makes an excellent case study for the historian, in large part because of its industrial and commercial origins. The cause of lead poisoning is clear—it is a plague of our own creation.[14]

A community in the grip of an epidemic seeks simultaneously to understand the reasons for the epidemic's existence and for its choice of victims. In practice, the community often seeks to blame forces outside and greater than itself as well as individuals (or groups) within itself, while not blaming the community as a whole. Biotic diseases easily fit such arguments. But environmental illnesses such as lead poisoning pose a challenge for a society seeking a superhuman scourge and subhuman scapegoats. Lead oxides are neither "natural" nor supernatural; neither nature nor God can be invoked as unmoved movers. Breathing is not a sin, so workers who must inhale the air of a lead-choked factory cannot be held accountable for getting sick.

Nevertheless, the culture found ways to blame lead-poisoning victims for their illness, by drawing from nineteenth-century beliefs about sin and about the strong and the weak. Charles Dickens has a fatalist landlady describe her tenant, who lay dying of lead poisoning acquired in the local lead mill. Some succumbed sooner, she said, some later: "'tis all according to the constitooshun." Lead poisoning was the wages of sin, as well. Shop foremen, industrialists, and reformers alike presumed that personal habits such as drinking and smoking predisposed to lead poisoning. Physicians and policymakers continue to see cultural backwardness and parental irresponsibility as "causes" of childhood plumbism.[15]

Anti-lead activists could not blame the victim, however—but neither could they blame God or germ for "the Lead Menace." Instead, those in search of a looming, impersonal, motive force standing above or outside society that could visit such a plague upon the land found the next best thing: corporate capitalism.[16] In critical analyses from Alice Hamilton to Samuel Hays, in the rhetoric of journalists from *Everybody's* to *20/20,* lead poisoning has stood as a metaphor for corporate greed in much the same way as syphilis represented God's judgment on the wicked in earlier years.[17]

The analogy between an all-powerful and inscrutable deity and corporate capitalism is strong. It has also proven fatal, to the extent that it has succeeded in elevating "the corporation" above society, placing the responsibility for lead poisoning on the incorporeal, corporate "other." By having a corporate villain, Americans could enjoy the villain's products without guilt. Americans painted their homes with superior-quality white lead paints in the days when an arguably inferior, but nontoxic, substitute was finding favor in other nations. We bought "ethyl" for more powerful gasoline, for more powerful engines, for larger automobiles, when other nations were extending or maintaining their rail systems. On repeated occasions, conditions favored the public's interest being heard and acted upon, but the public, or its representatives, failed to apply the necessary leverage.

Lead has been in use throughout America's history and Americans have been poisoned by it for as long. In 1786, Benjamin Franklin concluded a letter about his interest in lead poisoning with a frequently quoted statement: "This, my dear Friend, is all I can at present recollect on the Subject. You will see by it, that the Opinion of this mischievous Effect from Lead is at least above Sixty Years old; and you will observe with Concern how long a useful Truth may be known and exist, before it is generally receiv'd and practis'd on."[18] My decision to limit this book to the twentieth century follows from Franklin's observation. It was not until the late nineteenth century that lead poisoning was in any measure "generally receiv'd and practis'd upon," and lead did not attain popular notice or sustain professional concern in the United States before 1910.

A second limit is this study's focus on two lead-based consumer products: lead-based paints and tetraethyl lead in gasoline. There is very little here about miners, operatives at primary smelters, potters, plumbers, or many other workers whose tasks brought daily exposure to lead. I do not dwell on the history of lead arsenate in food or the increasing concern over lead in water supplies and drinking fountains. The lead in these sources was incidental to their manufacture and use. Leaded gasoline and lead paints, on the other hand, were consumer items, marketed explicitly on the strength of particular chemical properties of lead.

My interest in paint flows largely from its ubiquity in the home environment, its intractability, and its toxicity. The tons of lead paint in and on American homes remain a concentrated source of the potent neurotoxin.[19] Paint remains an obstinate occupational hazard as well. Welders who remove lead paints from bridges and steel structures consistently turn up with some of the highest reported rates of occupational lead poisoning.[20] The decision to concentrate on paint also stems from the fundamental irony of our deliberately coating homes, cars, and toys with toxic pigments. A fresh coat of paint, especially white paint, emblemizes purity, cleanliness, renewal. Extolling the benefits of fresh paint for public health, a physician in 1937 argued that "the dust which has gathered on unpainted walls is detrimental to health for various reasons. . . . One does not know who lived in the house previously, whether the former occupant was ill or perhaps died after contaminating the walls with disease germs expelled in coughing." Paint could bring moral uplift as well. A Depression-era report in a lead-industry journal praised a U.S. Department of Agriculture housing project in Alabama that employed lead paints for interiors and exteriors:

> *The construction of a new home for Willis and Julia Thurman, Negro small land owners . . . is a typical example of rural housing rehabilitation and the part white lead and oil plays in it. . . . The use of white painted homes is definite forward progress in rural habilitation and housing. By getting away from the dark and depressing unpainted cabins of past years, the new houses are more attractive, cheerful, and raise the pride and morale of families and communities.*[21]

The association between rural poverty, depression, and paintlessness persisted. Thirty years later, Senator Edward Kennedy, cosponsor of the first federal legislation to ban lead paints in public housing, agreed with a witness in a Senate hearing that perhaps lead-poisoning rates were lower in Appalachia because many homes there "are not even painted."[22]

The history of leaded gasoline is also critical for understanding the nation's experience. Beginning in the 1920s, questions about the safety of tetraethyl lead produced some of the century's most sharply defined public debates about lead as a public health issue, more than the issues surrounding lead paint would produce before the late 1960s. But these questions were only asked in the 1920s, when tetraethyl lead was introduced, and again in the 1970s, when the United States began to phase it out. Critics remained silent in the intervening years, during which tons of fine particles of lead belched from automobile exhausts, floated in the air that all urban Americans breathed, and settled in a fine dust in the playgrounds and homes of millions of urban children. This silence is itself a historic artifact in need of interpretation.

In order to integrate these often disparate themes, this study is constructed as a braided narrative. Rather than resorting to or unconsciously drifting into narratives of progress, "declension," rise and fall, and so on, I have tried to allow lead poisoning itself and the changes in the ways physicians have treated it to set the book's shape.[23] Lead poisoning takes many forms. But whether resulting from occupational, pediatric, or environmental exposure, those forms can be roughly classified as either acute or chronic manifestations. One may encounter a brief and intense exposure to lead and contract severe symptoms or, after months or years of low-dose exposure, one can accumulate enough lead to produce chronic symptoms or damage to internal organs without clinical symptoms of lead poisoning. Of course, both can happen over a person's lifetime—and that is precisely the case with America's bout with lead poisoning in the twentieth century.

Alice Hamilton told of a "foreman in a sheet lead works" in the late nineteenth century, who suddenly contracted severe lead poisoning.[24] Given the dearth of effective treatments, he was lucky to escape with his life. His doctor's "treatment" was to advise the man to quit work and rely on the body's "normal" metabolisms to sequester the offending lead in the bones, where he believed it would do no further harm. But, Hamilton continued, "Sixteen years later he had pneumonia and . . . his physician put him on iodide treatment." Whether caused by the acidosis that often accompanies serious infection or, as Hamilton believed, the prescribed iodine, the lead that had been deposited in his bones years before was drawn from its repository, and the patient developed "ankle drop," a symptom typical of severe lead poisoning.[25]

In the early twentieth century, lead-paint manufacturers in the United States had a severe lead-poisoning problem. Their "cure"—corporate-sponsored occupational medicine and privately administered workmen's compensation—made the problem disappear from public view by sequestering it in the workplace. In a similar fashion, pediatric lead poisoning, "discovered" in the 1930s, was portrayed as an intractable problem of poverty and race, and practically vanished from public concern for several decades.

In the same way that pneumonia (or its treatment) brought forth the lead from the bones of Alice Hamilton's factory foreman, in the late 1960s radicalized doctors, activist inner-city groups, enraged parents of poisoned children, and empowered community and occupational health programs all worked to force lead poisoning to the forefront of public concern. What was led to the bones in the 1930s was led out again by the 1970s.

The nation's bouts with the acute aspects of its lead poisoning thus make for a crisply outlined cyclical narrative.[26] But focusing on cycles ignores the steady drone of occupational and environmental lead exposure that persisted between crises. Keeping chronic exposure in mind helps avoid the extremes of whiggish triumphalism and hypercritical diatribe. It would be easy to set the history of pediatric plumbism in a triumphal narrative of medico-technological conquest—after all, medical science took only fifty years to pass from the "discovery" of childhood lead poisoning to its "cure." But by conceiving of this progress primarily as a shift from the acute to the chronic form of plumbism, both past progress and the magnitude of the obstacles yet to overcome remain objects of critical analysis.

This book is divided into three parts. The first three chapters serve as a prelude, setting the stage for the drama to come. Chapter 1 introduces lead's ancient history and surveys current definitions of its toxic effects. Chapters 2 and 3 tell of America's obliviousness to lead's dangers at the beginning of the century, of pediatricians who could not see childhood lead poisoning and of metals producers, paint manufacturers, and legislators who could not see past the popularity of lead paint in order to curb its clear health hazards. Chapters 4 through 7 trace the origins of the American system for preventing and compensating occupational sickness and some of the environmental consequences of this system. Industrial hygiene is shown to be both a medical subdiscipline arising from an increasingly powerful medical profession and an expansion of the rationalization of American industrial culture.

Pediatric and environmental lead poisoning take center stage in part three. Chapters 8 and 9 analyze the reasons for the silent epidemic's long gestation in isolated city health departments. Chapter 10 shows how community activists, pediatricians, and scientists in the late 1960s made pediatric lead poisoning a potent issue in a number of cities and then focused their efforts on pushing the federal government into action. Chapter 11 analyzes the debates over environmental pollution from leaded gasoline, which generated a scientific reappraisal of the lead industry's definitions of "natural" lead exposure and, indeed, of the very definition of lead poisoning. Chapter 12 follows changes in health policies for reducing pediatric lead poisoning since the 1970s, as advocates, health-care providers, and bureaucrats tried to reconcile policy, politics, and practice to a constantly shifting epidemiology. The final chapter demonstrates how new ideas about the dangers of low-level lead exposure transformed all aspects of regulation and complicated all efforts to further reduce exposures.

Over a century ago, physician John Keating wrote an article about childhood lead poisoning for his pediatric textbook. "The justification for the present article," he began, "is found not in the abundance but in the meagreness of the present accumulation of facts relating to children."[27] One hundred years later, Herbert Needleman, the world's foremost lead-poisoning researcher, speculated as to whether physicians would still be treating lead poisoning in the twenty-first century. He concluded, "We will not end this man-made epidemic until we understand the reasons for its curious persistence in the face of considerable data about what lead does, and what is needed to rid ourselves of it."[28] Keating's preface bespeaks his century's belief in redemption by facts; Needleman, whose reputation rests on his skillful interpretation of mountains of "facts relating to children," suggests that further progress in reducing lead poisoning will require historic understanding more than additional data. Journalists and public health activists in the late 1960s referred to the discovery of "the silent epidemic" of childhood lead poisoning. This study will argue that for most of the century lead poisoning, in all its guises, was silenced by design—and that since it was silenced once, it may be silenced once again.

ONE

Plumbing
the Depths

Five thousand years ago, ancient metallurgists separated lead from silver. At least two thousand years ago, careful observers identified the symptoms of poisoning by the more humble element, and yet still today, we surround ourselves with lead.

Greed sought after silver but did not fail to rejoice at having found minium.

PLINY,
Natural History

Lead-acid batteries start our cars, sheet lead and solder are integral parts of our buildings, lead oxides color our plastics and provide corrosion protection for our steel-hulled ships, lead-shielded cables help maintain clear, undistorted telecommunications. And lead continues in its ancient role as an incorruptible lining for coffins.

As lead is all around us, it is in us, too. Our lungs and intestines absorb the waste products of leaden consumer goods from bygone years—finely powdered lead carbonate from weathering lead paint on walls and woodwork, finer particles of lead monoxide left behind by cars burning leaded gasoline. Lead then makes its way to every part of our bodies: our skeletons absorb it into the matrix of the bone; it is sopped up by soft body tissues; our kidneys and livers carry its heavy burden. But it is in our brains that lead exacts its steepest toll. Lead-induced brain damage killed or maimed untold thousands in the twentieth century, and many health experts fear that every generation will continue to suffer subtle but demonstrable neurological damage, at an inestimable cost, until we cleanse the toxin from our surroundings. America's cities are veritable lead mines. Exhausting their deadly veins will cost hundreds of billions of dollars and will require a major shift in public consciousness. The lead will not go away by itself—after all, one of the element's chief attractions has always been its incorruptibility.

*   *   *

Lead poisoning shows many faces, prompting some to label it an "aping disease."[1] Its definition has been subject to nearly constant revision, though modern standards classify its many varieties according to two polarities: chronic versus acute, referring to the abruptness with which symptoms appear and their duration; and clinical versus subclinical, referring to the severity of the poisoning. The terms *subclinical* or *asymptomatic* describe cases in which measurable damage from lead evokes no outward symptoms.[2] *Clinical* lead poisoning manifests a wide range of indicators. The milder symptoms include fatigue and sleeplessness, pallor, loss of appetite, irritability or malaise, or other sudden behavior changes. Severe clinical symptoms include weakness, abdominal pain, constipation, persistent vomiting, clumsiness, extreme dizziness, and other signs of nerve damage—from paralysis of limbs to terminal symptoms such as convulsions and swelling of the brain (lead encephalopathy).[3]

At the beginning of the twentieth century, the relative lack of diagnostic tools meant that only acute, clinical plumbism was accurately diagnosed, and then only if a source of lead exposure was obvious. The first improvements in diagnostic tools enhanced physicians' ability to find symptoms inside the patient's body: microscopes to find signs of blood damage, x-rays to find signs of lead accumulating in the bones. Since the 1940s, however, diagnosis has relied more on the direct measurement of lead in body tissues and less on the patient's symptoms, hazardous behaviors, or environment. The resulting shift in the definition of plumbism has laid bare the association between low-level lead absorption and diseases of the nervous system, kidneys, and reproductive system, many chronic illnesses, and various developmental and behavioral concerns. This, in turn, has transformed the regulation of environmental lead.

Presumptions about the population at risk have undergone just as profound a change over the past century. Figure 1-1 is drawn from official reports of lead-poisoning deaths from city and state boards of health in the United States. As an index of actual deaths, the data are nearly useless, but they are useful indicators of the shift in the medical community's perception of lead poisoning—from a strictly occupational disease to a pediatric problem. Prior to the 1950s, most of the research assured the public that lead's victims fell neatly into two categories: workers whose dilatory employers left them beyond the pale of "modern" occupational hygiene; and extremely poor, neglected, or "backward" children who ate lead pigments. In the late 1950s, childhood lead poisoning was reforged as a racial disease, symptomatic of the "backwardness" of an African-American culture that

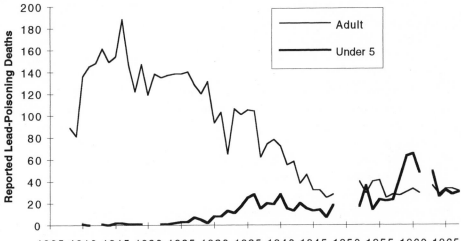

FIGURE 1-1

Reported Lead-Poisoning Deaths in the United States, 1908-67
The gap from 1949-51 is due to two changes in classification. After 1951 lead-poisoning deaths were not broken down by age but by place of poisoning. For the purposes of this figure, deaths occurring "in the home" were assumed to be pediatric cases, and all deaths under the remaining categories of "industrial" and "not specified" were assumed to be adults. Judging by the numbers of fatal childhood lead poisonings reported in just a handful of cities, as tabulated in chapter 8, a number of deaths "not specified" must, however, have been children. For example, in 1958, five cities reported a total of forty-four childhood lead-poisoning fatalities, but the U.S. Department of Health reported only forty-three deaths "in the home" for the entire nation; another twenty-six lead-poisoning fatalities were listed as "not specified."
Data for 1908-48, Bureau of the Census, *Mortality Statistics* (Washington: GPO, 1909-49); for 1950-67, U.S. Department of Health, Education and Welfare, U.S. Public Health Service, *Vital Statistics of the United States* (Washington: GPO, 1951-68).

was held to encourage dirt eating in its children. Environmentalists in the 1970s cast plumbism as an environmental pandemic. Since the late 1970s it has often been a source of anxiety for young, upwardly mobile parents as they renovate old homes.

Lead, in various compounds, begins its deadly course when absorbed into the blood, most often from the intestines or lungs. Much of the interloping lead is delivered to the body's waste stream for elimination. The blood's ability to delead itself depends on many factors, including the nature of the lead exposure and the age and health of the victim. Children—

with their rapid respiration and metabolisms designed for sponge-like absorption of nutrients—retain a higher percentage of ingested lead than do adults.

Over 95 percent of the retained lead is distributed to the bones. While bound in the matrix of the bones, lead ions are of little threat.[4] However, bone-bound lead can be a time bomb. Months or even years after exposure, metabolic changes can allow the deposited bone lead to leach back into the bloodstream, where it resumes its attack on the blood and soft body tissues (see Figure 1-2).

Lead damages the blood itself by shortening blood-cell life expectancy and sharply reducing the production of new blood.[5] Lead's effect on the kidneys depends on both level and duration of exposure. Lead can cause a variety of disorders that are almost indistinguishable from other renal diseases. With only brief exposure, most symptoms are reversible. Long-term exposure—even at levels that do not bring on clinical symptoms of plumbism—can, however, result in chronic or irreversible kidney damage.[6] The nervous system is particularly vulnerable to lead. At lower exposure levels, nerve conduction is reduced. As lead concentrations rise, so does nerve damage—to the point where neurons are killed, permanently replaced by nonfunctioning glial cells. Lead encephalopathy causes the convulsions, loss of consciousness, and coma that accompany fatal lead poisoning.[7]

Lead's impact on the reproductive system has long been cited to justify excluding women from lead-polluted workplaces, even though for years studies have shown that lead impedes conception by affecting sperm production just as it affects the ovaries. But conception is only the first hurdle; the fetus in a "leaded" womb faces additional, ongoing exposure. Lead crosses the placenta beginning in the first trimester, and although mother and fetus carry approximately the same lead burden per ounce, blood-lead levels that have no measurable effect on the former pose a grave risk to the latter. Higher rates of stillbirth and infant mortality among the children of highly leaded mothers were obvious to occupational hygienists a century ago. The subclinical, congenital poisoning of the children of lead workers today is nearly invisible—and highly contested.[8]

Although there is still no cure for lead poisoning, no elixir that will reverse the damage caused by years of exposure to the toxin, a number of treatments developed since World War II can reduce absorbed lead in the body. The protocols for these treatments are constantly being revised but in general, for both children and adults, they vary with severity of exposure. Mild cases are treated by environmental management, which includes

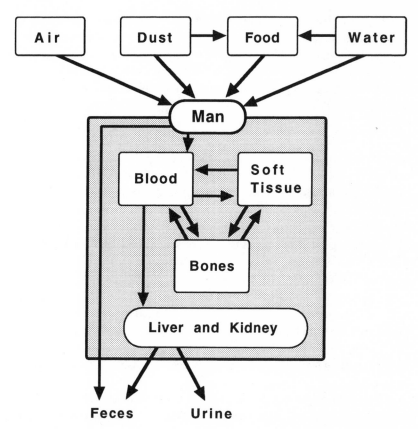

FIGURE 1-2
Pathways of Lead from the Environment to and within Man
After Environmental Protection Agency, *Air Quality Criteria for Lead* (Research Triangle Park,
N.C.: U.S. Environmental Protection Agency, Environmental Criteria and Assessment Office, 1986).

careful observation—usually on an out-patient basis—and the prevention of further exposure. Management of more severe plumbism involves
removing the patient from the environmental lead and then removing
the lead from the body by administering chelating chemicals, the most
common now employed being calcium disodium edatate (CaEDTA);
dimercaptopropanol, often called British anti-Lewisite (BAL); and 2,3-
dimercaptosuccinic acid (DMSA), or succimer. Chelation usually produces
a quick drop in blood lead, followed, once treatment is stopped, by a gradual rise beginning when newly deleaded blood absorbs additional lead
stored in body tissues. Further measurements determine whether another
course of treatment is required.[9]

The body absorbs lead slowly but releases it even more slowly. Where there is high exposure of short duration, timely treatments can readily cleanse the blood, but if lead has time to be absorbed into the tissues, the blood continues to be recontaminated and systematic poisoning results. A single, massive exposure to lead is more manageable than long-term, lower level exposure.[10]

Mankind has had a very long exposure. Humans evolved in the presence of minuscule traces of lead, which makes up significantly less than 1 percent of the earth's crust. Volcanic eruptions were the only probable sources of toxic exposures for early humans. According to geologist David Purves, modern human intolerance of lead is an artifact of early man's lead-free environment.[11] Early people most frequently encountered lead in the form of outcroppings of its most common ore, crystalline lead sulphide (galena). The gray rock was easily ground and mixed with other minerals for use in pigments. Galena-based eye makeup dates as far back as 4000 B.C. As recently as 1969, galena was still used in eye paints in India, and over a dozen cases of Indian women in London being poisoned by imported lead sulphide mascara were reported in 1968.[12]

Development of the technology necessary for lead smelting was encouraged because of lead ore's most valuable "contaminant"—silver, prized as early as 5000 B.C. for its beauty, malleability, and rarity. Lead and silver are mixed in the earth's veins like blood and water. It is natural that the fortunes of both metals became inextricably linked. Galena was readily available and easily smelted—strong inducements to exploit this new-found source of the precious metal. Indeed, the technology used to separate silver from lead may predate, or at least run parallel to, the development of bronze smelting.[13] The wealth of nations rose and fell with their sources of silver-bearing lead. Ancient Athens owed much of its wealth to the nearby Laurion mines, the richest source of silver in the Bronze Age Mediterranean. In the fifth century B.C., Athenians produced up to ten thousand tons of lead and silver each year.[14] It was more than a coincidence that Athens' power waned as its mines were played out.

Silver is extracted from argentiferous lead by a process called "cupellation." When lead ore is smelted, the lead and silver melt away commingled. This alloy is then heated in a shallow furnace. The lead on the surface reacts with air to form lead monoxide (litharge), which is skimmed off repeatedly. On average, 400,000 tons of litharge were skimmed away for every ton of cupelled silver produced. Although initially much of the lead was simply discarded, it could be resmelted or used to produce white lead pigment, and the close parallel between rising silver production and lead consumption suggests that markets and applications were quickly invented for the mountains of lead waste accumulating at the silver forges.[15] As Pliny declared, "Greed sought after silver but did not fail to rejoice at having found minium" (red lead).[16]

In time, the demand for lead made the smelting of ore with negligible silver content economically viable. Geochemist Clair Patterson estimates that around 200 B.C., 75 percent of Roman lead was a byproduct of silver mining but that by 50 B.C., the Roman Republic's demand was so great that about half the lead consumed was smelted from ore with negligible silver content.[17] By this time lead had become an intrinsic part of Roman life—an indispensable material for "civilized" living, a boon to civil engineers—and a ubiquitous poison.

Metallic lead's lackluster appearance made the metal undesirable for use in jewelry, but it was readily worked, cast, and alloyed, assuring its adoption for many utilitarian purposes. Bountiful archaeological evidence for the ancient world's love of lead remains—the element's incorruptibility guaranteed the survival of such artifacts. For example, "plumber" names both a trade and a historic connection with *plumbum* as a conveyer of water dating back at least as far as the city of Ur.[18] Many ancient cultures around the Mediterranean used lead pipes, but the Romans' enormous hydroengineering projects called for unprecedented quantities of lead. High-pressure pipes carried water over hills from reservoir to reservoir and through distribution networks in Rome.[19]

Lead is also linked to the history of writing. Of course, the "lead" now most closely associated with writing—pencil lead—is not lead at all but graphite, the misnomer stemming from the Middle Ages, when the newly discovered mineral was named "German lead." Early pharaohs and Assyrian kings inscribed messages on sheets of lead. Medieval scribes used gray lead styluses to draw fine, erasable lines on parchment to guide their writing. The ancient Chinese wrote on bamboo with white lead ink.[20] Lead became even more important for the written word after Gutenberg. Movable type made from lead-tin alloys was readily cast and easily reused, yet produced a crisp outline. Linotype machines, which automatically cast molten-lead type as an operator keys in text to be printed, are still used and continue to be a significant source of occupational plumbism.[21]

Lead is one of the densest elements, giving rise to its common applications in weapons, weights, and standard measures. Long before the advent of gunpowder, Greeks hurled leaden bullets from leather slings at Persians on the plain of Marathon. Adding insult to injury, these bullets sometimes bore inscribed messages, such as "Take this," or "Desist." Lead's mass and corrosion resistance made it the natural material for standard measures. Lead weights have been found in early Minoan sites.[22]

Lead's weight and permanence naturally linked the element to death, burial rites, and the gods. Lead coffins date back to 2200 B.C. Egypt. Phoenician tombs contained lead sheets inscribed with messages to the underworld. Roman colonists throughout Europe commonly buried their dead in lead coffins; Greeks and Romans frequently stored cremated remains in leaden urns. In North America, galena beads are often found in Native American burial sites.[23] The ancient Egyptians associated lead with Osiris, while the Greeks and Romans associated the metal with Cronus/Saturn. Both the Greeks and the ancient Chinese believed lead was the "father" of other metals, just as Cronus was the father of the Greek gods (another possible association: as Cronus ate his children, so lead seemed to consume its children, silver and gold, in the furnace). This connection continued in alchemy and astrology, and hence to modern medicine and science. Alchemists' symbolic language equated Saturn and lead. Scientists and physicians still refer to lead and its compounds as Saturn or extracts of Saturn; lead poisoning is still saturnism.[24]

The most widely used lead compound in history—found in the ruins of Ur and inside the abdomens of American children today—is lead carbonate, or ceruse: pure white, easily manufactured, easily mixed in oil, and easily applied to almost any surface, from canvas to wood to skin. The ancients quickly discovered several methods of manufacturing ceruse, establishing what Nriagu calls "the first chemical industry in the history of mankind."[25] And just as quickly, they discovered its toxicity. In the second century B.C., the Greek poet and physician Nicander condemned "the hateful brew compounded with gleaming, deadly WHITE LEAD whose fresh colour is like milk which foams all over when you milk it rich in the springtime." Nicander described most of the classic symptoms of saturnism— including hallucinations and paralysis—and recommended several purgative treatments.[26]

For at least 2500 years people painted their faces and bodies with white lead to cover pocks and to give the skin a fashionable pallor.[27] In Greece of the fourth century B.C., Xenophon jibed at ladies who applied a "plaster of ceruse and minium," and Chinese documents from the same era name various lead cosmetics mixed with rice powder.[28] The high foreheads of Elizabethan Englishwomen may have stemmed in part from long-term use of lead plasters, which caused patches of hair to drop from the scalp. Shaving eliminated unsightly bald patches. In the sixteenth century, ladies wore

Venetian ceruse, or spirits of Saturn—essentially white lead in a vinegar solution. Nineteenth-century women could lighten their skin with "Laird's Bloom of Youth," "Liebert's Cosmetique Infallible," "Eugenie's Favorite," or "Ali Ahmed's Treasure of the Desert," leaden face powders all. Ointments and powders containing white lead were sold in the United States until well into the 1900s. In the 1920s, physicians discovered that over the years, thousands of Chinese and Japanese infants had been killed or poisoned by ingesting white lead powder from their mothers' faces and breasts.[29]

Lead compounds lent their brilliant colors (and toxicity) to mankind's first ceramic glazes and they continue to be used throughout the world for this purpose, often with serious health consequences. Lead also made an ideal base for enamel. The enamel used in bathroom and kitchen fixtures at the beginning of the twentieth century contained between 10 and 40 percent lead.[30] While it is doubtful that many people have been poisoned by lead in the kitchen sink or bathtub, poisoning from lead-glazed ceramics and lead-enameled cooking utensils has been well documented from antiquity to today. Acids in food dissolve the lead in storage containers in a form readily absorbed by the alimentary tract. The resulting chronic plumbism is so slow in onset that establishing its cause is next to impossible. Mexican toxicologists have found traditional low-fire lead glazes to be a perennial source of serious pediatric plumbism.[31] We need not throw out the heirloom lead crystal—in most pieces the lead is tightly bound to the glass. However, acidic liquids stored for lengthy periods in leaded crystal decanters can absorb significant amounts of lead.[32]

Although lead's toxicity was well known, saturnine medications have been extremely popular since Nicander's time. Physicians have long noted this contradiction: Bernardino Ramazzini, an eighteenth-century pioneer in industrial medicine, observed, "How strange it is that lead—which . . . is commonly called the surgeon's mainstay . . . should . . . throw up merely by exhalation seeds so deadly that potters who need its aid are thus stricken."[33] Dentists used lead for fillings until the seventeenth century. The ancient Romans used lead for fillings as well, but in a procedure for extracting teeth, not preserving them: Roman dentists plugged decay-hollowed teeth with molten lead in order to prevent their shattering under the pressure of forceps.[34]

Lead salves and lotions were a part of the pharmacopoeia from West Africa to China, from the days of Mesopotamia to the twentieth century. Spanish explorers in the New World included lead plasters and ointments among their medications. When they returned to Europe afflicted with syphilis, their physicians often treated the new condition with ointments containing lead.[35] Seventeenth- and eighteenth-century physicians prescribed salts of lead for hemorrhages, epilepsy, dropsy, "female irregularities," and as an antidiarrheal. Seventeenth-century British "Professor of Physic" William Salmon prescribed swallowing gold or lead pills (bullets worked in a pinch) as a cure for "colic due to circumvolution of the Intestines through a wind," as did Ramazzini. The mass of lead was supposed to push the obstruction through the "twisted" bowels.[36]

Industrial physician Carey P. McCord found that apothecaries and physicians in early America "almost invariably mentioned red and white lead, 'litharge' and diachylon [lead oleate] for plasters and lead soaps." A writer in 1932 recommended applying "a milky solution of 'sugar of lead'" to treat poison ivy, and McCord, writing in 1954, noted that lead solutions were still widely used "in the treatment of poison ivy dermatitis and occasionally as a soak for sprains." In 1926, Columbia University medical professor Carter Wood claimed that of approximately 250 cancer patients he treated with injections of finely ground lead, one-fifth received some relief.[37]

Throughout human history, men and women have employed lead's devastating effect on reproduction in their perpetual attempts to control fertility. Aristotle discussed "anointing that part of the womb on which the seed falls with oil of cedar or with ointment of lead or with frankincense commingled with olive oil." In ancient Rome, men suffering from nocturnal emissions or wishing to produce temporary sterility fastened lead plates to their chests. When contraception failed, women swallowed lead compounds as an abortifacient, a function they served well into modern times. British physician Thomas Oliver noted in 1902 that "women of the lower classes when pregnant . . . resort to diachylon pills." As late as 1974, women in some parts of the world still did.[38]

Throughout history, one of the most common sources of lead exposure has been alcohol. From remote antiquity, when vintners first decanted their wares into leaden vessels, to the present, when backwoods moonshiners continue to extract "squeezin's" from lead-clad stills, millions of drinkers have sipped a little Saturn with their Bacchus. Prior to the twentieth century, lead-adulterated wines and spirits probably caused more plumbism than any other nonoccupational source. The first adulteration was most likely inadvertent. Greek and Roman vintners employed lead at every stage of wine production. In Greece, they also added lead to their wine with sea water, boiled down in lead pots and used as a flavoring. The most expensive Roman wines were sweetened with a fruit concoction called *sapa*, usually made by boiling down inferior grapes, almost always in lead pots. Wine was stored in sealed leaden vessels or glazed ceramic crocks, and drunk from tumblers made of lead or heavily leaded bronze. In recent years, speculation regarding the historic consequences of such widespread adulteration has produced a spirited scholarly debate.[39]

Gradually, experience and experimentation led to the intentional adulteration of wines. Vintners discovered that lead could retard fermentation. Wines stored in the presence of lead lasted longer and tasted sweeter than those stored without lead. Chemical tinkerers soon created saturnine recipes to alter the taste, even to mimic other producers' unadulterated wines. Charlemagne is reputed to have called for the banning of leaded wines in A.D. 802. By the Renaissance, wine-related lead poisoning prompted harsh countermeasures: the German city of Ulm executed its convicted wine adulterers.[40]

Nevertheless, the practice of wine adulteration spread, as did the health consequences. In 1639 an epidemic of severe colic racked the French town of Poitiers in the district of Poitou. François Citois, a local physician, described the symptoms, including constipation and twisted bowels, kidney stones, paralysis, and encephalopathy. Sixty years later, Eberhard Gockel, city physician of Ulm, contracted the illness himself while investigating an outbreak of colic of Poitou in the Wengen monastery. Gockel conducted a survey of his region, the Neckar valley, and found that wine adulteration and colic of Poitou were both widespread.[41]

Meanwhile, England had its own lead-poisoning epidemic. The Devonshire colic was known to be brought on by immodest consumption of cider. In 1767, after nearly one hundred years of perennial complaints of colic, George Baker tested Devonshire cider and found that it was heavily leaded. Baker's investigations showed that at every stage, some aspect of cider production—from the glazed earthenware rollers that expressed the juice, to the soldered joints and repairs in sheet-metal troughs, to the cisterns in which the cider aged—introduced more and more lead into the product. Making matters worse, many farmers and merchants added ceruse to correct cider's acidity or to check fermentation. The region's cider producers, on the other hand, insisted that the only lead to be found in Devonshire cider came from the shot farmers used when picking birds off the apple piles.[42]

Two factors stimulated Baker's curiosity: a review of Citois's work, and correspondence with a respected American scientist, Benjamin Franklin.[43] Franklin found out about lead poisoning while still a teenager in Boston, where he learned from "a General discourse . . . of a Complaint from

North Carolina against New England Rum, that it poisoned their People, giving them the Dry Bellyach." In 1723, an inquiry showed that lead in the stills caused the ailments. That same year, Massachusetts Bay Colony passed the first known anti-lead legislation in the New World, which banned the use of "leaden heads or worms" in the distillation of rum and other spirits.[44]

Despite a few nineteenth-century reforms, the adulteration and culinary poisoning continued. In 1820, British physician Fredrick Accum complained that the widespread practice of "leading" wines "adds the crime of murder to that of fraud." According to Richard P. Wedeen of the Veteran's Administration Medical Center in East Orange, New Jersey, saturnine gout from alcohol, well known and accurately described in the eighteenth century, was all but forgotten in the twentieth, erroneously considered "nonexistent outside of the moonshine belt." Adulteration continued unabated throughout the nineteenth century. But American physicians, steeped in Temperance culture, came to see alcohol as less a medical than a moral issue. In blaming the sinner and his "demon alcohol," they ignored the very real poison lurking in the sinner's drink. Symptoms of lead poisoning were frequently attributed to alcoholism—lead encephalopathy was misdiagnosed as delirium tremens, lead colic as pancreatitis or alcoholic gastritis. No one knows how many healthy appendixes and gallbladders were removed when physicians misdiagnosed lead colic, but Wedeen suspects such errors were common before the 1960s.[45]

In recent decades, as lead poisoning in the present became an increasingly powerful public issue, scholars began uncovering lead's secret past. The result is a growing body of fascinating fact and speculation about lead and lead poisoning, an expanding field of interest that cuts across disciplines. Art historians are examining painters' biographies for evidence that the long-standing tendency to associate "genius, madness, and melancholy" may be rooted in the widespread use of lead pigments over the centuries.[46] In 1713, physician Bernardino Ramazzini published his suspicions that Correggio and Raphael were likely victims of lead poisoning. Today, similar speculation runs to Carracci, Goya, and many others. Van Gogh was particularly fond of eating one pigment, which he sucked from his brushes. It is probable that this tasty paint was lead based, as many pigments were mixed into a base of white lead, noted for its sweetness.[47]

Bone-lead analysis is becoming an important tool for the physical anthropologist, the archaeologist, and the historian. Whether the subject is an ancient Nubian child or the crew of a nineteenth-century ship that explored the frozen Arctic, bone-lead analysis has produced lively and refreshing speculation.[48] Since the late 1970s, paleopathologist Arthur C. Aufderheide and his colleagues have conducted a series of exhumations and careful bone-lead content studies at colonial American sites to compare the health and diets of masters and slaves.[49]

Although lead poisoning is both an ancient and a global phenomenon, in the twentieth century it acquired a particularly American flavor. By 1900, America's robust lead industries were well on their way to becoming the world's largest suppliers of lead products. That year, the National Lead Company distributed a pamphlet entitled "Uncle Sam's Experience with Paints," narrated by Uncle Sam himself. The recent "scrap" with Spain had prompted some major repainting and now, Sam reported, "my navy looks better than it ever did. Besides, I have been furbishing up the new places that have come under my flag and feel that I have repaid the people for any inconvenience I may have put them to while bringing them into line and giving them their freedom." Havana and Santiago were now free not only from tyrants, but from "the smudge of yellow fever," because, Sam crowed, "wherever I go I introduce cleanliness. And for cleanliness there is nothing like paint—the *best* paint—Pure White Lead and Pure Linseed Oil."

If one century ago America's lead industry stood ready, in the words of another paint manufacturer, to "cover the earth" with lead paint, the United States in 1900 was still a largely disinterested consumer of foreign ideas about recognizing, preventing, and, in particular, regulating lead poisoning. By World War II, however, the United States had also achieved preeminence in the study of lead poisoning in the workplace, home, and general environment. Childhood plumbism was a major focus of postwar lead-poisoning research and thus appeared to the world as an American problem. The international medical community saw America's "silent epidemic" as a peculiar product of America's urban population, shaped by the nation's racial and class divides and aggravated by its comparatively backward system of health care for the poor.

At the century's end the nation had dramatically reduced its own population's exposure to lead, but it continues to market the poison to the rest of the world. The United States is now the world's leading supplier of both lead products and anti–lead-poisoning activism. Around the globe, American environmentalists and public health experts are engaged in the contentious business of convincing Third World countries to forego leaden products that American businesses enticed them into depending upon. The United States may, for the most part, have kicked its lead habit—

Americans finished the century with less lead in their bodies than their great-grandparents started it with. But for much of the rest of the world, the situation may be just the opposite at the beginning of the twenty-first century. The experience of America's leaden century may suggest what to expect.

TWO

Childhood
Lead Poisoning
before 1930

Near the end of August 1913, the Harriet Lane Home of the Johns Hopkins Hospital admitted a 5-year-old boy from the Baltimore Home for the Friendless.[1] The child was comatose. His legs and arms twitched and flexed in spasms, his head remained rigidly drawn back to his shoulders, and his body shook with general convulsions. Over the next two weeks, doctors performed four spinal taps to divine the cause of the child's malady and to relieve the cranial pressure that threatened fatal encephalopathy. As they found nothing to suggest tuberculosis, syphilis, or any kind of major infection, his condition was diagnosed as "a serious meningitis of unknown cause." A month later, he was returned to the Home of the Friendless. He seemed healthy until March, when a prolonged bout of headaches and vomiting gave rise to convulsions. He was readmitted to the hospital "in a condition which was almost identical to that of his first admission." Lumbar punctures once again revealed no clue to etiology.

Then someone noticed "a slight discoloration about a tooth." This was soon identified as a "lead line," a clinical sign of lead poisoning frequently found in occupational cases, the result of a deposit of lead sulphide on the victim's gums. The source of the lead became apparent when the child was found with his face besmirched with white lead paint gnawed off his hospital crib. The rails of his dormitory bedstead at the Home for the Friendless were chipped and denuded, and the staff confirmed that unless closely supervised "he would gnaw any painted object." Again the boy remained in the hospital for a month, and once again the doctors thought

The child lives in a lead world.

JOHN C. RUDDOCK,
"LEAD POISONING IN
CHILDREN WITH SPECIAL
REFERENCE TO PICA"

his recovery was complete. But three weeks after his release, the child suffered a sudden convulsion and died.

In both its near inscrutability and its fatal outcome, this child's case, one of the first published reports of pediatric lead-paint poisoning in the United States, was typical of serious lead poisoning in the years before tools for accurate diagnosis and effective treatment became available. It says much about medical science in 1913. The doctors' training and cultural backgrounds prepared them to expect germ-based diseases, not environmental toxins. Their black bags brimmed with equipment for administering Noguchi globulin tests, for testing for Wassermann reactions or reducing Fehling's solution. Similarly, their intellectual baggage was filled with the theories and magic bullets of Koch and Pasteur. Their notion of "environment" encompassed germ-bearing filth and the deleterious effects of bad air but not the commonplace, manmade toxins confronting every child, every person, in an increasingly urban America.

In the late 1960s, when public health researchers became aware of the true scale of childhood lead poisoning, journalists wrote of the discovery of a "silent epidemic."[2] Thirty years later, the name no longer applies. Lead poisoning is still a serious problem, but it is instructive to contrast the frequent and hyperbolic reporting of lead poisoning in recent years, given the small number of life-threatening cases reported, with the press's near silence on the subject forty—or ninety—years ago. Hundreds died then, and few reported it. Now, lead-poisoning deaths are blessedly few, but hundreds report it.[3]

The advances made in treating lead poisoning are reflected in the dropping blood-lead level considered "acceptable" by federal agencies. Before 1970, no pediatric standard existed; many doctors and hospitals used the threshold for toxicity established by lead company–sponsored research for diagnosing lead poisoning in adult lead workers. This standard defined levels higher than 80 μg/dL as dangerous.[4] As late as the early 1950s, Baltimore's aggressive pediatric lead-screening program defined cases in which the blood-lead level exceeded 70 μg/dL only as "possible" lead poisoning, even if lead paint was found in the child's home. In 1970, the Surgeon General defined 60 μg/dL as "undue lead absorption," but just one year later, the Public Health Service recommended deleading treatments for children with blood-lead levels over 40 μg/dL. The Centers for Disease Control lowered the threshold to 30 μg/dL in 1975, to 25 μg/dL in 1985, and to 15 μg/dL in 1990.[5] Today, most lead-poisoning research seeks to measure the subtle physiological damage brought on by these "low" blood-lead levels.[6]

Decreasing blood-lead levels have not reduced the concern expressed by environmentalists and child health advocates over the remaining dangers. The number of government hearings on childhood lead poisoning has increased, as has the number of landlords sued for negligence because of lead paint. The reason for this apparent irony is that the fight to end lead poisoning is part of a complete reshaping of the definitions of acceptable risk and wellness, and of how society views its duties to the poor. To understand the shifting perception of lead poisoning is to gain insight into some of the most significant changes in America's recent past.

This chapter examines the first shift in this awareness, the moment when childhood plumbism first began registering on American radar. Before 1900, many obstacles prevented the recognition of an age-specific environmental disease such as childhood lead-paint poisoning. Several of these obstacles gradually fell away between 1900 and 1930, though not through particular insight of visionary physicians or conscious policy changes implemented by progressive boards of health. Rather, they were eroded by fundamental changes in American society: changes in demographics and definitions of acceptable risk in the workplace and home, and a measurable shift in the valuation of children and childhood.

The effect of these changes upon awareness of childhood lead poisoning has not been adequately explored.[7] The standard approach follows the narrative scheme of "discovery," tracing the thread of citations on childhood lead-paint poisoning back to Australia, where around 1900, two pediatricians, A. J. Turner and J. Lockhart Gibson, established that the seasonally recurrent colics, vomiting, and paralysis of dozens of Queensland children were due entirely to their ingesting flaking and dusty lead paint from sun-weathered verandahs.[8] After this revelation, in the conventional view, awareness of pediatric lead poisoning grew organically—albeit in fits and starts—blossoming as a public health issue in Baltimore in the 1930s, later in other cities and towns, and exploding as a serious political issue in the 1960s.

But in fact the news from Australia had little, if any, immediate impact in the United States. This should not surprise anyone familiar with the way medical information is (or is not) transferred from one part of the world to another. Even if American pediatricians read Australian medical journals on a regular basis or knew something about Turner and Gibson's findings, they might have assumed that childhood lead-paint poisoning was specific to tropical, exotic Australia—an impression reinforced by Turner and Gibson themselves, whose reports implicated the hot, rainy Queensland summers and wooden frame bungalows on stilts typical in flood-prone Brisbane.[9] However, many of Brisbane's climatic and architectural characteristics were also common in locales such as Savannah, Georgia, or Tampa, Florida. Tampa was at the time similar in size to Brisbane, and is situated at almost exactly the same (albeit north) latitude. Furthermore, its homes were almost all white-painted wood frame. That American pediatricians did not respond to the Australians' findings seems consistent with Bruno Latour's position that scientists find new evidence convincing only if it allows them "to go faster in the direction that they wanted to go in any case."[10]

At the beginning of the twentieth century then, the name "silent epidemic" was entirely appropriate for childhood lead-paint poisoning. In all probability hundreds of children were being debilitated or killed by the paint in their homes every year.[11] For as long as infants have grown teeth, they have looked for something to cut them on. And as pediatrician John C. Ruddock observed in 1924, "a child lives in a lead world."[12] Lead paint crumbled from walls then just as it does today, and it tasted just as sweet—if not sweeter, since there was more lead in it. It defies logic to believe that children in 1900 were less likely to cut their teeth on lead than were the well-documented victims of lead poisoning in the 1960s. American physicians in the first three decades of the twentieth century must have fundamentally misunderstood the true cause of these children's illness.

In 1923, L. Emmett Holt, Jr., the most noted pediatrician of the decade, declared, "Poisoning by lead is a rare occurrence in early life, and is, therefore, seldom considered as a diagnostic possibility in young children."[13] Holt's conclusion was certainly correct: plumbism *was* seldom considered by physicians. But was his premise correct? Was lead poisoning rare in early life in 1923? This seems possible only if leaded gasoline, which was not introduced until the mid-1920s, was more of an agent in bringing about "lead-paint poisoning" than was paint itself. And indeed, such a case can be made. As leaded gasoline was withdrawn from American pumps in the 1970s, lead levels in the blood of urban dwellers began to fall. "Lead-paint poisoning" fatalities virtually disappeared in the decade of the Clean Air Act.[14] American children's blood-lead levels are much lower in the 1990s than twenty years ago, and the world's highest blood-lead levels are to be found in children living in dense urban centers of countries where leaded gasoline continues to be burned.[15]

To prove that introducing leaded gasoline initiated the pediatric lead-poisoning crisis would require baseline blood-lead data in order to show a symmetrical increase in those levels in the 1920s and 1930s, as leaded gas was introduced. Unfortunately, no such hard data exist to establish the status *ante* tetraethyl. Tracking the number of case reports over time is highly problematic. Reports of childhood plumbism are exceedingly rare in medical journals from the first quarter of the twentieth century, even at the height of the furor over occupational lead poisoning in the early 1910s. An examination of the admission records of the Children's Hospital in Boston from 1900 to 1920 did not turn up a single patient admitted with lead poisoning.[16]

In the late 1920s and in the 1930s, after leaded gasoline was marketed nationwide, the frequency of childhood lead poisoning reported in medical journals increased dramatically.[17] This increase might seem to be damning evidence of tetraethyl lead's effects, but the association between increased consumption of leaded gasoline and increases in reported lead poisoning is very weak. As Figure 2-1 suggests, the consumption of tetraethyl lead remained comparatively low prior to the 1950s, twenty years after pediatricians and public health experts became aware of the epidemic nature of childhood lead poisoning. It is possible that the newly interested researchers were simply finding the cases of lead poisoning caused largely by leaded gasoline, even before it entered into widespread use. However, if the small absolute level of increase in environmental lead contributed by leaded gasoline prior to the 1950s had that large an influence upon the increase in lead poisoning, then the explosion of tetraethyl consumption after World War II would have brought a similar geometric increase in lead poisoning, and this apparently did not occur. The rise in reports is probably merely an index of increased interest in the subject of childhood lead poisoning. The frequency of plumbism always seems highest wherever public health organizations mount aggressive programs for detection and prevention. It is almost as if searching for lead poisoning "causes" it, as Baltimore's vigorous and pioneering programs to detect and prevent pediatric cases would demonstrate.[18]

As with the apparent rise of lead-poisoning fatalities in the 1920s, the dramatic decline in fatal plumbism in the 1970s is not due simply to the phase-down of leaded gasoline but also to new case-finding programs and improved treatment after the mid-1960s. Doctors in the United States reported no childhood fatalities from lead poisoning from any source in 1993. In 1990, a lead-paint–poisoning fatality involving a Wisconsin child who died after ingesting chips of paint containing as much as 30 percent lead was reported, but prior to that, no deaths related to lead-based paints had been reported in the United States since the mid-1970s.[19]

Banning tetraethyl lead did not remove all the lead from the urban envi-

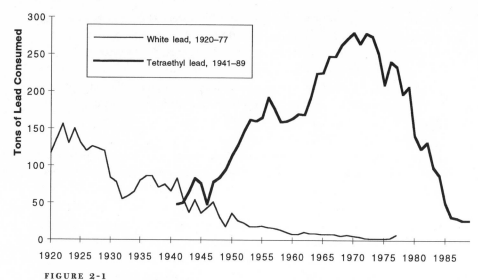

FIGURE 2-1
Falling Consumption of White Lead; Rising Consumption of Leaded Gasoline
U.S. Department of the Interior, Bureau of Mines, "Lead," in *Minerals Yearbook* 1920–89
(Washington: GPO, 1921–90). No statistics for tetraethyl lead were published prior to 1941.

ronment. The residue of fifty years of burning leaded gasoline remains in
the dust of every playground near a city street. Nonetheless, it seems unlikely that the average city child's exposure to lead is higher now than in
1920, given the drastic reductions in other environmental sources such as
lead arsenate in food, lead solder in tin cans, and lead salts in tap water.
Making this even more unlikely is the ubiquity of leaded paint seventy
years ago. The child in 1924 simply lived in more of a lead world than
today's child. And even if blood-lead levels seventy years ago had been
essentially the same as those of today, the lack of effective early detection
and treatment at that time would have resulted in many serious cases of
plumbism and more than a few fatalities each year.

By no means can it be argued that leaded gasoline was blameless. Burning leaded gasoline injected tons of absorbable lead into the environment,
elevating the blood-lead levels of most urban dwellers so that by the time
children ate their first paint chip, their bodies were already struggling to
cope with the low-grade fever of environmental plumbism. Still, it seems
improbable that there was as little lead-paint poisoning before 1930 as the
written record suggests. Unfortunately, there is no way to begin estimating
how many children's blood-lead levels then reached or exceeded today's
threshold for aggressive deleading. No publicity campaigns warned par-

ents of the dangers their infants faced when they gnawed on windowsills
or crib railings; no public authority put landlords on notice to provide
nontoxic paints in all apartments inhabited by young children; and few
doctors were equipped—whether through training, awareness, or technology—to detect or even suspect that a child's illness originated in an environmental toxin.

We cannot determine how many children were lead sickened or killed annually before 1930. On the other hand, we cannot assume that the scarcity
of recorded lead-poisoning fatalities reflects reality. But if we assume that
the most important risk factors for pediatric plumbism—lead paint, lead-
tainted food, and aerosolized lead dust in the urban environment—remained roughly the same from the 1920s until the late 1950s and early
1960s, when cities began lead-poisoning detection programs, then we may
use estimates from those early programs to project backward. During the
latter period, approximately thirty-seven children died each year from lead
poisoning in Baltimore, Chicago, New York, and Philadelphia, where approximately 15 percent of the nation's urban preschool population lived.
Assuming the same fatality rate in all urban areas produces a nationwide
estimate of 240 fatal plumbism cases each year. But the number of lead-
poisoning deaths reported between 1951 and 1960 never exceeded ninety-
eight, and the average was only sixty-seven. In other words, perhaps some
70 percent of all childhood lead-poisoning deaths went improperly reported.[20]

To see where these lead-poisoned children "disappeared," it is helpful
to reexamine how the doctors who reported the first cases of lead-paint
poisoning in the United States arrived at their diagnoses. In the case that
began this chapter, it was not until all biological causes were eliminated
that doctors sought clues in the child's physical environment. The physician's gaze was more intently fixed upon the microscope and test tube than
the patient. The clue that led to the environmental source of this disease
(the lead line on the child's gums) came by way of the only medical science
that routinely considered the toxicity of a patient's environment—occupational medicine. This case makes very plain how easily fatal lead poisoning
could be missed by even the most meticulous physicians.

The records of the Boston Children's Hospital from 1900 to 1920 provide ample evidence of the misdiagnosis of lead poisoning. The most powerful argument, given the likelihood that *some* children were dying of lead
poisoning in those years, is that of silence—the near total absence of diagnosed lead poisoning over two decades. If even a few cases were diagnosed

each year, one could conclude that physicians were at least aware of the disease. A low frequency of reported cases of plumbism might then be interpreted as indicating a low incidence. Total silence suggests complete misdiagnosis.[21] Semiquantitative examination of the records lends some statistical evidence for this, as will be shown below, but the most compelling data are also the least quantifiable—anecdotal or suggestive evidence from individual case reports.

Lead's toxic effects are so variable as to invite misdiagnosis. The symptoms of lead poisoning vary from gastrointestinal disorders to neurologic damage, and so many organs can be affected that it is tempting to go through medical records and label as "suspected lead poisoning" every case of summertime colic or vomiting, every paralysis or convulsion of unknown etiology.[22] For example, the records of the Department of Diseases of the Nervous Systems at Children's Hospital contain a remarkable number of cases in which a summertime bout with constipation or diarrhea led to convulsions, leaving a previously bright 2-year-old partially paralyzed or mentally incapacitated.[23] This course is entirely consistent with a severe case of lead poisoning.

For example, in the summer of 1905 an 11-month-old Boston girl developed convulsions shortly after her mother noticed her eating wallpaper. The examining physician noted that she did not play with toys, but put them in her mouth. The next summer she developed severe constipation and convulsions, after which her left arm was partially paralyzed. Her mother said that after the first bout of convulsions she did not seem as bright as before. The doctor's diagnosis: this child was "feeble minded."[24]

First-hand reports of living conditions in New York tenements offer some tantalizing environmental evidence. J. H. Morand, a public health nurse looking for the sources of polio infections in New York City in 1916, reported on the case of a 2-year-old boy who in early August became ill with pain, vomiting, and fever. The first two doctors who attended the boy in his home reported that he was teething and had stomach trouble, which they predicted would abate without treatment. Instead, less than a week later the child's left arm was found to be paralyzed, and he was hospitalized as a polio victim.[25]

Morand, searching for the dirt and filth expected in a germ-based disease, could find none. The sanitary conditions of the ground-floor tenement were good. In fact, she reported, "In this apartment the painters and plumbers were at work." How much lead dust had this child breathed as the painters dry-sanded the old paint, as was then the custom? Many parents that summer kept their children cooped up in their homes on account of the polio epidemic. The fact that the child was teething suggests even greater danger of exposure to environmental lead. Other nurses who visited children with symptoms suggesting plumbism as much as polio described tenements where plumbing problems caused chips of plaster and paint to fall from walls and ceilings.

A more systematic search strategy focusing on one disease—tuberculosis (TB) meningitis—reinforces the impression such anecdotal evidence provides, suggesting how frequently misdiagnosis of lead poisoning might have occurred. Like lead poisoning, TB meningitis is a "versatile disease" in that its symptoms are extremely variable and can "simulate" those of other diseases. Childhood lead-poisoning researchers beginning in the late 1920s suspected that the terminal stage of lead poisoning, lead encephalopathy, was often incorrectly diagnosed as tubercular meningitis.[26] The early stages of both diseases produce appetite loss, headache, vomiting, and acute constipation. Gradually, neurological symptoms arise: photophobia, rigidity of the neck, lethargy, and paralysis of the limbs or facial nerves. The most serious cases end in convulsions, coma, and death.

Today, the clinician has a number of tools to help discriminate between the two conditions. Blood-lead levels can easily be measured, and radiological examinations can determine most other sites of TB infection. But examining the spinal fluid, a mainstay of differential diagnosis at the turn of the twentieth century, remains important in distinguishing between tubercular and other forms of infectious meningitis. Three additional clinical markers are more practical for historical analysis. While of little use in diagnosing individual cases, aggregate data on fever profile, sixth-nerve palsy, and seasonality could distinguish groups in which tubercular and plumbic meningitis may have been confused. Tuberculosis meningitis tends to produce a gradual increase in temperature as the body fights off the infection, whereas lead encephalopathy may produce little or no fever, although temperatures do rise sharply just before death. Sixth-nerve palsy refers to facial paralysis, which may occur in both conditions, but appears more often in lead poisoning. Seasonality is the most accessible of these markers, as it can be gleaned from admission records. While TB meningitis and lead poisoning can occur any time in the year, summer is the peak season for lead poisoning because, it is now believed, lead that is stored in the bones most of the year gets drawn out with calcium, which is mobilized by increased levels of vitamin D, a product of the summer sun.[27] In contrast, TB meningitis has a much less pronounced seasonality and varies greatly from year to year and from place to place.[28]

Figure 2–2 is based on the TB meningitis cases admitted to Boston Children's Hospital between 1901 and 1919, compiled by age and month of death.[29] Disaggregating the data by age (Figure 2-2B) reveals a marked difference in susceptibility and seasonality for children under 3, who make up a little over half the cases. This age is also the group at highest risk for lead poisoning, for which the peak season is summer. The epidemiology of lead poisoning would suggest that few of the deaths occurring in the winter and spring were from lead, and tuberculosis is the most likely suspect in these cases. But the frequently reported confusion between the symptoms of tuberculosis and plumbism suggests that more than a few of the summertime cases probably originated in lead paint.

In 1915, physician Alfred Meyers examined the records of 105 TB meningitis cases reported at Boston Children's Hospital from 1910 to 1913.[30] Meyers was confident that the "diagnoses were unquestionable from the clinical picture and were made by competent men." His confidence was not diminished by the fact that tubercle bacilli were obtained in less than a quarter of the cases. Received clinical wisdom "proved" that these children were tubercular, despite the fact that in over 60 percent of the cases there was no pathological evidence of tuberculosis.

The evidence suggests that there was probably a good deal of lead poisoning around to talk about in the early twentieth century. How then do we explain the silence? First, a qualification: it is not quite accurate to say that American pediatricians "discovered" childhood lead poisoning in the 1910s. Isolated reports of pediatric plumbism appear in medical journals from the late nineteenth and into the early twentieth century. But in most cases the poisoning was seen as accidental, arising from unique and unpredictable conditions—food adulteration, contaminated water, lead-containing medications, or lead dust brought home in the clothes of the child's lead-worker parents.[31]

For example, in 1887 Philadelphia physician David Denison Stewart reported sixty-four cases of lead poisoning, eight of which were fatal. Most of the victims were children under 14 years of age; all of the deaths were attributable to eating buns colored with yellow lead chromate. Stewart tracked those cases to two bakers who replaced eggs in their batter with the inexpensive lead compound, sold as "extract of egg." Other physicians reported lead poisoning from tainted drinking water, canned or foil-wrapped foods, lead-based ointments on nursing mothers' breasts or faces, and lead nipple shields, or from sugar of lead taken as a quack medical remedy. Lead house paint was only implicated when children succumbed to vapors from wet paint or freshly painted walls. In most of these cases,

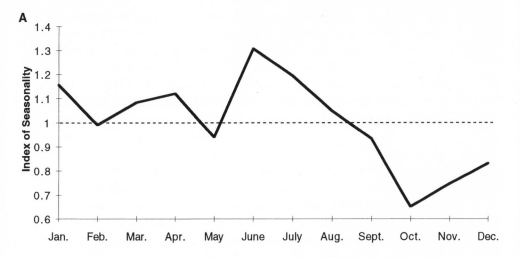

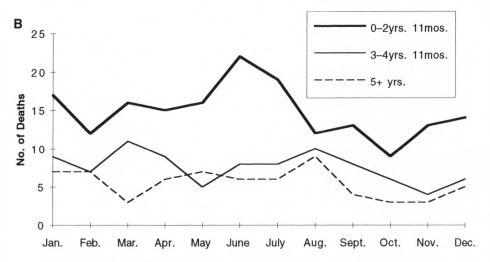

FIGURE 2-2

Fatal TB Meningitis, Boston Children's Hospital, 1901–19

A, seasonality index (all ages); B, number of deaths by age.

Admission logs 1869–1950, Boston Children's Hospital Archives, series 92–40.

the victims' symptoms probably arose from caustic chemicals in the paint, not the lead. In this period, no one reported cases of lead poisoning from breathing or eating lead-paint chips or dust.[32]

Childhood lead poisoning, then, was not seen as a public health issue as much as the result of parents' ignorance, the personal greed of corrupt

businessmen, or an inexplicable private tragedy. It seems that no one speculated that the most highly regarded and costliest paint product—white
lead, used as designed and manufactured—posed a serious public health
problem.[33] Indeed, childhood lead poisoning was not seen as much of a
problem at all. As late as 1911, lead poisoning did not even appear in the
index of the classic American pediatric reference of the time, Emmett
Holt's *The Diseases of Infancy and Childhood,* though the author does
mention that "the colic of lead and arsenic poisoning are both very rare
in children."[34]

Late in the second decade of the century, some physicians suspected
that they had underestimated the true magnitude of the problem. Baltimore
pediatrician Kenneth Blackfan suggested in 1917 that lead poisoning was a
probable cause of convulsions in cases where other sources were not clear.
Over the course of the 1920s, the scale of the problem grew increasingly
clear to physicians. Whereas in 1920, Robert Strong, clinical professor of
pediatrics at Tulane University, reported that "lead poisoning, so far as I
have been able to find in pediatric literature, does not seem to be common
in children," by 1926 Charles McKhann of Boston could declare, "Lead
poisoning is of relatively frequent occurrence in children."[35] Significantly,
by the time of McKhann's report, the primary source of pediatric plumbism
had been identified as pica, or a "perverted appetite, which causes the
ingestion of lead paint."[36]

When for their own reasons, American pediatricians began taking an
interest in lead poisoning, their literature searches turned up the Australian
experience but in no way did that research prompt the American.[37] So
why was the time right in the 1920s? What changes in American medicine,
or American culture, prompted the new interest in childhood lead poisoning?

At least three specific factors inhibited the discovery and study of pediatric plumbism in America before the 1920s: theories about the origins
of disease that emphasized germs and contagion, the dominance of biotic
diseases as a vector of childhood sickness, and the failure to study children's diseases apart from those affecting the general public. All three were
gradually eroded by long-term changes in medical theory, improvements
in nutrition and sanitation, and the enhanced valuation of children's lives.
As these inhibitory factors diminished, the necessary conditions developed to make pediatric lead poisoning visible, and interesting, to physicians.

It might reasonably be argued that childhood lead poisoning went
largely undetected before the 1920s because most physicians believed that

the germ theory, the dominant paradigm of disease causation of the time, provided an explanation for the symptoms they found in almost every case. This devotion to the germ theory explains why the diagnosis of tuberculosis may have often masked lead poisoning, an illness due to an environmental toxin. The dead hand of medical authority—even from nearly a century ago—can easily daunt the historical investigator. For example, pathologists' reports in Boston's Children's Hospital autopsies of patients diagnosed with TB meningitis almost always revealed the presence of TB, "proof" of a correct diagnosis. Such clear evidence might make all speculation as to the prevalence of misdiagnosis capricious, except that exposure to TB bacilli was nearly endemic in 1900. The correct pathologic observation—that is, the presence of TB—therefore does not prove the etiology of the brain damage that killed the patient.[38]

The ubiquity of endemic and epidemic germ-based diseases gave physicians good reason to think in terms of contagion. The shattering rates of infant mortality that persisted into the twentieth century stemmed in large part from infectious diseases.[39] The perception of great success with certain vaccines and pasteurization led physicians and public health officials to rely on further discoveries in microbiology to cure still more diseases. Lead poisoning killed perhaps three hundred children a year—a pittance, next to the toll of contagious disease. Environmental diseases might naturally be expected to get lost in the background.

It is now commonly understood that, by the end of the nineteenth century, the germ theory supplanted the previously dominant conceptions of disease origin. Before the bacteriologists' ascendancy, the public health community was divided over whether the best path to improved health lay in miasmic or contagion theory. The former can be described as proto-environmental, arguing that disease sprang spontaneously from filth and still air and advocating improvements in water supply, ventilation, and sewerage. Contagion theory—a forerunner of germ theory but largely confined to causation by direct human contact—urged quarantine and education in personal hygiene.[40]

Both nineteenth-century theories were put to work in the service of the new paradigm of germ theory. The changes in public health strategies that contributed to the reduction of mortality, especially in the cities, can easily be seen as a pragmatic employment of elements of each. Miasmists could cite the presence of germs in a city's water to increase the likelihood that more effective filtration or new sewage disposal techniques would be employed; contagionists' causes were certainly enhanced by the discovery of communicable germ-based diseases. What got lost in the mix that came to be known as the "New Public Health" were the nonbiotic, environmental disease agents.[41]

To see how this confluence inhibited the discovery of lead poisoning, a bit of counterfactual speculation is useful. Suppose that in 1900 the miasmists and not the germ theorists had carried the day—that the New Public Health, with its emphasis on laboratories and magic bullets, had not arrived. Paul Starr describes the New Public Health's "modernization of dirt." The miasmists had a "broad conception of dirt" that implied "a need for a correspondingly large investment in cleaning things up."[42] In this counterfactual America, the anticontagionists might have broadened the conception of dirt to include environmental dusts, peeling paint, or poisonous substances used as paint, and thus might have taken action against lead. In such a polity, federal laws banning lead paint might indeed have been enacted with the same force as the Pure Food and Drugs Act, instead of languishing in committee.[43] And as it turns out, a resurgence of environmental thinking—namely, a broadened conception of dirt—was required before pediatric lead poisoning would become visible.

But besides the prevalence of biotic diseases and the ideological blinders that followed from the successful application of the germ theory against them, an additional factor inhibited an increased awareness of lead poisoning as an age-specific disease of early childhood: the late development and relative immaturity of the American pediatric establishment. The previous century had seen the start of a profound transformation in the valuation of children and an appreciation of childhood as a unique phase of life. Beginning in the middle classes but gradually attaining near universality, children's place in the family changed—from a form of insurance against their parents' old age to the status of beloved but "economically worthless" treasures.[44] Despite this increased valuation, medical and public health practitioners were slow to realize to what degree children had their own unique ways of bearing diseases. Differential rates of change in mortality may have played a significant role in highlighting these differences. By the second half of the nineteenth century, overall mortality rates had dropped below 25 deaths per thousand, but childhood rates had merely inched downward (see Figure 2-3, which illustrates the figures for Massachusetts). Each passing year underscored the need to improve infant mortality, at least among those who had stopped seeing childhood diseases as nature's way of culling the weak.[45]

In its formative years, American pediatrics had as its primary goal the reduction of infant mortality, and the germ theory was its chief weapon. Richard Meckel argues that pediatricians' training, as well as their need for

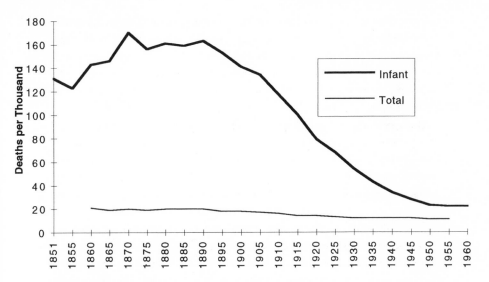

FIGURE 2-3
Infant Mortality versus Total Mortality in Massachusetts, 1851–1960
Historical Statistics of the United States, Series B148, "Infant Mortality Rate for Massachusetts: 1851
to 1970," 57; Series B193–200, "Death Rate, Massachusetts: 1860 to 1970," 63.

improved professional standing, allied them with germ theorists in pushing
for clean milk campaigns and infant feeding stations. Transforming infant
feeding into a "problem" to be handled by medical professionals allowed
pediatricians to garner authority by claiming expertise in what had formerly been chiefly a women's concern. Ironically, though, American pediatricians' employment of the germ theory inhibited the proper diagnosis of
poisoning from environmental toxins.[46]

Infant mortality rates plummeted in the first decades of the century,
although the contribution made by pediatricians and their milk stations is
debatable. Whatever the cause, decreased infant mortality was a necessary
precondition for the discovery of lead poisoning, as it left in place a healthier infant population less plagued by diarrheal diseases and tuberculosis.[47]
This reduction in the epidemiological background noise, together with the
establishment of an interventionist pediatrics industry, left only one major
ingredient to be added: an increased awareness of environmental poisons
in general and of lead in particular.

The Progressive Era fights for improved occupational hygiene, led by organizations such as the American Association for Labor Legislation and
by researchers such as Alice Hamilton and David Edsall, provided that

increased awareness. This first generation of occupational hygienists made dusty factories their laboratories and seldom apologized for prescribing drastic changes in working conditions on the basis of the most meager statistical and clinical data. Their highly publicized reports brought to the public an awareness of the effectiveness, in terms of both health and cost, of environmental remedies.[48]

But since occupational plumbism dominated the medical literature, the occupational symptoms of lead poisoning became normative. Progressive Era pediatricians would cite Tanquerel des Planches, the nineteenth-century pioneer in the study of occupational lead poisoning; in later decades, the voice of authority would be Robert Kehoe, an occupational hygienist in the employ of a large lead-using industry.[49] As long as pediatricians sought clinical signs of adult plumbism in their juvenile patients, they were bound to misdiagnose. For example, a reliable indication of high lead exposure in adults was the lead line, but lead lines appear infrequently in children—or in lead-poisoned adults with good oral hygiene.

The construction of childhood lead poisoning as a public health issue in the United States awaited three developments: a mature subdiscipline of pediatrics, greater sensitivity to environmental disease causation, and a reason to single out the particular toxin lead. The presence of these elements in Australia goes far in explaining why pediatricians there found lead paint-poisoning in abundance more than two decades before American pediatricians began studying it.

Both Gibson and Turner specialized in children's diseases from the late 1880s, when the American Pediatric Society was in its infancy. Both were regular contributors to the *Australian Medical Gazette,* where they reported studies of age-specific diseases. It is also relevant that a city as small as Brisbane in 1890 had a children's hospital. In addition to training and research experience in pediatrics, the Australian researchers exhibited a heightened awareness of environmental factors, a consequence of Brisbane's size and its status as capital to a frontier region. The significance of the frontier in Australian pediatric toxicology lies in the fact that in a town with few of the public health problems common to big cities, the annual outbreaks of debilitating pediatric lead poisoning stood out in sharp relief against a background of general salubrity.[50] Finally, there are two reasons to conclude that Turner and Gibson were familiar with occupational lead toxicology. First, silver-lead mining and smelting had been a growing industry in Queensland since 1870.[51] Second, Australian physicians would have been aware of the growing controversies in British occupational hygiene that had resulted in a flurry of research and regulation since the 1890s.

The prerequisites for the discovery of childhood lead-poisoning's epidemic nature were met in the United States by the 1920s. And indeed, over the next twenty years, isolated enclaves of interested researchers would continue to enhance knowledge about the disease's epidemiology and etiology. But until the 1970s, the medical and public health communities' responses remained sluggish, if not moribund. Much of this inertia came from the second redefinition of childhood plumbism in as many decades.

When John Ruddock wrote of the ubiquity of lead in the child's environment in 1924, he made no class or racial distinctions: "The child lives in a lead world," he wrote, not "the *poor* child," not "the *black* child."[52] But increasingly (partly for sound epidemiological reasons having to do with nutrition and housing conditions and partly to do with researchers' expectations and society's needs) lead poisoning came to be a ghettoized disease, in two senses. First, it was ghettoized in that those who studied it—occupational hygienists for the most part—stood outside the medical mainstream with its focus on polio, tuberculosis, and other biotic diseases; but, second, it was also a ghetto disease in that the ghettos of the inner cities would be the place where its victims were to be found. A line would be drawn separating the general public and those defined as "at risk" for lead poisoning: poor—especially poor *black*—children, as well as lead workers.

With the line drawn and those "at risk" securely within its brackets, America would feel free to continue its lead gluttony. Ironically, given the eventual association between lead paint and poor housing, the most popular lead product at the beginning of the century was white lead paint, a premium quality product of its time.

THREE

Toxic Purity

HOW THE UNITED STATES BECAME A NATION OF WHITE-LEADERS

The witness, Marion E. Rhodes of Missouri, had warmed to his subject. Committee members' constitutional objections could not dissuade him or tame his proselytizer's rhetoric. Science had proven his case; he had no use for statutory quibbling. "The most eminent scientists and doctors of Great Britain," he proclaimed, "reached the conclusion that white lead is a poison. I say it is a poison, it matters not in what form you approach it. . . . The small particles that result from chalking . . . when taken by inhalation into the lungs, are absorbed and become a poison to the system."[1]

There is no question of health involved in dealing with paint.

WILLIAM LUCAS OF LUCAS PAINTS, TESTIMONY BEFORE THE U.S. HOUSE COMMITTEE ON INTERSTATE AND FOREIGN COMMERCE

There would be nothing remarkable about Rhodes's testimony had it been given in one of the dozens of congressional committee hearings on lead poisoning held since 1970, when the federal government first sought a legislative remedy to the problem of lead paint.[2] His apocalyptic tone, his reliance upon scientific expertise, and his environmental sensibility would have fit right in. But Rhodes's testimony comes from a 1910 House of Representatives committee hearing on a bill that, had it been enacted, might have kept tons of lead paint off American walls and prevented thousands of cases of lead poisoning over the course of the century.

The Sixty-first Congress chose not to regulate lead paints. Doing so would have required settling issues far more ambiguous than those addressed by the complex rules set down in the Pure Food and Drugs Act, enacted only four years earlier. Distinguishing between butter and margarine or "rectified" and "straight" whiskey were simple tasks compared to defining "pure" paint. The public had demanded food regulation because they were enraged by horror stories of tainted and unhealthful foods; paint simply did not arouse the public's fervor. In fact, as it turns out, most Americans in 1910 agreed that whatever else went into it, the best paint contained lead.

The National Lead Company built upon this understanding in a 1910 advertising campaign that urged painters who used National Lead products to call themselves "white-leaders."

> *"I am a white-leader," said the painter.*
> *"What's a white-leader?" asked the property-owner.*
> *"A white-leader," replied the painter, "is first of all a man who believes in and uses pure white lead as a paint pigment.*
> *"A painter who uses white lead, a dealer who pushes white lead, a property-owner who specifies that white lead shall be used on his building— they are all entitled to be called white-leaders."*
> *"Why, I am a white-leader, then," said the property-owner.*[3]

The United States in 1910 was a nation of "white-leaders," upholding a toxic standard of purity that would stand for several decades, long after most European countries had begun switching to safer pigments.

This chapter offers an account of America's fatal attraction to lead paint. From the origins of the American paint industry to the Progressive Era, lead pigment manufacturers balanced a complex and uneasy set of relationships with the manufacturers of other pigments and paints. Makers of "mixed" paint tinkered and experimented, formulating the paints they hoped would guarantee them a lucrative place in the industrializing nation's expanding paint market. The public condemnations leveled by lead pigment manufacturers and mixed-paint makers against one another belied the symbiotic relationship they carried on in private, and they bought and sold each other's products as market and technical demands required. Matters came to a head in the times of the pure food and drugs movement with its strident antitrust rhetoric, but the paint industry, acting as a partnership among competing interests, muted its divisive voices in favor of cooperation and unified opposition to or coopting of state and federal regulation. The nation is still paying the cost of this peace.

From colonial days, Americans demanded white lead for paints, enamel, cosmetics, and medicines, and its manufacture, a process of "corroding" metallic lead in acidic vapors, was one of the first chemical industries in the United States. America's first large-scale producer of white lead was the Philadelphia chemical firm of Samuel Wetherill & Sons, which began corroding lead in 1804. Before 1812, most white lead consumed in the United States was imported from Britain, but exclusionary trading practices after the War of 1812 made domestic white lead manufacture profitable; with the lead mines acquired in the Louisiana Purchase, the industry soared. As more American production plants were built, domestic supplies grew, costs fell, and demand skyrocketed.[4]

In 1887 most of the largest white lead corroders combined to form the National Lead Trust, protecting themselves, paint industry spokesman George B. Heckel wryly explained, from competition that "had become so keen and its debasing consequences so disturbing to the conscience of moral-minded business men." The trust, headed by Colonel William P. Thompson, soon included all but three of the nation's largest white lead producers. In 1891, in the wake of the Sherman Anti-Trust Act, the National Lead Trust re-created itself as the National Lead Company, with Thompson as its first president. In 1907, remnants of the days of less formal association could still be seen in the nearly two dozen "varieties" of paint NLC affiliates sold, often in competition with one another. This ended when Orlando C. Harn, NLC's new advertising head (who had also masterminded the Heinz 57 Varieties campaign), conceived the familiar Dutch Boy logo, under which all NLC paints would be sold (Figure 3-1).[5]

White lead paint dominated the market in the early twentieth century because of its covering power and the lead industry's economic strength. Prior to the 1940s, when titanium dioxide pigments became standard, lead-based paint's superiority went largely unchallenged in the United States. Competitors vying for a share of the paint market had to contend with the highly organized and politically savvy lead industry. White lead was that industry's most important product: from 1900 to 1910, white lead pigments accounted for 30 to 40 percent of the United States' lead consumption, and they remained the largest single use of the metal until well into the 1920s, when lead for automotive batteries came to dominate.[6]

Of course, white lead companies were not the only paint producers, white lead was not the only paint base, and lead was not the only mineral being mined. As one paint manufacturer wrote in 1910, "there are many Pigments beside White Lead that are used to-day in making good paint, and for many purposes superior to white lead Paints."[7] Other materials and the companies that manufactured them frequently challenged lead's domination of the paint industry. The first commercial zinc oxide paints were developed in France during the 1840s. In the United States, Philadel-

FIGURE 3-1
The Dutch Boy Painter, Trademark for the National Lead Company
The Dutch Boy's portrait was painted by Lawrence Carmichael Earle, whom the
National Lead Company commissioned in 1907.
Warshaw Collection of Business Americana, Archives Center, National Museum of American History.

phia's Wetherill and Jones patented a process for producing zinc oxide in 1855. In 1895, several zinc manufacturers merged to form the New Jersey Zinc Company. By 1900, this combination had absorbed almost all zinc producers, making it, in the words of its publicist, George Heckel, "to all intents and purposes, a monopoly." And with its monopoly power, the company could hold the price of American zinc oxide at half a cent per pound less than the price of white lead.[8] Zinc was an especially strong competitor in the paint market at the beginning of the century: from 1900 to 1910 metallic zinc production increased almost three times as fast as lead production.[9]

A number of other mineral and chemical interests competed for respectability and market share in the late nineteenth and early twentieth centuries. Painters and paint manufacturers experimented with zinc oxide, zinc sulphide, barytes, and carbonates with mixed but largely encouraging success, producing a range of new products—some cheaper, some whiter than white lead, and some, they claimed, every bit as durable. All had their champions, who tended to ally with mixed paint and pigment manufacturers, those consumers of raw pigments and paint components of every variety.

The paint industry in 1910 was undergoing a profound transformation involving tremendous growth (both in number of producers and value of products), increasing concentration of ownership, and a deepening rift between the "modern" mixed-paint industry and the traditional and powerfully allied lead corroders over the very definition of their product. The value of paints and varnishes produced in the United States rose by 80 percent between 1899 and 1909, while the value of goods and services in the industry increased by almost 50 percent.[10] The decades-old trend toward concentration of ownership persisted, although the technical and improvisatory nature of the industry permitted a steady accretion of new businesses, so that both competition and a plurality of interests were maintained. The number of paint and varnish manufacturers grew by 30 percent from 1904 to 1919, while the value of goods produced by these firms grew by 400 percent (see Table 1). By 1919, corporate-owned companies accounted for 76 percent of manufacturers, who made 94 percent of the nation's paints and varnishes.[11]

At the beginning of the century, the structure of ownership and corporate management of the lead industry—and its huge scale of production—exemplified the new industrial age. But within white lead plants, the means of production, though varying widely, for the most part continued to reflect the industry's ancient origins. When health investigators entered Illinois's

TABLE 1 *Ownership and Value of Products of Paint and Varnish Manufacturers, 1904-19*

	1904	*1909*	*1914*	*1919*
Number of establishments	639	791	800	830
Number corporate owned (%)	360 (56.34)	526 (66.50)	558 (69.75)	628 (75.66)
Value of products (in millions of dollars)	90.84	124.89	145.62	340.35
Value of corporate products (in millions of dollars) (%)	75.47 (83.08)	106.35 (85.15)	133.24 (91.50)	320.83 (94.26)

SOURCE: Bureau of the Census, *Historical Statistics of the United States, Colonial Times to 1970,* Part I (Washington: GPO, 1975), 224.

white lead factories in 1910, the production methods they found in most—even those that were new—had changed little in a hundred years.[12] Anyone with passing knowledge of the paint industry knew that making and using lead pigments was potentially deadly. At every stage and with every process—from transforming metallic lead into a bucket of paint, to applying that paint to a wall or to sanding it away—workers were sickened, workers died.

The process of making the "pure" lead paint that white-leaders demanded began cleanly enough, as workmen unloaded hundred-pound ingots, or "pigs," of dull, bluish-gray metallic lead from barges, boxcars, or trucks onto drays or wheelbarrows to be hauled to the smelter.[13] Paid less than skilled workers in the factory, the largely unskilled laborers whose only contact with lead was in its metallic state were less likely to succumb to severe sickness, although many contracted the milder or chronic forms of plumbism. Only a few months after leaving farm labor in his native Poland, 22-year-old Thomas O———was working on the docks of a New York white lead company unloading pigs. Despite his limited exposure to the more hazardous processes within the plant, three months after he started work he complained of stomach pains and appetite loss, for which the company doctor prescribed a nostrum.[14]

At the smelter, the pigs were melted and recast in the shape of large, thin, belt buckles. Smelter workers faced considerable risk, especially in factories with inadequate ventilation. Those who possessed a variety of marketable skills arranged to "alternate lead and non-lead employments" on a seasonal basis, but even this could not always prevent their acquiring chronic lead poisoning. Tonio M————, a Polish carpenter, worked as a "caster" in a New York white lead factory. When he first became incapacitated with lead colic, he left to ply his former craft in Connecticut. Poor pay and the realities of seasonal employment pushed him back to the paint factory, where he soon contracted partial paralysis and colic and was forced once again to leave.[15]

After smelting began "corroding" of the pure lead buckles with acid, in the presence of carbon dioxide. Most plants in the United States used the "Dutch" or "stack" process developed in the seventeenth century. Typically, the corrosion of "blue" metallic lead began in a small room or shed. Working in the "blue beds" was considered less hazardous than most jobs in the paint factory, and one could find native-born Americans doing this labor. In many European paint factories, women, banned by law from the most hazardous jobs, were permitted to build blue stacks. In the United States, custom—not law—forbade their employment in any lead-using departments.[16]

First, workers covered the floor with spent tanbark brought from tanneries (or sometimes with a thick layer of manure).[17] In another room, members of the "setting crew" stacked the lead buckles inside specially designed ceramic pots, with small reservoirs in their bases containing dilute acetic acid, usually vinegar. Once they had covered the floor with a layer of buckle-filled pots, the stackers laid boards over the assemblage, forming a platform for the next layer of tannin and pots of lead. Layer upon layer was set down until the stack nearly touched the ceiling. After six to fourteen weeks, the chemical action was complete, transmuting the "blue bed" into a "white bed." The pots now overflowed with frothy ceruse. The frosting of white lead encrusting the buckles often grew so thick that the vessels were cracked.[18]

Disassembling the white beds began the riskier phase of lead production and was largely carried out by men who, if they spoke at all while they worked, did so in Polish, Czech, or Italian. In the worst factories, workers dumped the contents of the pots onto separating tables where they hand-scraped and pounded the flaky lead carbonate from what remained of the buckles (see Figure 3-2). The more modern plants often employed mechanical separators, although the workers who fed these machines still stood in clouds of toxic dust with no protection beyond the bandanna they might wear over their mouths if the day were not too hot.[19] Other workers

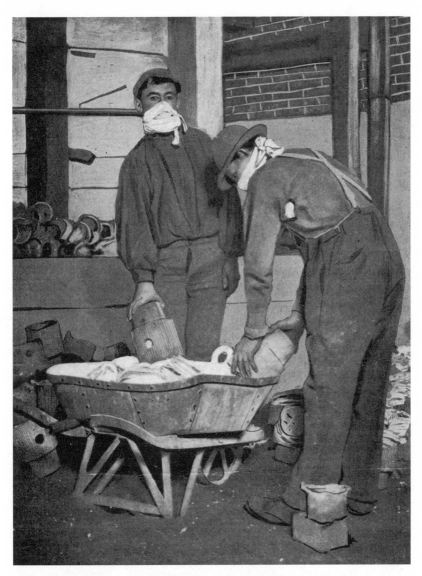

FIGURE 3-2
The Dutch or "Stack" Process for Producing White Lead
Workers remove white lead in fine powder form from ceramic pots.
American Labor Legislation Review 2 (June 12): plate facing 368.

recovered the oxide from the crockery or swept it from the strips of tannin, the boards that separated the stacks, and the floor. The corroded buckles were returned to the blue beds or the smelter, depending upon how much metallic lead remained.

After the "stripping" process, laborers washed the white powder to separate any tannin and particles of blue lead and shoveled the wet white lead into crocks to be dried in ovens. The fine, deadly, powdered ceruse was then dumped into wooden casks for shipment or mixed with linseed oil to make paint. Both processes were dusty and dangerous. Aleck S———, a 31-year-old Russian Pole who had been in the United States less than a year, contracted severe colic, cramps, and partial paralysis in his wrists and arms after only ten months of packing white lead; Nathan G———, Polish and a cobbler by trade, worked for three months in a white lead plant as a "mixer," blending white lead powder with oil, until he was hospitalized with cramps, colic, vomiting, and loss of appetite.[20]

The Pullman Palace Car Company purchased hundreds of these casks to make paints for their train carriages. In Pullman's paint department, unskilled workers pried off lids, scooped powder into buckets, and blended linseed oil, pigments, and turpentine. Inside the train cars—within a cloud of turpentine vapors and dust—workmen, "some of whom had never held a paint brush before," spread the paint on ceilings and in tiny dressing rooms. Meanwhile, experienced painters brushed the oily suspension on the cars' exteriors in a series of thin coats. These workers almost never undertook the most hazardous job in the shop—"rubbing down," or sanding each coat in preparation for the next. At this work, factory investigator Alice Hamilton found "an Italian of twenty years of age, two Lithuanian brothers, a Greek, and two Russians, none of whom were painters by trade." Neither the company nor their skilled coworkers had informed them of the dangers, despite the fact that the plant's medical department, "which consisted of one old doctor who had been a surgeon in the Civil War," frequently sent lead-poisoned painters to the Cook County Hospital, sometimes within the first month of their employment.[21]

Thousands of barrels went to skilled house painters, who preferred lead paint despite the risks they knew it entailed. With house painting, as in industrial paint shops such as Pullman's, the dusty job of opening the casks of white lead and mixing the powder with oil usually fell to painters' helpers, as did sandpapering. Weather permitting, skilled painters preferred outside jobs, because they knew the fresh air reduced their chances of contracting "painter's colic"—a common name for the gastrointestinal form of plumbism. Others tried to vary the kind of work they did, alternat-

ing between painting and varnishing jobs. When Russian-born house-
painter Morris R——— was younger, he took only outside work, hoping
to reduce his exposure to lead. But at age 47 in 1911, he was restricted to
inside work because vision problems that gave him "a certain droop to the
eyelids" prevented him from working on scaffolds. He began to be plagued
with bouts of colic and obstinate constipation, the only periods of relief
coming when there was no work.[22] That painters consistently eclipsed
workers in other trades in statistical reports of lead poisoning reveals how
often their strategies for self-preservation failed.

Once on the wall of a house or apartment, the paint need cause no
serious sickness for decades—unless, of course, a young child occupying
the room nibbled at it. Eventually time, the dictates of fashion, or public
health mandates would bring painters back with scrapers and sandpaper.
Skilled painters would put their young apprentices to work removing the
weathered paint. How often were they visited by neighborhood children,
curious about the workmen, fascinated with the appearance of their familiar surroundings now shrouded in drop cloths and coated with a fine white
dust where they could leave footprints and that swirled in clouds when
they ran?

Despite the known hazards, there can be no denying how American painters loved lead paint. John Dewar, master painter and executive officer in
the International Association of Master House Painters and Decorators,
was an indefatigable white-leader. Testifying in support of a 1910 law that
would mandate ingredient labels on any paint other than "pure" white
lead, Dewar invoked the Pure Food and Drugs Act: "These, gentlemen,
are the days when the art and science of adulteration is rampant all over
this country. In the days of the pure food and drug exposure people stood
back and gasped at the results produced. I want to tell you that in the paint
industry of this day . . . the same adulteration, the same mislabeling, are
rampant, and we have the proof of it."[23]

The Progressive Era debates over defining the best paints focused less
on health than on the definition of paint, the freedom to experiment, and
the obligation either to respect tradition or to abandon old standbys for
new, improved standards—in short, on the future of the industry. Neither
side was particularly consistent. Lead companies stressed their role as pur-
veyors of ancient, time-honored traditions in paint technology (the words
"Old Dutch Process" appeared in the NLC logo and Sherwin-Williams
advertisements) but sold a substantial portion of their product to those
who manufactured mixed paints.[24] Mixed-paint manufacturers by defini-

tion turned their backs on tradition, experimenting with various minerals and hiring chemists in unprecedented numbers to concoct an improved, or cheaper, product. Many, however, passed their new-fangled mixed paints off as old-fashioned "pure" lead. This deception was the bone of contention that legislators were asked to resolve.

The campaign for legislation against "adulterated" paint ignited at the state level around 1900 and fizzled out at the federal level in 1910. Nebraska passed the first pro-lead paint-labeling law in 1902. Each year for several years thereafter, the lead corroders introduced a flurry of bills in state legislatures. Members of the master painters' association, led by John Dewar, wrote letters of support, testifying about their costly experiences with adulterated paints.[25] But the Paint Manufacturers' Association, representing the mixed-paint trade, maintained a network of paid informants at most state capitals. As soon as a bill was proposed, it deluged the appropriate committee chairman with correspondence decrying formula labeling as ineffective, and advocating the association's alternative "model bill" that required no label but punished manufacturers of fraudulently labeled paints. So successful was this intelligence campaign that, according to the association's secretary, "until North Dakota stepped into the limelight not a single paint law in any State got past the Legislature."[26]

The Paint Manufacturers' Association had no spy in Bismarck; after all, North Dakota had no paint manufacturing companies to speak of, nor any large mineral interests. The state did have an aggressively Progressive legislature, though, and it also had Dr. Edwin F. Ladd, of the State Agricultural College, who had played an important role in passing the federal Pure Food and Drugs Act.[27] The bill Ladd drafted for North Dakota defined "unadulterated" paints as those containing only lead, zinc (the two are frequently commingled in ores), "pure linseed oil," "japan drier," and "pure colors." Paints containing any other ingredients had to bear labels disclosing what "adulterants" had been used, and in what proportion.[28] North Dakota's law withstood challenges brought by the Paint Manufacturers' Association, which took the case to the federal Supreme Court.[29]

Following North Dakota's example, Minnesota, South Dakota, and Iowa passed similar laws in 1907. Ladd freely distributed his inflammatory "bulletin no. 70," a sort of *The Jungle* for the paint industry, in which the blatant mislabeling of paint products sold as pure lead was revealed. North Dakota analysts had found, for example, that Climax White Lead contained absolutely no lead, nor did Wier's (Improved) Bavarian White Lead; Anti-trust White Lead had less than 15 percent.[30]

The Paint Manufacturers' Association's network of spies was no longer able to control this prairie fire of state-level legislation. A number of mixed-paint manufacturers realized that one moderate federal labeling law might well be superior to a hodgepodge of state laws and bring to a halt the internecine battles that were making the paint industry a joke among businessmen.[31] A brief alliance between manufacturers' and painters' groups produced a federal bill, similar in language to the Pure Food and Drugs Act of 1906, to require paint-labeling with detailed content analysis.[32]

When Weldon Heyburn of Idaho introduced the bill in the Senate in 1907, retailers and painters from around the country wrote letters and came to Washington to support it.[33] Retailers generally supported paint-labeling laws, because regardless of whether they were aware that the paints their suppliers provided were mislabeled or "adulterated," it was they who dealt with angry customers. The Retail Hardware Association of the Carolinas declared that "prepared Paint is subject to the most dishonest adulterations in its manufacture, and detection of its adulteration is almost impossible to the ordinary consumer. . . . As it is now he often gets water for oil, and cheap minerals for lead, as the maker can name anything that can be spread on with a brush 'pure paint' regardless of its composition."[34]

Of all the so-called adulterations of paint products, none raised the ire of painters as much as the substitution of "cheap minerals for lead."[35] John Dewar complained that the prospective purchaser of white lead paint—professional or homeowner alike—was "buying a pig in a poke; that he is swapping, as it were, his jackknife sight unseen. . . . It is only when his paint washes off the building or cracks with fissures that resemble that of the alligator skin that he knows he has been defeated in his attempt to produce in his work the best results possible."[36] Apparently, few paint companies in 1910 actually sold pure white lead. Bureau of Chemistry analysis of paints purchased in the District of Columbia nearly replicated North Dakota's earlier findings.[37]

The bill's opponents focused on the public's right to purchase less expensive mixed paints and on the potential hardships involved in formula labeling. William Lucas, president of the Paint Manufacturers' Association (which came to oppose the bill it helped frame), asked whether a farmer should have to use premium paint to cover his barn or chicken shed. Some manufacturers worried that labeling would compel them "to disclose everything for the benefit of their competitors."[38] Others objected to the costs of analyzing their products—and to the possibility that mistaken analyses or changes in a paint's composition during its shelf-life could result in prosecution.[39]

Many of the larger firms were already labeling their paints by 1910, primarily as a response to state laws, but also as a measure to forestall further legislation. Manufacturers sent copies of the formula labels their products proudly bore.[40] One mixed-paint maker boasted that quality manufacturers had no fear of labeling: "Can any piano maker produce a Steinway? No! But all piano makers know the materials in the Steinway." Baryte producer William A. Buddecke was chauvinistic about his product and thought that paint manufacturers should be proud of what they put in their cans.[41]

Proud souls like Buddecke aside, most mixed-paint producers opposed the Heyburn bill and attacked its supporters' conspiratorial and monopolistic intentions. Bade Brothers of New York wrote that the bill was "playing right into the hands of the Lead and Oil Trusts." The D. T. Wier White Lead Company of St. Louis agreed. When Wier's premium product, Bavarian White Lead, was found by North Dakota's chemists to contain no lead, the company pulled its products out of North Dakota, complaining that "we can not see that the law in question has benefited anyone except the two trusts who control two pigments [lead and zinc] specified by the law."[42]

Some paint manufacturers implicated the country's largest painters' union in the lead trust conspiracy. A. Burdsal, an Indianapolis paint manufacturer and member of the Paint Grinders' Association, argued that the master painters' association and the paint grinders (who favored access to a wide variety of pigment bases) were locked in "an irrepressible conflict" over mixed paints, the paint grinders' chief product. Master painters ground and mixed their own pigments and preferred lead compounds. Mixed paints, Burdsal claimed, took much of the science and mystery out of house painting, allowing "hundreds of thousands of men all over the United States . . . to follow the profession of painting, and compete with [the master painters] in their business."[43]

The two sides debating paint-labeling legislation were most clearly defined by their relations to competing raw materials—lead, zinc, barytes, and mineral pigments. The white lead manufacturers and the master painters stood as conservators of a long-standing tradition of quality. The mixed-paint manufacturers, most of whose products contained some white lead, wanted to preserve the freedom to formulate new, better, or cheaper paints. Never did the debate over the Heyburn bill concern the toxicity of lead paint. Indeed, the bill's staunchest critics questioned only the economic power of the lead corroders, not the threat to health their product posed.

Missouri Representative Richard Bartholdt's bill, introduced in 1910 "in the interest of the public health," put lead poisoning at the center of the dispute.[44] The House never seriously considered his bill, and he never introduced similar legislation. The testimony presented on the last day of May 1910, however, included the strongest warnings about lead-paint poisoning that Congress would hear for the next sixty years.

Bartholdt introduced his legislation just days after the Heyburn hearings ended, and although much of its language was borrowed from the Senate bill, Bartholdt's bill went well beyond it—indeed, beyond the Pure Food and Drugs Act—by calling for federal intervention in regulating the manufacture, sale, and use of any paint containing white lead.[45] In addition to content labeling, Bartholdt wanted all lead-containing paints to be "labeled with a skull and crossbones and the words 'Poison; white lead.'" The bill spelled out detailed rules for the manufacture and application of lead paints. Employers would have to provide painters with protective clothing and printed instructions warning them of the dangers of using lead paint and would be required to post warning signs wherever such paint was being applied. And under the new law, the federal government would prohibit painters from drinking on the job.

Painters Magazine immediately labeled the bill a "piece of freak legislation," both for its scope and because of its intent to reduce lead poisoning. The magazine reminded painters that those who followed simple precautions when using lead paint "need have no fear of lead poisoning."[46] Painters knew that lead poisoning was a real danger; after all, one of the disease's most frequent manifestations was called "painter's colic." Just the same, the magazine argued, "every one familiar with American journeymen painters knows that they are, as a rule, as healthy as mechanics in other branches of the building trades."[47] The painter could avoid lead poisoning "if he leaves alcohol, or anything sour out of his stomach, at least during working hours, and washes his hands with water acidulated with sulfuric acid after working time, before using soap for cleanliness. The greatest trouble is indulgence in liquor drinking."[48] Such complacence pervaded the lead-using trades of America, squelching any chance for strict government regulation.

This was not the case in Europe, however, where many of the bill's key provisions were already in place. Bartholdt lifted the occupational hygiene provisions from German laws enacted in 1905.[49] He might have looked to France for even heavier restrictions: where German law sought only to make the manufacture and use of lead paint safer, France led its neighbors in partly banning white lead for interior uses. In 1909, after prolonged efforts begun in the mid–nineteenth century, the French Chamber of Deputies declared that in five years, lead paints "shall be forbidden in all painting of whatsoever nature it may be . . . as well upon the exterior as upon the interior of buildings."[50] Lead was still widely available in Europe, but government regulations made its use safer for those who manufactured it and those who applied it. In the United States, no federal legislation mandated occupational hygiene, let alone the kind of intervention called for in Bartholdt's bill.

Bartholdt's chief supporter was Marion E. Rhodes, a state representative from Missouri who had tried to pass similar legislation. Although he repeatedly asserted that Bartholdt's bill did "not seek to prohibit the use of white lead in this country," Rhodes waxed evangelical when he described the dangers of lead paint and the courageous nature of the legislation enacted by European nations. He presented several reports of occupational lead poisoning from hospitals and health departments in St. Louis, Boston, and New York.[51] "You will observe, gentlemen," he pointed out while passing around a small box of white powder purchased from a druggist in Washington, D.C., "that in that receptacle, this article which is designated 'carbonated lead' appears under the skull and crossbones and a poison label." To buy the white lead, Rhodes had had to give the druggist his name and residence, and declare the purpose for which it was to be used. If the lead in the small box required a poison label, Rhodes reasoned, then "a hogshead, a carload, or a house full of the same substance would likewise be poison."[52]

Rhodes argued that lead paint was a public health hazard as well. In language as strong as any uttered by environmentalists in lead-paint poisoning hearings in the 1970s, he claimed that toxic dust, flaking from doors and windows, was filling the air in the legislative chamber. Rhodes warned the committee that if "you closed down every window, hermetically sealed every aperture through which air enters, we would finally die."[53] Bartholdt and Rhodes offered no empirical evidence of lead's threat to the public, but then, the bill's primary aim was occupational hygiene, not public health.

Bartholdt mustered little support for his bill. John Dewar and his International Association of Master House Painters and Decorators did not support it, despite its provisions for safer working conditions and a section that would have allowed only "experienced or union men" to use white lead. Dewar, in fact, would not back a bill that restricted the use of white lead in any way.[54] No paint manufacturers testified in the hearings on the bill, and the House record includes only one letter of support from a paint manufacturer, H. V. Kent, president of the Kent and Purdy Paint Company ("Makers of Good Paint"). Kent, who like many of Bartholdt's supporters was from Missouri, wrote that this bill was superior to others, which were blatantly "in the interest of the White Lead Trust."[55]

Bartholdt tried to enlist Dr. Harvey Wiley, head of the Department of Agriculture's Bureau of Chemistry, who had been instrumental in passing and administering the Pure Food and Drugs Act. Bartholdt chose the wrong ally. Wiley's definitions of "purity" often favored tradition over chemistry: white lead in oil was "pure"; paints containing zinc or barytes were "adulterated." The bill should not pick out white lead for special health warnings, he advised. Instead, the law should treat all pigments as if toxic.[56] Committee members shared Wiley's prejudice against "adulterants": one legislator asked Bartholdt whether any lead paint was poisonous or only those that had been adulterated, despite the commonsensical conclusion that "adulterating" a toxic paint by blending in harmless pigments reduces its toxicity.[57] A "pure" toxin was seen as superior to a healthful "adulterant."

House files on the hearings contain no direct evidence of lead industry opposition to Bartholdt's bill, although the bill's advocates frequently alluded to the impact of the "white lead trust."[58] Despite the stormy reception in the industry press, only one witness spoke against the bill. Eugene A. Philbin, a lawyer representing the National Paint, Oil and Varnish Association, dismissed as unnecessary the bill's provisions for occupational hygiene, noting that the industry was already cleaning its own house. The section requiring skull and crossbones on labels and signs, he argued, made the bill intolerable.[59] In personal correspondence to the committee chair, Philbin revealed that the bill's sponsors had a financial interest in Buddecke's baryte company and that Rhodes was "a close political friend of Mr. Buddecke."[60]

If the lead industry seemed nonchalant about Bartholdt's bill, it had good reason. The opposition was so strong and diverse that the lead-paint industry did not need to resort to monopolist machinations. Paint manufacturers lent no support to the bill because they needed white lead. In 1910 most mixed paints contained at least some white lead to increase covering power and act as a hardening agent (sixty-six years later, paint makers were still fighting federal legislation that would ban lead hardeners from paints). And many pigments, such as bright yellows and certain reds, were oxides of lead. Restricting white lead might result in the prohibition of all lead paints, and at the very least radically alter the way paints were made and used.[61] Even Rhodes, when asked by committee chair James Mann whether lead was not "the best material known as a pigment," replied that "we concede that is true, and concede it is an important industry in this country."[62] Bartholdt's bill, with its anemic support, its obviously "interested" parties, and its "socialistic" attitudes toward regulation of the workplace, posed no threat either to the lead corroders or to the mixed-paint companies who relied on their product.

Both paint bills introduced during the Sixty-first Congress died without a vote, though long-time paint industry insider George Heckel recalled twenty years later that "in every session of Congress since that day," a bill similar to Heyburn's "has appeared and been argued to death."[63] Bartholdt's bill would not come back to haunt the halls of Congress. Deemed too intrusive and expansive for consideration beyond the Mann committee, it descended, unreported, into oblivion.

The campaign for state and federal legislation against "adulterated" paint was only one front on which the opposing sides fought. Years before the legislative campaigns began, the mixed-paint and lead industries did battle in the pages of the paint-trade press. Ironically, by the time Heyburn and Bartholdt offered their legislation, paint advertisements reflected a relatively peaceful coexistence. Had potential legislators examined this self-imposed peace, they may have been dissuaded from mounting their doomed campaigns in the first place.

In the late nineteenth century, lead corroders had belittled mixed paints and, in both the trade journals and in pamphlets for the general public, had warned of "adulterated" lead paint being sold by unscrupulous dealers. In turn, the mixed-paint faction, in addition to employing antitrust rhetoric, made frequent use of their "trump" card—lead poisoning (see Figure 3-3). George Heckel recalled that in his early days in the zinc industry, the topic of lead poisoning "was the weapon both of offense and defense of the mixed paint man. . . . When the lead man shouted 'water'—'barytes'—'coal oil'—'benzine'—the paint man whispered 'poison.'"[64]

Ironically, as legislative fervor swelled, most of the rancor vanished from paint-industry advertisements. Heckel gave credit for this armistice to the cool heads of public relations men such as himself. Prior to the arrival of Heckel's NLC counterpart, Orlando Harn, relations between New Jersey Zinc and the NLC were "an eye for an eye." But Harn and Heckel made peace: "We wiped our blades, hung up our armor and sat down comfortably, each to his own knitting." In order to protect their mutual interests, manufacturers and distributors of zinc oxide, lithopone, and barytes would promote their products' quality and economy and seldom allude to the

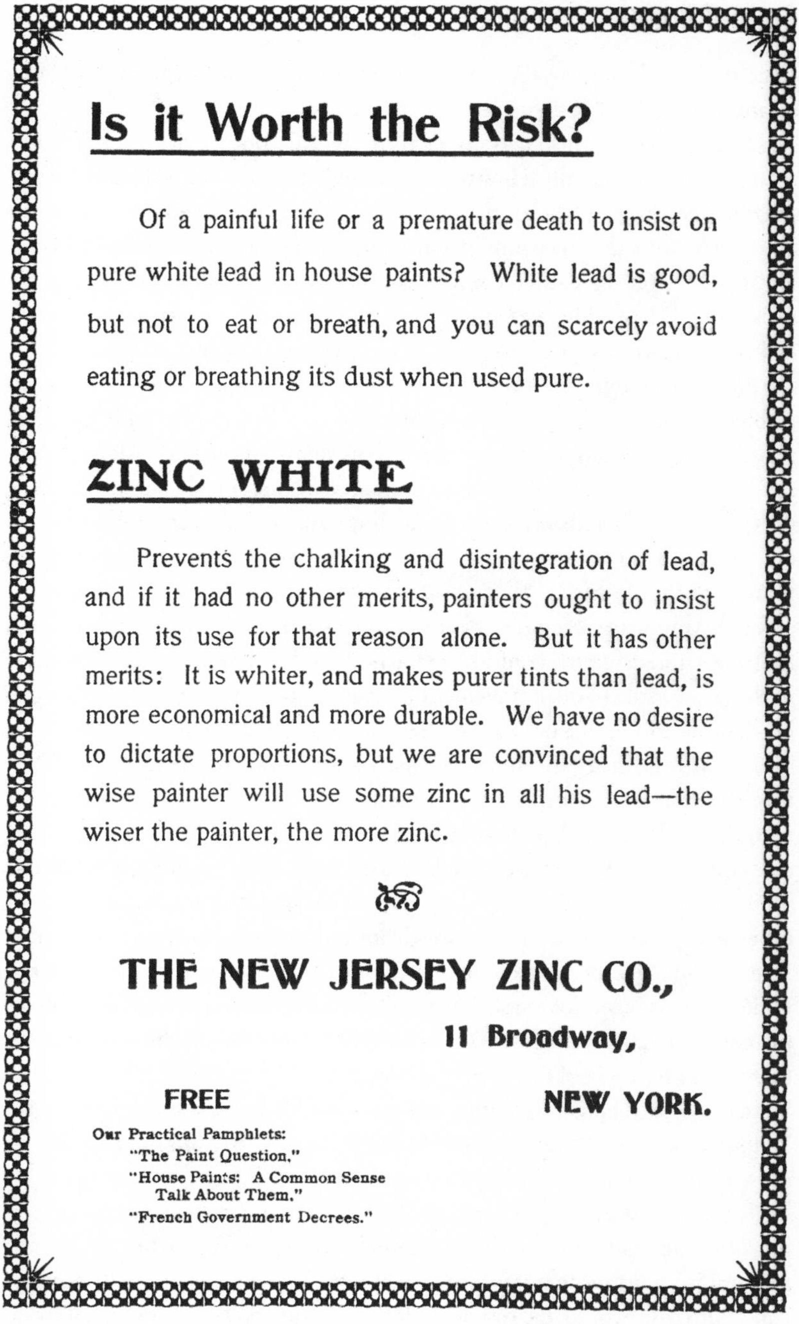

Is it Worth the Risk?

Of a painful life or a premature death to insist on pure white lead in house paints? White lead is good, but not to eat or breath, and you can scarcely avoid eating or breathing its dust when used pure.

ZINC WHITE

Prevents the chalking and disintegration of lead, and if it had no other merits, painters ought to insist upon its use for that reason alone. But it has other merits: It is whiter, and makes purer tints than lead, is more economical and more durable. We have no desire to dictate proportions, but we are convinced that the wise painter will use some zinc in all his lead—the wiser the painter, the more zinc.

THE NEW JERSEY ZINC CO.,

11 Broadway,

FREE

NEW YORK.

Our Practical Pamphlets:
"The Paint Question."
"House Paints: A Common Sense
Talk About Them."
"French Government Decrees."

FIGURE 3-3
1903 New Jersey Zinc Company Advertisement Warning Painters of the Dangers of Lead Paint
Official Journal of the Brotherhood of Painters, Decorators and Paperhangers of America, 16th Annual, Sept 1903.

toxicity of lead; in turn, the lead industry toned down its rhetoric of adulteration.[65]

Despite this accord, certain factions within both sides of the lead-mixed paint divide still felt strongly enough to push the legislative campaigns forward. But where advocates of food and drug regulation had been able to stir up and direct public animus to their cause, there was no popular support for regulating paints, whether for protection from fraud or poisoning. The public was simply not interested or, if it was, that interest did not extend beyond finding a good paint at an affordable price. Had Bartholdt overcome the prevailing skepticism about lead paint's threat to public health, perhaps the story would have turned out differently. But that was not going to happen when even the president of the Paint Manufacturers' Association, ostensibly a lead-industry opponent, declared "there is no question of health involved in dealing with paint, and we do not see any more reason for singling it out . . . as the subject of drastic legislation than any other industry."[66]

Had Americans been less devoted to lead paint, it is much more likely that the United States would have participated—in spirit if not directly—in the post–World War I movement by European signatories to the League of Nations to restrict or ban white lead from most interior applications. Twice in three years the International Labor Office of the League of Nations tried in vain to involve the United States in its sweeping reforms. In 1919 the ILO met in Washington, a meeting that Alice Hamilton described as "a pitifully futile attempt to enlist American interest in at least one activity of the League of Nations."[67] America reiterated its intransigence in October 1921. Four hundred delegates from forty nations attended the Third International Labor Conference in Geneva, but representatives from the United States were not among them. Regulating the lead trades was one of the conference's key goals. Most significant among the resolutions was one that would have saved the lives of thousands of American children had the United States been a signatory—a ban on lead paint for interior use. Within a decade, countries from Czechoslovakia to Cuba had agreed to the Geneva resolutions.[68]

Trade unions and big business alike feared internationalism. Smug feelings of American superiority were largely to blame. According to Hamilton, unions worried that if international regulations were adopted, "American standards would be dragged down to the level of 'pauper labor of Europe.'" It is difficult to see how pauperization would have resulted from the ILO's recommendations for the lead industry, which included banning children and women from ore reduction and smelting, setting strict requirements for reporting sickness, and requiring protective clothes and clean eating places.[69]

Two painters' associations sent wildly different resolutions to the conference. In direct opposition to the line they would take a decade later, the Brotherhood of Painters, Decorators and Paperhangers, then about 125,000 strong, stated that "the use of white lead is a grave menace to the health and lives of painters and should be forbidden." But the International Association of Master House Painters and Decorators, whose Progressive Era leader, John Dewar, had so vehemently defended the purity of white lead in the halls of Congress, maintained its preference: "Any general prohibition of the use of white lead in painting would result in less permanent and satisfactory work being done."[70]

The United States government agreed. Prior to the 1970s, no federal law prohibited the use of lead paint for most residential applications.[71] Between 1910 and 1977 over four thousand tons of lead pigments were applied to homes and products in the United States.[72] Without mandatory labeling, the myth of lead's superiority continued, since most paint buyers could not know how little lead some of the best paints contained. Without federal requirements for the most basic occupational hygiene measures, thousands of painters and laborers in the lead-using industries suffered from acute plumbism, while untold numbers suffered from lead poisoning's more insidious chronic forms. Under the Occupational Safety and Health Act of 1970, federal controls on worker safety in the lead industries were not mandated until late in that decade.[73]

Instead, white-leaders continued to "cover the earth"—at least the American corner of it—with lead. One of lead paint's most endearing characteristics, from the painter's point of view, was that it made repainting so easy. According to a lead-industry publication from the 1930s, "white lead paint slowly wears away, leaving an even, slightly chalky surface which is excellent for repainting."[74] What wears away, however—falling to the floor, settling in windowsills—is a fine dust of a potent neurotoxin that painters employed at their peril.

Of course, lead paint's toxicity did not stop by affecting only those who sanded away the dusty residue of old coats in order to apply fresh, new, pure lead paint to the walls and woodwork of America's homes. The fine dust on the floor often made its way into the mouths of babes. Many young children nibbled a flake or two from weathered railings or chewed upon a convenient wooden windowsill or stair tread and found the taste not unpleasant—somewhat sweet, in fact. Those who ate enough to get sick, got sick. Those who ate more often died.

FOUR

Occupational Lead Poisoning in the Progressive Era

In the summer of 1910, the Illinois Commission on Occupational Diseases sent Alice Hamilton—its "special investigator"—to Brussels for the Fourth International Congress on Occupational Accidents and Diseases. Hamilton's survey of lead-using industries in Illinois was well underway, and in Europe she hoped to visit factories to learn all she could of "preventive hygiene."

It is drop by drop, it is little by little, day after day for weeks and months, and finally enough is accumulated to produce symptoms.

JUDGE J. STONE
Adams v Acme White Lead & Color Works

She found more than data. As she wrote in her autobiography, the visit changed her life, setting her on the course she would follow through a thirty-year career "exploring the dangerous trades."[1] Americans presented only two papers at the congress, and when the discussion turned to what progress the United States was making, one Dr. Glibert, a Belgian Labor Department official, dismissed American industrial hygiene: "It is well known that there is no industrial hygiene in the United States. *Ça n'existe pas.*"[2]

Hamilton agreed that it was a "deplorable impression our country made." She failed to define, however, what "*ça*" it was they were talking about. If the haughty Belgian and the chastened American delegate meant that the profession of industrial medicine in America was ill defined and disreputable, there could be little argument. If they meant that compared with European governments, the American state was powerless to regulate factory conditions, they were essentially correct. But if Glibert and Hamilton meant that American industry, workers, and doctors were unaware of the occupational hazards in the nation's industrial plants, they were surely mistaken.

A long tradition of historical writing on occupational medicine has taken Glibert's view, rightfully perceiving that "industrial medicine's 'great decade'" was not until the years from 1910 to 1920. But in exalting their torchbearers with whiggish enthusiasm, these internalist histories often exaggerate the depth of the prevailing darkness.[3] Some historians have depicted the industrial sanitarians as either bold pioneers—the avant-garde of social reform, the muckrakers' factota—or as agents of middle-class machinations to exert social control over the laboring classes; others have criticized the industrial hygienists for their uncritical scientism.[4] Few, if any, question the originality of the hygienists' explorations or the audacity of their criticisms; nor do they question Hamilton's assessment that industrial hygiene did not exist. The tablet of occupational medicine in the United States is assumed to have been blank.[5]

This chapter challenges that assumption. While I credit Hamilton and her generation of pioneers for the light they shed on conditions in American factories, I also examine the Progressive hygienists in light of their times and discuss what Europeans and Americans knew about occupational lead poisoning in the decades prior to Hamilton's investigations. I compare nineteenth-century European efforts at regulating "the dangerous trades," then assess how the European experience influenced American efforts at occupational reform, and conclude with American industry's eventual turn from the European path of state regulation to embrace a privately administered regime of workmen's compensation.

Lead's toxicity was well understood years before industrial hygiene came into its own in the United States.[6] Anyone in the lead trades was unlikely to remain wholly ignorant of the dangers. Management and workers put in place mechanisms to "deal" with lead poisoning. Some of these methods did reduce individuals' exposure, but only by redistributing the risk. Most, however, were mere coping mechanisms for controlling who would be most exposed and for denying the human costs of unhygienic conditions. Plant owners put recent immigrants in the most hazardous jobs. Native-born Americans took the majority of jobs in less hazardous areas and kept their sisters and daughters out of the lead works completely.[7] Judging by these measures and by the consistency—across time and space—of received wisdom about how best to avoid occupational lead poisoning, there existed in America in 1910 a culture of plumbism, a culture built from the stuff of nineteenth-century industrial ideology: denial, fatalism, Social Darwinism, sexism, racism, and "workers' control."[8]

Against this array of complacent traditions, America's industrial culture was changing with a glacier's speed and imperturbability. But once the changes arrived, they would force massive reassessment of the relations between manufacturers and laborers. Although this chapter is about occupational hygiene in the lead industries, on another level it seeks to identify the cultural factors in the late nineteenth and early twentieth centuries that pushed aside long-standing traditions about accepting risks and sidestepping responsibility. While local, state, and federal governments were often willing participants in reform, vested industrial interests easily muted or directed their power. Whether the role of government regulation in this period shifted from paternalism to maternalism had far less impact on the health and safety of workers in the lead industries than industry's own search for the most efficient and orderly use of natural, economic, and human resources.

Similarly, the experience of the lead industry suggests that organized labor played only a minor role in the birth of professionalized occupational hygiene. This was not merely a new branch of medicine but a new way of handling workplace dangers, one that was as fundamentally different from obeisance to the "fellow-servant rule" as vertical integration was from horizontal, a development that would lengthen and enhance workers' lives in ways that "mere" pay raises could not approach.

The ascension of corporate-sponsored industrial hygiene may well represent, as recent critics have argued, a decline from a Progressive golden era, but the principle at the heart of occupational medicine—an enlarged notion of responsibility—would not go away. In 1948 Alice Hamilton recalled with an air of bemused incredulity the comment of a lead company spokesman whom she interviewed in 1912. "Do you mean to say," the manager had gasped, "that if a man gets lead poisoning in my plant I am responsible?"[9] That this attitude seemed quaint or shocking to Hamilton thirty-five years later suggests the fundamental changes in the expected relationship between worker and employer.

The conditions under which lead compounds were manufactured and employed at the turn of the century reveal three important facts: lead was deadly, workers and management knew it, but the culture of the workplace inhibited change and guaranteed widespread occupational lead poisoning. The origins of this saturnine culture predate the beginnings of American industrialization, for although it was occupational lead poisoning that first grabbed popular attention, lead had been a "consumer" poison since colonial days—dissolved in drops from leaden water pipes or in adulterated rum, blended with spices, or leached into acidic foods prepared or stored in pewter vessels.[10] Industrial hygienist and historian Carey McCord, who scoured the American colonial medical *oeuvre* for evidence of recognized lead-poisoning epidemics, accurately appraised the most serious problem historians of early lead poisoning must deal with: "Nowhere are there statistics—just allusions." Yet a growing body of research on lead poisoning in Europe and the Americas makes it certain that from the seventeenth century, Americans frequently suffered lead's acute forms as well as some of its more chronic manifestations.[11]

While it seems likely that most plumbism in the eighteenth and nineteenth centuries originated in food and drink, that does not mean that there was no occupational lead poisoning. Each new lead-using enterprise—every pottery, glass maker, and printer—meant additional occupational exposure.[12] A growing number of American physicians in the nineteenth century published reports of painters, miners, and potters crippled by lead. One bibliography of pre-1900 American publications on occupational health includes two hundred articles and books, and judging by the articles specifically about lead poisoning that were omitted, the true number ought to be far more.[13] In 1826, for example, the *Boston Medical Intelligencer* reported a case of a painter, who "had the habit of putting his painting pencil in his mouth, after using it in different colors, which contained more or less lead." The painter developed colic, constipation, and nausea, but was cured by drinking dilute vinegar and avoiding the source of poisoning.[14] In his early nineteenth-century treatise "On the Influence of Trades, Professions, and Occupations in the United States, in the Production of Disease," Benjamin McCready generalized the association between American painters' choice of pigment and their health: "Painters are in the habit of constantly employing a mineral which has long been known for its poisonous qualities, and it is to this that the peculiarities of their complaints are owing." Significantly, McCready noted that painters in his age already employed one of the tricks of the twentieth-century trade: "The men seldom work for any length of time at this department of the business" (i.e., painting with "dead white"), "alternating it with others that are comparatively innoxious."[15]

In the second half of the nineteenth century, the pace of American exploration into occupational conditions accelerated considerably. Unprecedented industrialization and immigration transformed cities into crowded and filthy centers, stimulating a search for new ways of dealing with the rapidly changing conditions of Americans' increasingly urban world.[16] Newly formed city health departments, ever on the lookout for more authority, inspected factories. An 1865 report on New York City's health conditions noted the occupational diseases suffered by that city's workers.

Eight years later, the New York City Board of Health published physician Allen Hamilton's survey of lead-using factories in the Lower East Side. The Massachusetts Bureau of Statistics included safety conditions in surveys beginning in the early 1870s. The bureau's late 1880s *The Working Girls of Boston* noted that in type foundries "the workroom is always filled with a fine lead dust, caused by 'rubbing'" and reported that employees had described cases "of young girls who have died from the effects of the work." In 1904 the Massachusetts State Federation of Women's Clubs noted lead poisoning in rubber factories surveyed in New England.[17] State laws also reflected a growing awareness of occupational diseases. As early as 1894 nine states required factory inspections. By 1904 seventeen states required ventilation in workplaces, eleven defined adequate lighting, and sixteen mandated fans and blowers.[18]

This record of state intervention and medical research would seem to give the lie to the Belgian hygienist who asserted that occupational hygiene did not exist in America. But for all this investigation and regulatory action and despite considerable progress toward improving occupational safety, the United States in 1910 remained behind most European nations. An examination of nineteenth-century industrial hygiene efforts in Germany, Britain, and France—Europe's three largest lead-using countries—shows the size of the gap, suggests the reasons for it, and partly accounts for the sudden shift in American industries' attitudes toward occupational hygiene during the Progressive Era.[19]

Europe's powerful early twentieth-century laws for factory inspection and regulation of the manufacture and use of lead were largely a product of a wider movement for social legislation that had dominated politics from the late nineteenth century. Hence, any argument about the origins of European lead laws can quickly get mired in the murky backwaters of comparative "welfare-state formation" analysis.[20] Laying aside general and theoretical comparisons, however, several specific factors did favor Europe's early recognition of occupational lead poisoning and spurred some lead-using manufacturers to modify their processes to protect workers. The professionalization and institutionalization of European medicine is one of the most important of these.[21] The role of the European public hospital can be seen in the lead studies conducted at La Charité Hospital in Paris, commencing in the late eighteenth century and culminating in the influential work of Louis Tanquerel des Planches in the 1830s. Tanquerel examined the case histories of over 1,200 patients, and in 1839 published a two-volume study of lead's neurotoxic effects. His data showed that lead compounds—especially pigments—posed the most dangerous occupational hazard.[22]

During the second half of the nineteenth century Tanquerel's work proved invaluable for France's growing zinc industry in its campaigns against lead paint.[23] In order to rebuff this challenge and forestall an outright ban, many French lead-compound manufacturers armed themselves with Tanquerel's findings and adopted innovative processes that made the final phases of stripping, drying, and packing less hazardous, including enclosing or automating dusty operations to reduce workers' exposure. An early twentieth-century observer reported that at the factory of French paint manufacturer Besançon et Cie, "while the old Dutch method of manufacture was in use there was an immunity from plumbism among the employés." This observer attributed the firm's success to banning "female labor"; reducing workers' exposure to dry white lead, largely by combining the processes of drying and mixing pigments in oil; close supervision of workers, including regular medical inspection; and providing alternate jobs if workers became "leaded."[24] The improvements in French lead factories were not initially state mandated; rather, they were voluntarily (but by no means universally) adopted by management.

In Germany, state-sponsored social reforms, not economic pressure from competing industries, spurred the adoption of more hygienic conditions. After 1878, government officials inspected lead-using industries on a regular basis. The Reich's 1884 workmen's compensation law declared occupational diseases (including lead poisoning) no different from compensable industrial accidents, putting financial pressure on German manufacturers to clean up their processes and facilities. Regulations enacted two years later required protective clothing and breathing apparatus and banned especially hazardous practices such as dry sanding of white lead paint. By the early twentieth century German ordinances regulated painters and others who worked with lead compounds and mandated regular medical examinations. In short, from the last quarter of the nineteenth century, Germany's active state imposed a high cost for sick workers.[25]

In this setting of economic incentive and statutory rigor, hygiene in Germany's lead industries improved dramatically. From 1886 to 1907 the number of lead-poisoning cases in one plant in Tarnowitz fell from about 50 percent of workers to 7 percent (of course, not all German plants matched this record). State-mandated factory reports and publication of the resulting statistics stimulated both medical interest and industrialists' awareness. As a result, according to the British Principal Lady Inspector of Factories, German industrial hygiene in 1902 "far surpassed the standards obtaining until recently in England."[26]

The path Britain took in reducing its industrial lead poisoning followed neither the German nor the French model, although progress in these nations served as both a fillip and a source of information for British reformers. Britain's initial hygiene laws, paternalistic and weak, seemed predicated on the notion that information and moral authority would change the hearts, and subsequently the business practices, of the industrialists. Reports in the British press that industrial conditions were bringing on a gradual "physical degeneracy" shaped early efforts to clean up factories. Charles Dickens, writing as the "Uncommercial Traveller" in his weekly periodical *All the Year Round,* visited the home of an unemployed laborer in a "squalid maze of streets, courts, and alleys of miserable houses" east of London. In a corner lay a young woman, moaning quietly: "'Tis the poor craythur that stays here, sur," the man's wife explained, "and 'tis very bad she is, and 'tis very bad she's been this long time, and 'tis better she'll never be, and 'tis slape she does all day, and 'tis wake she does all night, and 'tis the lead, sur." She explained that their tenant worked in the local lead mill but because she had a weak "constitooshun" she got "leadpisoned, bad as can be . . . and her brain is coming out at her ear, and it hurts her dreadful."[27]

In 1861, no government regulations protected lead workers like Dickens's "poor craythur." Between 1833 and 1864 Britain enacted a series of laws that limited the working day of most industrial laborers to ten hours, though the conditions under which those hours were spent received only toothless regulatory edicts.[28] The turning point for British occupational hygiene came in the 1890s, as physicians, with government sponsorship, initiated medical surveys of "dangerous trades." In 1895, the Home Secretary established the Dangerous Trades Committee, which over the next three years studied twenty-one British industries. Although the committee members could not enter factories without the owners' consent, rarely were they barred. Physician Thomas Oliver argued that although many manufacturers welcomed the inspectors' advice for improved hygiene and constructed new and safer factories, only much stricter laws would bring the majority into compliance.[29]

British laws enacted between 1895 and 1906 went far toward turning investigative knowledge into regulatory power. The Factory and Workshop Act of 1895 required physicians to report all cases of occupational poisoning by lead, phosphorus, or arsenic, as well as any cases of anthrax contracted in factories or workshops. The Workmen's Compensation Act of 1906 defined twenty-five occupational diseases as compensable, provided that a physician certified that the disease was acquired on the job.[30] British white lead manufacturers, under pressure from the Home Office to reduce poisoning, began employing many of Oliver's suggestions in the early 1900s. Improved ventilation, replacement of dry sanding with wet, and automated ovens brought dramatic improvements in Great Britain's industrial lead-poisoning rates, from 1,058 cases in 1900 to 576 in the 1909–1910 recording period.

Oliver edited the Dangerous Trades Committee's report, published in 1902. *The Dangerous Trades: The Historical, Social, and Legal Aspects of Industrial Occupations as Affecting Health, by a Number of Experts* became a model for occupational health researchers in both Europe and the United States. One American reader, Alice Hamilton, already a committed champion of Chicago's working classes, found in Oliver's reports a way to put into action her concerns about sick workers and their families.

Medical historian Lloyd Taylor wrote that "the emergence of the practice of industrial medicine in America owed everything to the remarkable career of Dr. Alice Hamilton."[31] Recent historiography has tended to canonize her; she has been credited with setting the course for American industrial medicine, with personally embodying a paradigm shift in toxicology from domination by germ theory and clinical biology to greater reliance upon the tools of the social sciences, and with inventing modern urban environmentalism. Still, it does not exaggerate Hamilton's impact to say that she helped change American medicine's attitude toward occupational hazards from blind acceptance of the way things were to open-eyed determination to see what could be done.[32]

Alice Hamilton was born in 1869, the second of five children in the well-to-do family of Montgomery and Gertrude Hamilton of Fort Wayne, Indiana.[33] The Hamiltons were among the town's social elite. Montgomery was active in local politics and the Presbyterian church, and was an outspoken supporter of home rule for Ireland and social and economic reform in the United States. Schooled at home, the Hamilton children learned French, German, and Latin, and their parents expected them to extemporize on a wide variety of subjects. At 14, Alice went to Miss Porter's School for Young Ladies, where she dreamed of becoming "a medical missionary to Persia, or maybe in the slums of New York, or both."[34]

Alice followed her young girl's dream and chose a career in medicine, entering the University of Michigan in 1889 and receiving her medical degree in 1893. After interning at Boston's New England Hospital for Women, she studied pathology in Munich and Leipzig, where she was the only woman in her class. From 1896 to 1897 she completed graduate work at the Johns Hopkins University, an institution that encouraged its students to specialize at a time when most members of the American Medical Association were general practitioners. Hamilton came to know William H. Welch, Johns Hopkins's professor of pathology, whose pioneering work in the interactions of disease and environment influenced a generation of medical professionals. She jumped at her first job offer—teaching pathology at the Women's Medical School of Northwestern University—"not only because it meant employment in my own field, but because it was in Chicago. At last I could realize the dream I had had for years, of going to live in Hull-House."[35]

Hull House was the most famous of over one hundred Progressive Era settlement houses—urban missions where well educated, concerned men and women came to live, creating what they saw as oases amid urban industrial squalor. Jane Addams established her settlement house in 1889, taking over the run-down Hull mansion in Chicago's overcrowded nineteenth ward. By 1897, Hull House was a model for settlement houses around the country, with twenty-five permanent residents trained to lecture, counsel, and act upon issues ranging from factory conditions to juvenile delinquency to birth control. Hamilton stayed at Hull House for twenty-two years.[36]

She had been teaching at Northwestern University and living at Hull House for about ten years when she first read Thomas Oliver's *Dangerous Trades*. From her years at the settlement house she knew well the inhuman conditions, shortened lives, and debilitating diseases suffered by Chicago's industrial workers. She also believed—based on Jane Addams's example—that through persistent caring, education of both the oppressed and the oppressor, and political agitation, a small group of dedicated activists could ameliorate the effects of industrial conditions. But amelioration was not enough. In her autobiography, *Exploring the Dangerous Trades,* she wrote that reading Oliver's book and a "muck-raking" article castigating the nation's lack of compensation laws sent her "to the Crerar Library to read everything [she] could find on the dangers to industrial workers, and what could be done to protect them."[37]

In America in 1910, a great deal "could be done to protect them," but substantial improvement required the conjunction of a number of factors. Activists in the medical and social sciences organized to take action on the part of workers. The government displayed new political will and started to build an infrastructure sufficient to execute it. And, for a variety of reasons, a critical mass of key industries permitted and even sought government investigation, and appeared willing to implement investigators' nonbinding recommendations. In short, America's new age of industrial hygiene stemmed from the complex interplay of progressive activists seeking reform, industrialists seeking efficiency and quiescent relations with labor, and government seeking domestic tranquility and industrial progress.

At the twentieth century's opening, the convulsions of a liability system in its death throes threatened both tranquility and progress. The old ways of handling workplace injuries and deaths were no longer viable. For much of the nineteenth century, three common law precepts, as interpreted by American jurists and codified in state liability laws, dictated the terms by which responsibility for sick and injured workers was negotiated. Assumption of risk, contributory negligence, and the fellow-servant rule forged an iron restraint that robbed most industrial workers and their families of any hope of recovering compensation for crippling or fatal injuries. In essence, if a worker complained after being hit on the head by a falling hammer, the three rules allowed the employer to answer, "If you can't keep your eye on what careless workers around you are doing, then you shouldn't work here." Workers could sue their employers over liability, but as muckraker William Hard sarcastically advised, "Go ahead and sue the U.S. Steel Corporation. The courts are open to you just as they are open to the U.S. Steel Corporation. . . . You are at liberty to try to starve out the U.S. Steel Corporation just as the U.S. Steel Corporation is at liberty to try to starve you out." Litigation, wrote Hard, was "a rich man's game, like automobiling and yachting." But from the Progressive Era industrialists' standpoint, winning contributory negligence cases—a game in which they held all the trumps—was expensive and wasteful.[38]

The liability system had made some sense generations earlier, when factories and workshops bore more similarities than differences. Owners and the courts could with some justification apportion responsibility for accidents to workers, but only to the degree that "workers' control" characterized the labor relations within a workplace.[39] But times had changed. "The Factory Age had come," wrote Frank W. Lewis in 1909; "great inventions and the application of steam to machinery were transforming the industrial world. It was gradually dawning upon the minds of thoughtful men that these great changes had made imperative new standards of law as related to workmen."[40] As factories grew larger—as processes became more automated and the scale and scope of processes and machinery diverged from human proportions—the ground beneath these precepts gave way, and gradually the thorny path to a new system appeared.

A number of factors helped clear this path. Labor militancy in key large-scale industries forced the issue as never before.[41] Gradually, juries demonstrated increasing sympathy for injured industrial workers, and with every liability reform law, the number and size of awards to plaintiffs, though never high, grew enough to alarm employers and make the prospects of compensation insurance more palatable. Presiding over these court settlements, a judiciary increasingly willing to depart from formal adherence to the cold calculus of precedent entertained more pragmatic, or "realistic," arguments in reaching its decisions.[42]

But because change in the courts might be slow in coming, reformers also sought legislative solutions to occupational hazards. The American Association for Labor Legislation (AALL), established in 1906 as a branch of the International Association for Labor Legislation, was instrumental in drafting and promoting state labor laws before World War I.[43] The European connection was critical for the American association. Under the leadership of social scientists Richard Ely, John Commons, and John B. Andrews, the AALL sought to replicate in the United States the two-tiered approach of many European hygiene laws: state-supervised hygiene programs to prevent illness, and mandatory insurance to compensate for industrial accidents and disease.[44]

For two years the AALL did little more than conduct letter-writing campaigns to build support and elicit funding. Then, revelations of a ghastly occupational disease, phosphorus necrosis ("phossy jaw"), in the match industry gave the AALL an issue around which to marshal support for its legislative programs—a focused situation with dramatic potential for igniting public support, and a problem that could be remedied relatively inexpensively through legislation.[45]

Before 1912, workers in match factories often suffered from phossy jaw—a painful and disfiguring disease commonly resulting in the loss of an eye or the deterioration of the jaw. Europe had wrestled with the condition since 1845. In France, where the state produced all the matches and paid the costs of phossy jaw, government-subsidized scientists developed sesquisulphide, a substitute for white phosphorus. By 1908 European match factories were all but free of the disease.[46]

Although doctors reported cases of phossy jaw in the United States as early as 1851, the American medical and scientific community paid little heed. According to Alice Hamilton, "practically nothing was published in

American medical journals from 1851 to 1909," the year that the Bureau of Labor published the results of its investigation of phossy jaw.[47] The study, conducted in large part by the AALL's Irene Osgood and John Andrews, turned up 150 cases, including four fatalities. The AALL then coordinated a lobbying and public awareness campaign against phossy jaw. Boycotts, letter-writing campaigns, and articles such as Gordon Thayer's "Matches or Men?" brought swift action. Diamond Match company, which owned the U.S. rights to sesquisulphide, made its process available to all manufacturers, free of royalties. In 1912 Congress passed the Esch-Hughes bill, which taxed white phosphorus out of the competition. In just a few years, a relatively small group of activists had educated and mobilized the public and Congress to take effective action that almost eliminated an industrial disease.[48] Eliminating lead poisoning from the workplace was not going to be as simple, but the AALL's victory over phossy jaw supplied needed confidence in the early stages of the fight.

Pressure for legislative action on occupational health mounted. Labor organizations clamored for liability reform; muckrakers, social scientists, and organizations such as the AALL and the Russell Sage Foundation advocated enlarged state involvement in occupational safety; industrialists sought to minimize the potential for punitive reforms. Given the limited powers of the federal government, it fell to the industrial states to find a way to deflect and channel the tremendous pressure mounting on all sides.

Illinois, with its booming industries, core of activist scholars from Chicago, and progressive state government, took up the challenge. The state's voters elected Democrat Charles S. Deneen governor in 1904 on the basis of his promises of political and industrial reform. In 1907, the Illinois Commission on Workingmen's Insurance declared, "Undoubtedly the time is not distant when industrial states must take up the problem of legislation upon sickness insurance. To provide a scientific basis for such legislation we recommend the appointment of a competent commission . . . to make a thorough study of . . . diseases among workpeople." The resulting commission comprised five physicians, two members of the state's Labor Department, University of Chicago sociologist Charles Henderson, and one employer. Henderson, no stranger at Hull House, enlisted Alice Hamilton. But since all agreed she was the only member qualified to conduct the study, she resigned from the commission to become its managing director.[49]

The commission decided that the survey should focus on manufacturing trades that dealt directly with known poisons such as lead, arsenic, brass, zinc, and cyanide. Hamilton limited her own work to the most insidious and pervasive of these: lead. In 1910 and 1911 she and her staff visited 304 plants: paint factories, lead pipe and ceramics manufacturers, enamelers—any lead-using business she could discover. Hunting for unknown sources of plumbism, assistants scoured hospital records and interviewed company physicians and labor leaders. They found several sources not mentioned in the voluminous European literature: "making freight car seals, coffin 'trim' . . . pottery decoration; polishing cut glass; brass founding; wrapping cigars in so-called tinfoil, which is really lead." When a lead-using trade was found, Hamilton herself visited the plants, braving the heat of smelters, climbing catwalks, walking through clouds of white lead powder, peering into cauldrons. She also examined workers and conducted extensive interviews with employees, supervisors, and owners.[50]

The commission's findings, published in 1911, helped shake loose the complacency of governments from Springfield to Washington. The study uncovered in abundance forms of plumbism that had been all but eliminated from European factories decades before. No longer could legislators and industrialists blithely delude themselves that American technology and democratic ideals guaranteed greater safety for workers than that afforded by Europe's crumbling monarchies. The situation demanded swift action—and more study.

Initially, there was far more of the latter than the former. In New York City, Edwin Pratt, an economist at the New York School of Philanthropy, conducted a hasty study of lead-paint manufacturers. In St. Louis, students at the Washington University School of Social Economy surveyed Missouri's lead industries, turning up over a thousand cases of lead poisoning. In 1912, David Edsall, a professor at Harvard Medical School, enlarged the occupational medicine program at Massachusetts General Hospital's seven-year-old Social Services Department, focusing his research efforts on occupational lead poisoning. In addition to collecting data and delivering treatment, the department enlisted workers to distribute pamphlets such as "Advice to Persons Working with Lead" and encouraged them to start safety committees at their plants. Hamilton went on to direct a number of surveys of lead industries for the federal Department of Labor.[51]

Almost all of these early studies relied on similar strategies for collecting and analyzing data. With no equipment capable of measuring environmental exposures and in the absence of standards and technology with which to assess the effects of lead exposure in workers' bodies, Hamilton and her peers followed the model of nineteenth-century labor bureau surveys: describing processes and conditions; collecting impressions and anecdotal evidence from workers; and comparing conditions in one factory with those found in others in the United States and Europe. They reinforced their largely impressionistic evidence with simple statistics garnered from local hospital and physicians' records. Gordon Thayer's muckraking 1913 exposé in *Everybody's Magazine* employed the technique brilliantly:

> *In a Düsseldorf Factory, employing 150 men, two cases of lead-poisoning were discovered in 1910 by the examining physician. In an American factory, employing 142 men, twenty-five cases were sent to the doctor with lead-poisoning. In an English white and red lead factory employing ninety men, there was not one case of lead poisoning in five successive years. In an American white and red lead factory employing eighty-five men, the doctor's record for six months showed thirty-five men leaded. Another English factory employs 182 men and did not have a case of lead-poisoning last year. An American factory with 170 men had sixty poisoned during the same year.*[52]

More significant than these studies' stylistic similarities was their agreement on legislative remedies. They argued for laws like Germany's, laws that would mandate state inspection of factories, regular medical examinations, and strict requirements for reporting all cases of occupational lead poisoning. The report of the Illinois survey concluded with detailed recommendations for a hygiene bill, which would require remedial action as well as specific precautions—from maintaining adequate ventilation to providing washrooms and work clothes for all workers. Pratt recommended an enlarged, more active role for the State of New York's Medical Inspector of Factories, vigorous surveillance of factories, and state sponsorship of prevention programs.[53]

Several states did take legislative action, but the pace was not as swift as Hamilton and other progressive reformers would have liked. In 1911, six states passed occupational disease bills, but these merely required manufacturers to report all cases of lead poisoning. Only the Illinois bill went beyond the reporting requirement. In accordance with the Occupational Disease Commission's suggestions, the Illinois law required safety measures to reduce plumbism, including monthly medical examinations and "approved respirators, overalls, dressing rooms, lavatory and bath facilities."[54] To facilitate passage of similar laws in other states, the AALL drafted a model occupational disease bill and lobbied legislatures in the biggest lead-producing states to pass their own versions. As they had done during the phossy jaw campaign, they enlisted Gordon Thayer to write the scathing summary of the various lead surveys quoted above.[55]

The AALL's bill, like the Illinois law, did not require or even address

ILLUSTRATIONS
BY HARRY TOWNSEND

The Lead
Menace
in
Gordon Thayer

FIGURE 4-1
Illustration from Gordon Thayer's "The Lead Menace"
Everybody's Magazine, 1 March 1913.

financial compensation for sickness, injury, or death. It sought prevention,
not redress. But by the time Thayer's article appeared, industry-supported
workmen's compensation laws were sweeping through the state capitals,
overshadowing and even invalidating proactive, preventive labor protection bills like the association's. By 1911, ten states had enacted workmen's
compensation laws; by 1920, forty-two states had; and by 1948 all states
had such laws on the books.[56]

Historian Robert Asher characterized the ascendancy of workmen's
compensation as the ironic fulfillment of workers' failed efforts to achieve
equitable liability law reform: "Workmen's compensation, with its assurance of relief, regardless of the fault of the worker or the employer, had
supplanted the common law of industrial torts."[57] Early in the campaign,
the AALL's John Andrews predicted that workers in the lead industries
would benefit from compensation requirements, because "placing the
financial cost of lead poisoning upon the lead industry will promote greater
cleanliness in the lead trades. It will pay to clean up. A considerable part
of the money now paid to employers' liability companies and to ambulance

chasers could, under a just system of compensation, go where it belongs—
to the injured workman or his family."[58]

But most workmen's compensation laws, designed to compensate for
specific injuries suffered on the job, sidestepped occupational diseases,
which might develop over years and have origins and causes that often
remained obscure. For most of the 1910s and 1920s fewer than ten states
required manufacturers to compensate industrial diseases in the same way
as accidents, and as late as 1946, only thirteen states did. Even in those
states, legal provisos by no means assured compensation for industrial lead
poisoning. The courts remained the chief venue for lead-poisoned workers
seeking compensation.[59]

Defining the boundary between accidents and industrial disease was critical for industry to control the awards paid to injured workers under state
compensation laws. At one level the distinction is obvious. The accident
occurs in an instant—a moment's distraction, a hidden flaw in a cable, and
the surging machinery swallows the hand, the falling girder crushes bones
and flesh. But occupational disease, Michigan Supreme Court Judge J.
Stone declared in 1914, "is a matter of weeks or months or years. . . . It
is drop by drop, it is little by little, day after day for weeks and months,
and finally enough is accumulated to produce symptoms."[60] And over the
course of these years, as the poisoned worker's appetite fades, as increasing
neurological dysfunction distort his personality and behavior, he is left not
with one unequivocal "disease" but a set of seemingly unrelated health
problems, including contagious illness and opportunistic infections his
weakened body can no longer fight off. For such workers and their families,
the end result differed little from that of an accident: a palsied hand as
useless as the mangled one; a disabled or dead worker. But to the employer,
his insurer, the state compensation boards, and the courts, the distinction
was essential.

No one at the Acme White Lead & Color Works denied that Augustus
Adams had died of lead poisoning. Through the winter and spring of 1913,
Adams tended a sifting machine in the Detroit paint manufacturer's red
lead department. It took five months for the lead to make him sick enough
to stop working. It took another month to kill him. Sarah Adams, his
widow, applied for compensation under Michigan's recently enacted compensation law, and the arbitration committee found in her favor. Acme
appealed, arguing that lead poisoning was a disease, not an accident—and
hence excluded from Michigan's compensation law. Further, Acme argued
that "if the act does apply to industrial diseases, it is so far unconstitu-

tional." The board upheld the award. The language of the act, the board argued, "clearly includes all personal injuries arising out of and in the course of the employment, whether the same are caused 'by accident' or otherwise."[61]

The Michigan Supreme Court reversed the state board's findings, basing its decision in common law. Black's law dictionary's definition—"Accident: An unforeseen event, occurring without the will or design of the person whose mere act causes it"—left no room for occupational illness, since in a lead-paint company "such poisoning" was "something that is contracted by a fairly certain percentage of those working." Further, the court narrowly interpreted the purpose of workmen's compensation as "more just and humane laws" to replace assumption of risk, contributory negligence, and the fellow-servant rule.[62]

Judge Stone foresaw that compensation laws might negate "any liability of the employer for injury resulting from occupational disease." Indeed, just six years later, a metal company in Illinois successfully argued that because of its mandatory participation in the workmen's compensation program, it could not also be penalized for plant conditions that took only five months to reduce John Labanoski, a healthy 187-pound worker, to a sallow, anemic, and delirious 144-pound inmate of the State Hospital for the Insane. Since the 1911 Illinois Compensation Act dictated the appropriate compensation for occupational diseases, the court found that Labanoski was not entitled to sue his employer for gross violations of the Occupational Disease Act.[63]

State-enforced compensation laws presented problems for workers who moved in search of jobs. In the autumn of 1923, Marion Winter left Kentucky for Dayton, Ohio, where he found work in the paint department at the Maxwell Motor Company. Through the fall, he touched up auto bodies in preparation for finishing coats, filling small imperfections with a putty containing 50 percent lead by weight and sanding to feather out the edges. Suddenly, in late November, he became seriously ill with lead poisoning. Perhaps cold weather had prompted the closing of the factory's windows and doors. For almost a week Winter tried unsuccessfully to complete a day's work, but then, on November 26, he went home to recover, after signing a paper his employer later claimed was an application for compensation. Through December and January, Winter was bedridden. The company sent food and two checks, totaling less than $75, which Winter assumed were "gratuities" but which the company maintained were his total compensation. Having discovered that Winter had not lived in Ohio for ninety days prior to "filing" for workmen's compensation, the company claimed it had discharged its obligation.[64]

The duration of illness and confusion over the definitions of various degrees of lead intoxication also muddied the legal waters. One February afternoon in 1922, a ship painter, identified in the court records only as Gerald, walked toward the boat where he was employed stripping red lead from steel plates. Suddenly, his painting kit and turpentine dropped from his left hand.[65] The twelve-year veteran of the Boston shipyards picked up his tools, but before he got to the ship, his palsied hand dropped its load once again. That evening his doctor confirmed what Gerald must have suspected—he had lead neuropathy, or "wrist-drop." When Gerald's case came before the state Industrial Accident Board, the shipyard's insurers argued that his condition was due to either kidney or heart disease, not lead poisoning. Strong evidence of Gerald's chronic exposure to lead, and medical testimony to the effect that Gerald's kidney damage might be a product of that exposure, convinced the board that Gerald was due compensation. The state Supreme Court noted that the evidence for heart disease was strong, but supported the board's findings of fact and affirmed the ruling.[66]

The vagueness of plumbism's symptoms and the ability of the disease to go undetected for years continued to complicate matters long after states began compensating occupational illness. For example, in the early 1940s, a paint factory laborer in New York who worked for a number of companies over the course of about ten years, suddenly developed lead colic and was sent home. Tests confirmed his private doctor's diagnosis that he was lead poisoned. The company concurred, but still avoided compensation under state law. The worker and his doctor testified that he had suffered from mild forms of plumbism for years—never bad enough to keep him off the job, but sufficient for him to have sought treatment. Because years had passed since the onset of initial symptoms, the company argued successfully that the worker had not met the law's requirement that compensation claims be filed in a timely manner. Such cases hinged on the distinction between "lead absorption" and "lead poisoning." The worker argued that he had suffered "only" from lead absorption until the recent bout of colic, at which time he "contracted" lead poisoning. The company argued that his debilitating condition was merely the worsening of a chronic illness the worker and his doctor had long known about but had failed to report.[67]

For all the holes in the compensatory armor, the mere presence of mandatory compensation laws of any kind helped work a transformation in industrial culture. Two contingents of middle-management experts concerned specifically with the health of workers came to the fore. The expanded role of industrial physicians who increasingly presented a coherent and convincing argument for hygiene, will be taken up in the next chapter. But by definition, workmen's compensation also bolstered the power and influence of insurance carriers in setting and supervising health conditions in their customers' plants. It brought insurance companies in to encourage the preventive hygiene programs that permanently changed factory culture and allowed the state—in the form of preventive, supervisory regulations—to stay out. Organized labor and social reformers generally responded favorably to the new compensation regime, reducing pressure on state legislatures to consider additional, more interventionist regulations and weakening the impetus to enforce those already on the books. In 1913, Hamilton admitted, "I have been very much discouraged by the results of the Illinois law," which, because of lax enforcement and poor framing, had encouraged company doctors to define out of existence the lead poisoning in their factories. It was evident, she went on, "that the policy they have adopted is to first protect the company which employs them."[68] Over the next two decades, the absence of state or federal enforcement left industries in charge of setting the goals for occupational hygiene. "Protecting the company" would remain chief among those goals.

The activists of Hamilton's generation ensured that lead poisoning would remain for decades the most researched environmental disease. But in depicting them as explorers who boldly tore the scales off industry's eyes, we risk ignoring their more mundane role as facilitators of, and participants in, an emerging industrial culture in which management acknowledged its responsibility to protect its workers from hazards it had long known even as it sought to control the terms and minimize the costs of that acknowledgment. This transition began long before Alice Hamilton scaled her first catwalk in a lead plant. As she studied germ theory in Leipzig, state courts were chipping away at the fellow-servant rule and city and state officials were conducting factory inspections. While she was convincing the Chicago health community that flies in open privies were the vector for transmission of typhoid, the National Lead Company was building a white lead factory that she would later praise as the finest she had seen in Europe or America.[69]

If Alice Hamilton did not single-handedly create the discipline, in the years after her investigations it could no longer be said that "there is no industrial hygiene in the United States."[70] But increased knowledge about conditions in American lead factories did not bring the increased role for the state that similar revelations had prompted in Europe. Unlike Edwardian Britain or Germany under Bismarck, the United States lacked a strong state that could force its industries to implement paternalistic workplace regulations, whether for "statist" or compassionate reasons. In France, commercial interests in zinc could pressure reform, both from within industry and from the state, and painters' unions were willing to go on strike to avoid the toxic ceruse. But painters in the United States remained fervent white-leaders, and American paint manufacturers, including zinc producers who depended upon lead in their formulations, would not fight for—or even desire—overturning lead's traditional status as the best and "purest" paint. Doing so would only threaten their access to "the useful metal."[71] So instead of following the European course of regulation by inspection, prevention, and insurance under the aegis of a central authority, the United States took a different road, putting into place a system of insurance that only came into play after the worker was injured and only involved the state when the system's private mechanisms failed.

The knowledge garnered by state and federal investigators did not lead to state and federal power. "I knew," Hamilton recalled, "that I had no power to cause the managers the slightest discomfort." On the other hand, she added, "I was thankful they thought I had."[72] Industry granted Hamilton and other governmental specialists regulatory powers via moral suasion. Under the Progressives' moral regulatory regime, acknowledgment equaled confession and small improvements made by penitent industrialists brought atonement—though not compliance with published regulations. But by assuming only a revelatory and confessional function, government relinquished any claim to regulatory power when the basis for it shifted from morality to scientific knowledge. In the two decades after Hamilton's surveys, acquiring scientific knowledge about lead's toxicity and controlling its interpretation became an imperative for the lead industry. Armed with "the facts," the industry would be able to convince its workers, the scientific community, and the public in general that its products posed no threat.

FIVE

Protecting Workers and Profits in the Lead Industries

Alice Hamilton retained "some bitter memories" from her early surveys. "Still," she insisted, "they are few compared to my pleasant memories, for the arrogant and cold-blooded employer has, in my experience, made up a very small minority." She saw in the industrialists' cooperation a reflection of her own championing spirit, her upper–middle-class values infused with her Lutheranism and the Progressivism of the time. "The iniquitous

As you look down on them you see so many fathers and husbands and brothers and sons, real men, individuals. . . . All I see is a lot of my hands, a part, and a bothersome part, of my machinery, and that is the way most of them feel.

PLANT MANAGER TO
ALICE HAMILTON, CA. 1911

conditions I so often found," she recalled many years later, "were not a proof of deliberate greed or even of actual indifference, but rather of ignorance and an indolent acceptance of things as they are." Enlightenment and close supervision were all industry needed.[1]

Hamilton's optimism was not totally misplaced. Something of an industrial epiphany had occurred. From the mid-1910s through the 1930s, even after the furor stirred up by state and federal surveys had faded, the industry of industrial hygiene continued to grow as deaths from occupational lead poisoning fell. In a period of weak unions and lax regulation, when almost no workmen's compensation laws covered occupational illnesses, the levels of hazardous exposure for most workers in the lead industry were sharply reduced. What prompted and sustained this prolonged period of self-improvement?

For the lead industry the answer lay in its acquisition and strategic use of scientific—and especially medical—knowledge. The Progressive Era surveys forced many in the lead industries to acknowledge what they had known all along: that their product was toxic; that most manufacturers failed to incorporate existing cheap and effective means of reducing lead hazards; and that, hence, they were routinely killing their workers. They also learned that—as Winthrop Talbot, editor of *Human Engineering*, proclaimed—"sanitation is a means of saving dollars and cents."[2]

On the other hand, the manner in which the Progressive reformers publicized the economic and human value of workplace safety besmirched the industry's reputation and threatened more irksome government interference. Knowledge was good; publicity, bad. Increasingly, the lead industry sought to control both the production and distribution of knowledge, promoting and directing research into lead's toxicity and the diagnosis and treatment of lead poisoning while limiting its promulgation to medical and industrial professionals.

This chapter and chapter six trace the development of industry-sponsored occupational medicine in the years between the public outcry over "the lead menace" in American factories and World War II. Over the course of a generation, industrial medicine grew beyond its origins as a disreputable trade practiced by ill-trained and poorly equipped company doctors to become a full-fledged specialty with ties to the academy. Individual doctors working largely in isolation and with little influence in corporate affairs became empowered through professional affiliation and a record of considerable medical success. The lead industry's vertical integration of the production, interpretation, and distribution of research proved invaluable in controlling the costs associated with sick workers, both by lowering the rates and severity of illness and by reducing the manufacturers' perceived liability through their near monopoly on the "objective" data upon which courts and compensation boards increasingly relied.

Although most lead-using workplaces reduced their rates of occupational lead poisoning, the most dramatic changes took place in larger factories. There, economies of scale permitted ongoing investments in hygiene, and greater participation by middle-management specialists such as company doctors encouraged those investments. In these chapters I concentrate on the larger companies and the hygiene programs of the Lead Industries Association. In doing so, however, I do not ignore the fact that many small plants, LIA affiliates included, often did not embrace the new approach to industrial hygiene. To underline the uneven nature of the improvements, the next chapter concludes with a discussion of lead poisoning among members of the painters' trades, whose continued use of lead paint guaranteed high rates of lead poisoning from which neither unions, reformers, nor government offered protection.

* * *

Speaking to an international audience of public health experts in 1927, Frederick Hoffman, a consulting statistician for Prudential Insurance and a long-time student of lead poisoning in the United States, enthused, "It is always a gratifying experience to be able to present facts and figures suggestive of progress in the vast field of industrial hygiene." After reporting statistics for lead-poisoning fatalities, he turned to specific industries. "In the manufacture of white lead particularly, in which cases of chronic lead-poisoning now are very rare, far-reaching improvements have been introduced to adequately safeguard the health and the interests of the workers."[3]

In light of the absolute numbers of Americans whose deaths were attributed to lead poisoning each year from 1910 to 1940, Hoffman's statement rings false. In all but five of these years, lead poisoning was listed as the cause of death for over 100 men, women, and children (see Appendix, Table A-1). On the other hand, the death rate from lead poisoning in those years declined by two-thirds (Figure 5-1). Of course, official records do not reflect the actual numbers of lead-poisoning fatalities. As Chapter 2 argued, dozens—if not hundreds—of children died from lead-paint poisoning each year and the cause mistakenly recorded as infectious disease such as TB meningitis. Still, the data do gauge the level of awareness and perceptions of lead poisoning and allow for rough comparison over time. And it is significant that this decrease occurred over a period when lead consumption in the United States nearly tripled. The rate of deaths per ton of lead consumed fell by 81 percent; where one American died for every 2.9 tons of lead consumed in 1910, that rate had fallen to one death for every 14 tons by 1941.[4]

It is impossible to state with any certainty what percentage of these deaths were industrial in nature. Census Bureau mortality reports for most of the century are extremely vague in distinguishing between occupational and nonoccupational fatalities. But in these years, adults accounted for an ever-shrinking number of lead-poisoning deaths, supporting Hoffman's proposition that the lead industry was considerably less fatal than in the past.[5] (This sanguine appraisal must be qualified by healthy skepticism prompted by the crude diagnostic procedures used and the tendency of company doctors to under-report lead poisoning. "There are many plants in New York City," expounded Dr. E. F. Smith in 1927, "from which no lead cases are reported, except those failing to 'get by' the coroner."[6])

Prevention, not treatment, was in large measure responsible for these dramatic declines. Medical researchers between 1910 and 1940 vastly in-

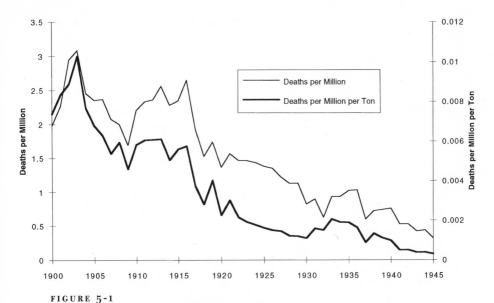

FIGURE 5-1
Rate of Deaths from Lead Poisoning, 1900-45
Bureau of the Census, *Mortality Statistics* (Washington: GPO, 1901-46); U.S. Department of the
Interior, Bureau of Mines, "Lead," in *Minerals Yearbook* (Washington: GPO, 1901-46).

creased their understanding of the physiology and pathology of lead poisoning, which enhanced physicians' ability to diagnose sick workers (and,
as became increasingly important, children). But breakthroughs in treatment would not occur until World War II. Unable to purge the mineral
from the victim's body without allowing its toxic effects to continue while
the lead was "in transit," physicians relied on "deposition and excretion":
by manipulating the body's metabolism through diet and chemical injections, they altered the worker's blood chemistry in order to encourage excretion through the digestive system or deposition in the bones.[7]

The only way to avoid serious occupational lead poisoning was through
prevention, for which lead-using industries had only two methods. The old
way—learned from years of experience with plentiful supplies of pliable, unskilled labor—was "rotation." As one paint factory manager explained to
Hamilton: "We make no effort to keep a steady force of laborers, encouraging the unskilled men to leave after a few weeks because we think it prevents them from becoming poisoned." Since customary wisdom held that
many white lead workers would get sick after several months, Hamilton
noted that "most foremen feel that they are only acting humanely if they
let their men go after a short time and engage new ones." Another operator

told her "that there was an advantage in employing colored labor, for the Negroes seldom work continuously more than a few months at a time."[8]

Hamilton agreed that prevention was the only solution, but argued that rotation did not prevent plumbism. Some workers seemed particularly susceptible, contracting it long before managers predicted. And nothing prevented a worker fired by one paint manufacturer on Monday from starting work for another on Tuesday. High turnover perpetuated an unskilled and undertrained labor force, increasing the danger for everyone. Hamilton advocated a new means of prevention: simple improvements in processes and equipment that would reduce exposure and permit the plant manager to replace "a gang of non-English-speaking foreigners whom he expects to lose in a month's time" with a well-trained and closely supervised permanent workforce. The hygienist's method of prevention produced healthy workers and enhanced productivity.[9] But without legal pressure, would arguments of efficiency and humanity impel lead manufacturers to clean up their factories?

Experiences with two other occupational diseases—phosphorus necrosis (phossy jaw) and silicosis—suggest two alternatives open to lead manufacturers hoping to quiet outrage over the lead menace. As we saw in Chapter 4, the case of phossy jaw showed that public pressure could spur implementation of new technology to eliminate a specific industrial hazard. But two factors made phossy jaw readily amenable to both public outrage and a quick technical fix.[10] The suddenness, specificity, and horrible consequences of phosphorus necrosis made it a fine candidate for reformers' fervor. Unlike lead poisoning, phossy jaw was unlikely to be confused with other sicknesses or to be blamed on intemperance or diet. And unlike lead poisoning, which often improved with removal from the toxic workplace, phossy jaw was incurable, leaving the victims it did not kill permanently disfigured. The second factor was the availability of a superior substitute for a product whose chief manufacturer owned the rights to the replacement and stood to benefit from an outright ban on phosphorous. As the debates over labeling laws for paints showed, the paint industry—not limited to the lead corroders—was not about to accede to a ban of the "useful metal."[11]

If lead poisoning was not to be eliminated by replacing lead with a nontoxic substitute, might it be ignored—as was silicosis, an even greater occupational killer? Known by a variety of names ("grinder's rot," "potter's rot," "dust phtyisis," "pneumoconiosis,") and often subsumed with black lung, brown lung, and other pulmonary diseases, silicosis is the degeneration of the lungs caused by prolonged exposure to nuisance dusts. Historians David Rosner and Gerald Markowitz have shown that silicosis was largely ignored in the microbe-oriented world of Progressive Era hygiene, its symptoms confused with tuberculosis. Not until the Depression, when increasing numbers of disabled or unemployed workers successfully sued their former employers for compensation, did silicosis receive serious attention as a separate disease entity.[12]

The relation of silicosis to lead poisoning is analogous to that of lead poisoning to phossy jaw: silicosis took longer to develop than lead poisoning, was even more subject to confusion with other chronic ailments, and occurred in a wider array of industries. But lead poisoning could not easily be ignored. For although silicosis may be the older disease, knowledge of lead's potency as an occupational hazard came centuries ahead of studies correctly diagnosing the complex of illnesses related to nuisance dusts.[13] Lead was a notorious poison, even if the level of its toxicity was imperfectly understood.[14]

Another reason lead poisoning could not be ignored was its economic importance. Silica is a hazard wherever men and women work in an atmosphere of dust and grit; it is not a product in its own right. Lead, on the other hand, *was* the product its poisoned workers made. More important, it was a key component in thousands of other products. "To list all the modern uses of lead," wrote a commentator in 1932, "would be as dull as Homer's catalogue of the ships." Bureau of Mines figures from the 1920s only hint at the ubiquity of the element and its compounds, which were used in products from storage batteries to the "tin" wrapping on cigars. White lead was the base material for almost all paints and ceramic glazes; red lead protected the hulls of ships and the cables of suspension bridges; electric cables were shrouded in lead; and many Americans were buried in lead-lined coffins.[15] The association between an infamous occupational poison and so many products would encourage the industry to protect both its workers and its reputation.

Paint industry executive George Heckel, like many contemporary commentators, argued that the lead industry's investments in hygiene were prompted—in part—by a gendered moralism. Labor economics had begun the process in the late nineteenth century, but "with the coming of the lady doctors into the political picture, there was 'something doing' right speedily." Heckel was explicit about the "feminine" basis for the reforms: "A woman just naturally loves a 'cause,'" he argued. "It has been woman, woman, woman, wherever she has been permitted to express herself, who has forced society to regard the rights and welfare of the individual."[16]

One celluloid-plant manager told Alice Hamilton as they viewed "his" workers, "Now take this crowd. As you look down on them you see so many fathers and husbands and brothers and sons, real men, individuals. . . . All I see is a lot of my hands, a part, and a bothersome part, of my machinery, and that is the way most of them feel. Until we get into industry the woman's way of looking at people we shall never run it as it ought to be run." Hamilton agreed. Although she consistently argued the economic benefits of proper industrial hygiene, she felt her true mission was to awaken "womanly" compassion nascent in the male-dominated industry: to employers and industry doctors "it seemed natural and right that a woman should put the care of the producing workman ahead of the value of the thing he was producing; in a man it would have been sentimentalism or radicalism."[17]

The more cynical view of Progressive Era industry's willingness to make workplaces safer is that it was motivated not by maternal insights or moral rectitude and responsibility, but by material concerns and the sense of impending labor crisis. The efficacy and cost-effectiveness of industrial hygiene proved that a little preventive maintenance went a long way to keep "bothersome machinery" running smoothly. As Dr. Otto Geier, an occupational physician in Cincinnati, reminded his peers: "You don't need to use the cry of humanity at all. [Managers] prefer to have you come in with a business proposition. . . . As a trimming on the side we may speak of the humanitarian work, but we must approach the business man with a business proposition."[18] In a similar vein, Alice Hamilton once argued that "it is of course, far better that the improvements should come in response to the demands of economical management, not of philanthropy, for they are on a much surer basis."[19]

But Heckel reserved some credit as well for the moral muscle flexed by male industrialists. "Let us be just to American industry," he pleaded, "it has acquitted itself well, at great cost to its invested capital, without a whimper." And although he recognized that "the cold-blooded economic proposition" played its part, more than efficiency was involved: "It may not be religion but it has a psychological affinity thereto. It is the general feeling, not born of reason or of calculation, that every one of us is, in a greater or lesser measure, responsible for the welfare and well being of his fellows."[20]

To some degree, this new "affinity" was generational. Many of the younger men who took the reins in lead industries brought with them a spirit of reform.[21] Hamilton told of her efforts to gain the cooperation of the management of a white-lead works she surveyed on the East Coast in

1911—"a truly terrible place." She went to visit the owner, "an old gentle-man" who dominated the meeting. Telling her of the plant's history, he "beamed with pride over its phenomenal growth." Whether overcome by his pride or inhibited by deference to his advanced age, Hamilton de-spaired of confronting him or gaining his cooperation: "I could as soon have told him that a beloved child of his was a criminal as to tell him that his plant was unfit to work in." But as it turned out, the man's son was now in charge of plant operations, and when she met him, he was eager to hear of the problems she had identified, requested Hamilton's recommenda-tions in writing, and "carried out far-reaching changes in equipment and method so that the place was unrecognizable." In another instance, to avoid a similar run-in with an aging plant owner, Hamilton wrote to the patriarch's daughter—the friend of an old boarding-school classmate—who quickly saw to it that a company representative would meet with Ham-ilton and arrange improvements.[22]

If there was a strong generational shift, a moral transformation that would be labeled "the Progressive spirit," conditions were also right for Heckel's "cold-blooded economic proposition" to push American indus-try toward a more "forward-looking" stance on the preservation of its workers' health. Whether part of a larger "search for order" or inspired by a "gospel of efficiency," American industries in the first quarter of the twentieth century self-consciously pursued economic rationalism, prompt-ing many of the business practices characteristic of the Progressive Era: trade associations and mergers, Taylorism, toleration (and skillful manipu-lation) of moderate government regulation, and the rational use of natural resources, or "conservation for use."[23]

Labor, in this new view, was a useful resource to be conserved. Cleve-land industrialist Winthrop Talbot stated in 1912 that a new conviction had arisen among "the employing world . . . that labor to be cheap must be efficient." And efficiency, as Frederick Hoffman had written in 1909, "im-plies a maximum of disease resistance, physical strength, and a long trade life free from serious interruptions caused by preventable illness."[24]

But workers made a most troublesome "resource." Coal and lumber might vary in quality, price, and availability, but they did not strike; water and electricity did not threaten capitalism with violent overthrow.[25] Not only did labor pose a threat of strikes and social unrest, but workers seemed to resist employers' efforts at enhancing efficiency, even when their health was at stake. Investigators such as Hamilton and Pratt frequently derided the "unhygienic" practices of workers, who persisted in eating, smoking, and chewing tobacco in dusty workplaces. George Heckel told of one white lead plant that provided workers with two lockers, one for work clothes and one for street clothes. To get from one locker to the other, it was necessary to pass through showers. Defiantly, the workers took to bringing umbrellas. Another plant offered unlimited supplies of milk to all workers who bathed—milk was assumed to help reduce absorption of lead, while bathing, it was hoped, ensured that exposure ended with the factory whistle. Heckel describes the outcome, in a passage that conveys a great deal about the owners' attitude toward regulation, philanthropy, and the laboring "other":

> *One day a husky, hairy "Hunky," evidently untouched by the cleansing stream, demanded his ration. It was refused, because he was "dirty." The big fellow slapped his dusty chest and shouted indignantly, "My dirt!" And, sentimentally inclined philanthropists take note, the state regulations required daily bathing. You can take a horse to the water but you can't make him drink it if he prefers something else."*[26]

If the hygienists' recommendations and government regulations were to be carried out successfully, industrialists had to show their sincerity and enhance their moral authority over workers. Then occupational hygiene would be seen as a benefit and could take its place in the employers' tool-box for containing labor unrest and acculturating workers to the new industrial order, along with other paternalist (and maternalist) examples of "industrial welfarism" or "welfare capitalism."[27]

Heckel's observation reminds us that the state was not entirely out of the picture—that government regulation, though largely toothless, increased industries' incentives to improve hygiene, if only to forestall more stringent regulation. Throughout the Progressive Era and well into the 1920s, state agencies concerned with factory hygiene continued to multiply, and government medical inspectors played ever larger roles in collecting data and enforcing state ventilation laws.[28] And despite the fact that winning compensation was often a Sisyphean task for the poisoned worker, every successful appeal added a costly incentive for industries to clean their workplaces and ally with industrial physicians who could convincingly, "scientifically" discriminate between true and false claims.[29]

Many Progressives may have feared that World War I would bring the end of reform, but others believed that war could prove instructive. Observed J. W. Schereschewsky of the Public Health Service in 1916, "The gigantic destruction now going on daily in the vast conflict of European nations only emphasizes the urgency for methods of conservation of life and health."[30] The war brought thousands of new workers into industry and produced a tremendous demand for labor, but the horrific conditions typical in many of the new factories suggest that this increased demand did not increase the value placed on workers' lives. Hamilton recalled that in war "the industrial world seemed to be given over to a sort of joyous ruthlessness which, I think, came not so much from greed for profits as from the intoxication of bigness."[31]

The race to produce arms and materiel for the war brought unprecedented state and federal involvement in the daily workings of the plants. The Public Health Service stepped up the investigatory activities of its Division of Industrial Hygiene, formed in 1912. Wartime measures strengthened hygiene programs in federal bureaus and agencies such as the Labor Department, the Ordnance Department, and the Shipbuilding Board.[32] Many Progressive Era reformers, sensing in this spurt of government activism an opportunity to enlarge the state's role in social welfare, pushed for new federal programs in the areas of child labor, working hours, the minimum wage, collective bargaining, and housing reform. National health insurance received spirited support (even the Surgeon General of the U.S. Public Health Service declared that "Health Insurance is the Next Great Step in Social Legislation"), until insurance carriers and doctors stigmatized the whole idea as "Prussianism."[33]

Other reformers moved from private agencies into wartime government positions and those who, like Alice Hamilton, were already government investigators, gained regulatory clout.[34] Although a vocal opponent of the war (she was a member of Jane Addams's delegation to the pacifist International Congress of Women at the Hague in 1915), Hamilton worked tirelessly to prevent death and illness among the workers who manufactured the weapons of war. She found it "profoundly depressing work. It was not only the sight of men sickening and dying in the effort to produce something that would wound or kill other men like themselves . . . but it was also my helplessness to protect them against quite unnecessary dangers." Besides conducting Bureau of Labor surveys of wartime industries and new chemical factories, Hamilton struggled with the frustrating standard-setting process on the War Labor Board.[35]

Before the exigencies of war could bring American industry, American reformers, and American government into full accord on workplace safety, the Armistice was signed, Wilson was dead, Harding was in the White House, and the United States seemed ready to get back to business as usual. Yet, in many parts of American industry, new technologies, labor shortages, and health mandates from state departments of labor or private insurance carriers had fundamentally altered the rules. And for all the nation's official demurs and protestations of detachment from the League of Nations, America's industries had taken a leading place in international economic affairs. The lead industry experienced both increased efficiency and expansion into world markets in the postwar years. Annual exports of refined lead rose from a five-year average of 1,164 short tons in 1901–1905 to 87,006 short tons for 1926–1930. In 1899, 8,751 workers produced 208,466 tons of lead; in 1925, only 6,115 lead workers were needed to produce 651,000 tons.[36] In short, for the American lead industries there was no "as usual" to get back to.

Alice Hamilton noted a drastic change in manufacturers' attitudes toward government supervision immediately after the war. "For the first time I found that my connection with the Federal Government, instead of opening all doors as it always had before, now only made me an object of resentment and distrust."[37] It is easy to ascribe this simply to an unchastened industry flexing its muscle in business-oriented times. But perhaps such headstrong resistance proved that the only regulatory power came from law, and that in the 1920s, industry was essentially above what little law existed.

A more thorough explanation includes what lead manufacturers had learned from Progressive Era surveys, the "compensation crisis," and World War I. Few questioned the economic benefits of improving plant hygiene: increased productivity and cost savings through reduced turnover, lower insurance, and decreased liability awards. The industry had learned both the benefits and the potential danger of government regulation, and that it was something to control, not fight. Furthermore, a large segment of the lead industry had accepted, to an unprecedented level, responsibility for its workers' health. Regardless of whether this change originated in the humanitarian impulses of enlightened industrialists, once such moral arguments were publicly declared as corporate philosophy, the new commitments carried considerable power in the proper institutional setting. The enlarged field of industry-sponsored occupational medicine provided that setting.[38]

From 1900 to 1930, the number of physicians principally involved in industrial medicine increased dramatically. The vagaries of medical practice make it nearly impossible to give firm numbers—for many physicians, industrial medicine took up only part of their time, and others examined and treated workers for more than one company. The growth of new professional associations, journals, and training programs in medical schools reflects the increased importance of industrial medicine. The industry-sponsored National Safety Council began in 1912, as the National Council for Industrial Safety. In 1914, the American Public Health Association created its Section on Industrial Hygiene. And in 1916 about two dozen industrial physicians from around the country established the American Association of Industrial Physicians and Surgeons. From its charter membership of 125, it grew to about six hundred members in three years.[39] Researchers for the Rockefeller Foundation estimated that in 1919, nine hundred corporations employed more than 1,500 industrial physicians, most on a full-time basis.[40]

The American Medical Association had long opposed physicians doing "contract work" with companies, fearing that such arrangements eroded professional autonomy. But the AMA recognized the importance of the growing field in 1915 when Otto Geier, chairman of the AMA's Preventive Medicine Section, directed a symposium on industrial hygiene at the association's annual meeting. In 1922, the AMA formally charged its public health section with responsibility for industrial medicine. The *Journal of Industrial Hygiene,* started by members of Harvard's Department of Industrial Medicine, published its first number in 1919.[41] The practice of occupational medicine had come a long way from the days when Alice Hamilton could find a smelter that hired one doctor for its six thousand workers, each of whom paid $1.75 per month for the doctor's services.[42]

Harry Mock, onetime chief surgeon of Sears and Roebuck, in 1917 contrasted the industrial physician of his generation with the company doctor of twenty years earlier. His predecessor's "standing in the profession was of a very low average and the character of his work was of a low standard. He was a company surgeon in word and deed and too often was only on the side of the employer as represented by the insurance company."[43] Historians of public and industrial health tend to share Mock's appraisal, but extend it to his generation of industrial physicians—and beyond. In doing so, they fail to see past the obvious strategic advantage of occupational medicine, casting company doctors as simple tools of capitalist exploitation. Certainly many workers may have seen the company physician in this light. Their unions consistently opposed the practice of medical examinations because of their potential for abuse by employers seeking additional power over hiring and firing.[44] But even if the industrialists' intentions were as monomial as some historians' arguments are monocausal, the results of their decision to enlist the physician in their war against labor were manifold.

Many medical schools created or enhanced training in occupational medicine in the first two decades of the twentieth century. In 1919, Alice Hamilton's preeminence was so widely recognized that when the Harvard Medical School decided to hire a specialist in industrial medicine, it could find no one more qualified. Harvard had never hired a woman, but after the Harvard Corporation addressed the sticky issues of whether to allow her into the Harvard Club (not even professors' wives were admitted) or to attend commencements ("under no circumstances may a woman sit on the platform"), she was hired.[45]

Industry had embraced the medical profession, the profession had accepted the occupational physician, and universities and medical schools had begun providing the scientific underpinnings of the new field. The stage was thus set for the relationships that would prevail in the lead industry until the 1960s. Privately employed occupational specialists supervised employees' health in the workplace, serving both as shields and gadflies. They kept workers well while protecting the company from what their science considered unwarranted liability of workers. Further, their presence in the factory constantly reminded management of the costs of sick workers. With statistical analyses provided by members of their professional organizations, they reassured management of the cost-effectiveness of their programs.

These professionals took their standards and procedures from research conducted at respected, ostensibly "objective" academic centers—research paid for largely by the industry itself. The lead-poisoning studies conducted in the 1920s by a team of Harvard Medical School researchers, funded almost entirely by the white lead industry, show clearly the symbiotic relationship that developed in the 1920s between industry, the academy, and industrial physicians. They also reveal the potential for abuse.[46]

In 1905, Richard Clarke Cabot established a Social Services Department at Massachusetts General Hospital (MGH). The department, run by social worker Ida Cannon, served as an out-patient clinic, treating chronic and minor illness and amassing data on industrial diseases.[47] In 1913, David Edsall, the Jackson Professor of Clinical Medicine at Harvard, began a program in the Social Services Department to increase the amount of occupation-related data collected from MGH patients.[48] Alice Hamilton reported that MGH was one of only two hospitals she had visited whose records were complete enough to be of any use in her investigations.[49]

In 1916 Edsall opened a second clinic at the hospital, operated by physician Wade Wright, specifically for industrial diseases. Edsall and Wright concentrated on lead poisoning and, as has become a commonplace in lead research, found that plumbism appeared when you looked for it. "In a period of five years before the industrial clinic was established," Wright reported, "147 cases were diagnosed as lead poisoning in the various departments of the hospital. In the first year, the industrial clinic diagnosed 148 cases of lead poisoning."[50] In that year alone, the clinic treated more than 5,100 of Boston's industrial workers. World War I temporarily closed the clinic's doors, but after the war it remained in operation until 1924.[51]

After the war, the clinic's primary focus shifted from collecting data and treating workers to helping industries improve hygiene through the application of "objectively" obtained scientific knowledge. This change in focus brought a new funding regime. Harvard's Division of Industrial Hygiene, which Edsall established in 1918, relied for its basic operating costs on a constant influx of funding provided by manufacturers through a "subscription" program.[52]

Early in 1920, Alice Hamilton, then teaching at Harvard Medical School, met with Frank Hammar, president of a large white lead company in St. Louis, to discuss the possibility of paint makers funding basic research in lead toxicity. Hamilton then asked Edsall to pursue the matter more formally. Edsall wrote to Hammar about the Medical School program's activities and goals, and suggested that a mutually beneficial arrangement might be possible.[53] Hammar secured the approval of the other members of the American Institute of Lead Manufacturers, a loose affiliation that would become the Lead Industries Association. Next Hammar met with Edsall, Hamilton, and other researchers to set the basic terms for a research program funded by the lead industry. Edsall insisted that the program should not be limited to occupational concerns, "since any sort of restriction implies the acceptance of conventional ideas relative to lead poisoning." He also demanded that the findings be published "in medical periodicals which carry authority."[54] If the American Institute of Lead Manufacturers could assure this degree of freedom—and $51,000 for a three-year program—the Harvard researchers could get to work. In February 1921, Hammar gave a tentative go-ahead.[55]

The initial study, directed by Edsall protege Dr. Joseph Aub, produced a number of papers published in the *Journal of Industrial Hygiene* between 1922 and 1925. Subsequently collected in the succinctly titled monograph *Lead Poisoning*, published in 1926, the studies formed the most important single body of research on lead poisoning since Tanquerel's 1839 *Traité des Maladies de Plomb ou Saturnines*. At the end of the three years, Aub secured funds from the lead industry to carry on smaller studies and grants of between five and ten thousand dollars annually, "for the care of patients and for the incidental expenses attendant upon the clinical and laboratory work here at the Hospital." As late as 1934, the LIA was making regular contributions.[56]

In an annual review of progress for the Medical School, Edsall reported that "the investigation resulted in an intimate relation with the lead industry, most profitable both to the manufacturers and to medical science."[57] He did not mean that the research was designed to please or profit lead makers. He did not intend to study lead poisoning "in the light of preventive medicine, but with the desire to understand more fully the chemistry, physiology, and clinical aspects of the disease itself." The Harvard researchers' most significant findings related to the chemistry of lead absorption and excretion in blood, bone, and soft tissues.

The group argued that the funding scheme afforded them complete academic freedom. The study's third annual report claimed that the sponsor "has not at any time offered suggestions as to any policy or point of view, but has allowed the work to run its course and to demonstrate the truth without regard for a possible economic effect upon the lead industries." Aub wrote to a lead manufacturer that "it is a great pleasure to be able to state that no one connected with the Lead Institute has in any way tried to influence our investigations, conclusions, or publications."[58]

But while the lead manufacturers were deferential to the doctors at the prestigious university, they were anything but disinterested. Early in the study, Edward Cornish, president of the National Lead Company, wrote to Edsall. He began with a disclaimer: "It would be unbecoming for us to make even a suggestion as to your line and methods of research." But he did have a suggestion: his peers would find it useful to have a scientific answer to the question of whether lead is more dangerous when breathed or swallowed.[59] Alice Hamilton had convinced Cornish of the importance of ventilation and respirators back in 1911, and by her account he had succeeded in getting his workers to cooperate. Hence, it was not his employees as much as his customers to whom he hoped to offer the scientific confirmation. Two years later, well into the research, Cornish reported: "Your finding that lead is relatively more dangerous when absorbed through the lungs will assist us in inducing painters to adopt wet sandpapering." Cornish's benign meddling was consistent with Alice Hamilton's highly favorable appraisal of his motives and actions.[60]

Another lead maker offered free advice, along with his company's check for $3,141.13, urging the researchers to remain skeptical of false claims

because "every time an employ of a lead factory has any sort of indisposi-
tion, no matter of what character, it is charged to the fault of the lead. . . .
I trust you will pardon my suggestion to you, therefore, that a very great
deal of the published records of lead poisoning cases are entirely unde-
pendable and not properly to be considered in arriving at conclusions in
the investigation which you are carrying on."[61]

Most of the correspondence between sponsors and the Harvard
researchers echoes this tone. The manufacturers displayed obvious pride
in "their" research program, but their letters do not suggest that they
wielded significant personal influence. Still, it seems unlikely that the
source of funding had no effect on the progress of the studies. Joseph Aub
kept in close contact with the lead industry for the rest of his career, and
while it would be unfair to characterize him merely as an apologist for the
LIA, his relationship with the manufacturers over forty years was warm
and, it would appear, mutually beneficial. In the 1940s, as the nature and
scope of pediatric plumbism was becoming better understood, physicians
at Boston Children's Hospital occasionally had to deal with both the findings and the reputations of Aub and his associates at Harvard Medical
School.[62]

In time, the major lead producers would take more direct control of
medical research that affected the production and use of their products.
As we will see, the manufacturers of leaded gasoline, the Ethyl Corporation
and E. I. du Pont de Nemours, would sponsor hundreds of specific medical, chemical, and epidemiological studies of lead toxicology and establish
a global network of industrial physicians and managers to promulgate standards for the manufacture and safe handling of leaded gasoline. In the
absence of significant government funding, lead-industry–sponsored
research became the world's most important source of both theoretical and
practical knowledge about the toxicity of lead.[63]

Alice Hamilton proudly noted that at the International Congress on
Occupational Accidents and Diseases held in 1925, the United States was
widely hailed as the world's leader in research into occupational medicine,
in stark contrast to 1911, when conferees snickered at the very notion of
occupational hygiene in America.[64] A large measure of the credit for American industrial hygiene's improved standing no doubt belongs to the
government-sponsored surveys of the Progressive Era. But by the time of
the 1925 conference, occupational hygiene research in the United States
had more than grown; it had metamorphosed. The union of professional
medicine, modern industry, and a weak state had produced a uniquely
American institution of industrial hygiene. Lead-industry sponsorship of

academic research, as well as primary research conducted by industrial hygienists themselves, served both the hygienists and the manufacturers by providing the scientific basis for industry's claims that it could protect its own. And if serious occupational lead poisoning continued in the smaller plants whose dilatory owners refused to follow the LIA's lead, that merely underlined the claim.

SIX

Company Doctors
on the Job

By the mid-1930s, lead-industry physicians formed a loosely knit professional community with a body of shared assumptions, medical and technological know-how, and managerial influence unimaginable to their nineteenth-century counterparts. They attended medical conferences where the published findings of Robert Kehoe or Joseph Aub were discussed, debated, and picked apart in terms of the impact they might have on plant operations.[1] They pitched their hygiene programs to management. And they wielded tremendous power over employees' daily lives: they decided who worked and who was fired, set strict hygiene regulations and mandated regular physical exams, took specimens of blood and body wastes, and fed or injected workers with various minerals and chemicals to control their absorption of lead. The modern plant physician alloyed medical authority with managerial power, forging a new set of relationships among worker, boss, and body.

You let us have your guinea pigs, and we will let you have our men.

DR. K. A. KOERBER, NLC
PLANT PHYSICIAN, TO
HARVARD MEDICAL SCHOOL
RESEARCHERS, 1937

Industrial physicians in lead plants could take a measure of professional validation from their industry's affiliation with important primary research in prestigious institutions. But the theoretical breakthroughs this research produced were of limited value in their daily business of keeping employees on the job. Increased understanding of the physiology of lead poisoning was less significant in reducing it than was the simple increase in the number of plants that employed elementary, old-fashioned methods of prevention and exposure control. Until effective treatments became available after World War II, company doctors continued to rely on basic ventilation and cleaner processes, preemployment interviews and medical screenings for "susceptible" workers, and prompt removal at the first sign of a worker's intoxication.

In the 1930s, to get a job in a lead plant that adhered to modern standards of hygiene, a worker first had to pass "a stringent preemployment examination by the physician himself" that included the taking of a detailed history to uncover past infections such as tuberculosis and syphilis.[2] K. A. Koerber, the physician at the NLC's Philadelphia plant, claimed that he administered "the same examination exactly that the United States Army usually gives its officers, plus the blood count and the clinic test." Elston Belknap, a noted industrial hygienist for a large battery manufacturer, advised physicians to turn away applicants who might already have absorbed lead and recommended probing their past employment for signs of previous exposure.[3]

G. H. Gehrmann, medical director of DuPont's tetraethyl plant, counseled plant doctors to consider intellectual and psychological factors as well. Those "decidedly below normal intelligence" did not qualify; nor did the "wise guy" who would "invariably fail to follow instructions and often take keen delight in disobeying them." Gehrmann recommended describing in full detail to the prospective employee the dangers of the work and the precautions to be followed. "He is dismissed from further consideration if . . . he expresses fear. Employees who are afraid of lead work are very liable to be constantly presenting subjective symptoms with no objective findings, and soon become a problem to the examiner."[4]

After the interview, the doctor's probing turned to the worker's body, starting with the mouth, where the tell-tale sign of a lead line might appear "in the gum near teeth which are carious or surrounded by active pyorrhea." The absence of a lead line did not mean a clean bill of health, however. Simple oral hygiene could prevent it, so lead-poisoned workers with good teeth might show no dots of lead sulphide on the gums. Therefore, the preemployment examination included detailed laboratory studies: "blood Wassermann test, routine urine examination, complete blood count and a rigorous search for stippled blood cells." In the latter test, a stained blood smear was examined for cells containing dark granules, or stipples. A high stipple count indicated anemia—not unique to lead poisoning, but a crude indicator of lead absorption in subjects known to have had significant exposure.[5]

From the employer's perspective, the obvious benefit of rigorous preemployment exams was the guarantee of a staff of fairly healthy workers—at the outset, at least. But the blood tests did not end with employment. Only careful surveillance by way of periodic reexaminations would give early warnings of hazardous conditions and point out workers whose "constitutions" predisposed them to lead poisoning. Dr. Koerber advised monthly workups in order to "pick it up before they get lead colic."[6]

Keeping workers healthy required the cooperation of engineers, management, physicians, and the workers themselves. Engineering probably played the greatest role in prevention. What Hamilton found in the 1910s remained true decades later: the factories with the lowest levels of lead poisoning controlled the amount of lead dust circulating in the factory air. A smelter manager told a group of industry physicians in 1937 that "we're spending about nine dollars for prevention to one dollar for cure." He gave three examples of simple improvements—filters, vacuum-operated barrel fillers, and ventilated hoods. He argued that "there is no great increase in the expense of the operation of the plant and it is reducing the hazard." This he determined by his physicians' reports of workers' stipple counts—an early example of using the workers as "biological monitors."[7]

The efficacy of mechanical improvements in hygiene, explained by Aub's findings that lead was more readily absorbed into the bloodstream if it entered the body through the respiratory rather than the digestive system, put to rest many of the beliefs that had laid blame on workers' personal habits such as chewing tobacco, eating on the job, or drinking (on or off the job). Ventilation and filtering were morally neutral, protecting the "slovenly" and the clean alike. Moving the center of responsibility to factory owners encouraged greater vigilance on their part. The Progressive Era paint manufacturer, after conferring with his lawyer, might have gotten away with posting safety rules whose primary purpose seemed to be to provide a jury with evidence that the workers should have known and followed them, permitting the owner to blame scofflaws for poisoning themselves.[8] His counterpart a generation later, after listening to his company doctor and his insurance agent, would direct his shop foremen to enforce greater compliance with company hygiene rules.

Under the regime of occupational hygiene, controlling workers went far beyond requiring them to wear protective clothing or take showers. At regular intervals—twice a year in some plants, once a month in others—each worker went to the company clinic to give blood or to turn in urine or stool samples. Determining the amount of lead in small samples of blood—the standard measure in recent times—was not routinely done, as this required costly equipment. Most smaller plants had to rely on microscopic observation in which technicians scanned blood smears for signs of cell damage, including stippling and decreased hemoglobin. Many employed a blood test that identified the precursors of stippled cells, giving physicians a limited opportunity to act before the onset of blood damage and to prevent lead poisoning.[9]

Diagnosis and classification of poisoned workers varied with the policies of the plant, the physician's training, and precedents set in workmen's compensation cases. In general, physicians classified saturnine workers by a rough tripartition. Almost all workers showed some signs of lead "absorption"—a few stippled cells, a trace of lead in the urine, perhaps a bit of a lead line on the gums. Usually this level of exposure produced no further symptoms. The second level included those workers whose laboratory tests and physical symptoms agreed that they were poisoned, but who were not sick enough to require removal from the plant. The third group exhibited dangerously high lead levels or blood damage and were incapacitated to some degree by abdominal or neurologic symptoms, such as wrist drop or palsies. The worst of these cases involved the brain itself. Lead encephalitis was often fatal and almost always resulted in permanent brain damage. By the 1930s, it was very rare in occupational settings.[10]

Most problematic, both for plant operations and in determining workmen's compensation, was distinguishing between "normal" lead absorption and "abnormal" intoxication, or poisoning. As Koerber exclaimed to his peers, "Lead absorption and lead intoxication,—I don't know just how you fellows do your analyses and so on—but I don't think you can take a blood on anyone that has been around lead without finding stippling. . . . What I'd like to know is where does absorption begin and where does intoxication begin?"[11]

Individual susceptibility complicated the definition. Some workers whose blood and urine showed only slight evidence of lead still contracted gastrointestinal symptoms such as colic or appetite disorders. From the plant physician's perspective, these workers seemed remarkably sensitive. "They taste lead from the day they get in there and come to you and tell you they always taste lead." At the other extreme were those whose laboratory results suggested massive absorption, but who made no complaints and appeared perfectly normal on examination. Obviously, setting standards based on the most or least susceptible had serious economic consequences. Koerber warned that since almost all workers had some degree of blood stippling, "if you are going to call that absorption you'll have to shut all the plants down."[12] But setting standards based on the seemingly impervious lead worker or limiting intervention to those who exhibited clinical signs of more serious absorption invited higher compensation or liability costs: compensation boards or juries could easily disagree with the company's definition.

These uncertainties reinforced the lead industry's need for "scientific" and "objective" standards. In April 1937, Felix Wormser, secretary-treasurer of the LIA, called a conference of physicians employed in member companies. He invited Joseph Aub of Harvard and Lawrence Fairhall, formerly of Harvard and now principal industrial toxicologist at the Public Health Service, to speak to the physicians who gathered at Palmer House in Chicago. Both dwelt on laboratory measures and refinements in diagnosis and chemical analysis but offered little that would clear up the problem of definition. At the conclusion of the conference, Dr. Koerber recommended that the association form a committee to conduct inspections and set standards: "My feeling is that if we had our own policemen going around, and put some teeth in the [inspections], maybe we could clean it up ourselves." Then he turned to the speakers. "I'd like to see Dr. Fairhall and Dr. Aub get together. They monkey with guinea pigs, but we are working with human beings. You let us have your guinea pigs, and we will let you have our men. You come and visit our plant, whenever you want to, and we'll work with you."[13]

At the time of the conference, the fastest growing lead industry was the manufacture of tetraethyl lead, and it already had its own policemen. Robert Kehoe was its J. Edgar Hoover. The LIA would gradually follow suit, but as a voluntary trade association with no rigid patent controls such as those the Ethyl Corporation held on tetraethyl lead, it had no teeth to put into inspections. Nor, of course, did Fairhall's agency. "The United States Public Health Service has no police power," Fairhall explained to the conferees, repeating almost verbatim the words Alice Hamilton had used two decades before: "They enter an industry and examine it only at the invitation of the industry." A PHS assistant surgeon despaired of finding absolute standards, and agreed with Aub that "in the last analysis the diagnosis of plumbism must depend on keen judgment . . . no fast rules can be adopted."[14]

Once physicians identified, by whatever standards, a case of lead poisoning, treating it involved another set of controversial choices. Before the advent of safe and effective chelating agents, treatment was a dangerous game. Aub argued that since "lead in the bones is inert" and "doesn't do any harm," the quickest and safest approach was to force its immediate deposition in the bones. "When a person has been relieved of pain," he asserted, "practically all of the lead will be found in the bones." Once there, the lead could be drawn out by injection of chemicals and dietary changes, with medical staff monitoring the process by frequent analysis of blood and urine. Deleading was a delicate process; if done too quickly, acute symptoms or death could result. Aub admitted to a group of medical officers that in developing the protocol he had "precipitated five lead colics, and one case of encephalopathy."[15]

This treatment followed two long and painful steps. One sixteen-year veteran of a New York lead smelter, identified as S. M., went to the Jewish Hospital in Brooklyn two weeks after his long-familiar symptoms of chronic plumbism flared up into full-blown lead poisoning.[16] Every day for three weeks, the 45-year-old man ate a calcium-rich diet, swallowed calcium lactate and magnesium sulfate, and endured daily injections of calcium gluconate. When his colic subsided, his doctor began the deleading, which took another three weeks in the hospital. He was switched to a low-calcium diet and given phosphoric acid and magnesium sulfate to leach the lead from his bones. The amount of lead found in his urine and feces nearly doubled during this process. After an additional three weeks in the hospital, he was sent home to rest for one month, following which he was readmitted for more treatment. When his doctor reported this case to a panel of the American Medical Association several months later, S. M. had been through three treatment cycles and it was anticipated that he would have to endure at least one more trip to the hospital.

Less severe cases presented occupational physicians with more options. Removing the worker from exposure was one—easily the most popular in Hamilton's day, when it was accomplished simply by firing the sickened worker. But modern labor policy discouraged high labor turnover; besides, management wanted to minimize the number of men dismissed for health reasons because, as a National Lead Company physician explained, "the minute he is laid off, he is going to run to the lawyer's office."[17] Another option was to offer poisoned workers temporary assignments in safe areas. Some companies farmed susceptible workers out to different factories or, as another doctor recommended, formed "sunshine gangs" of at-risk employees who would do jobs outside or around the plant, "at the same salary or wage" as their normal jobs paid.[18]

A stint on the "sunshine gang" was meant to end with return to the stackhouse. Often, however, having shown once their "susceptibility" to plumbism, these workers would again turn up at the clinic with stippled blood cells and lead colic. Laying them off remained a liability, but allowing them to live in an endless cycle of deleading, removal from work, and illness was an affront to many physicians' sensibilities. "Your concern is the health of the man," G. E. Brockway of the NLC entreated. "You are merely performing your duty as a physician to get rid of that man in this stage rather than wait." But the doctor was also a company employee, and physicians complained that even advising workers that they should find safer jobs could land them on the manager's carpet. In an opinion that tells a great deal about the industrial physicians' divided loyalties, Dr. D. R. Johns of the Anaconda Sales Company declared at the 1937 LIA conference, "Where a man has more than one attack I think our company will agree that he is a poor hazard and they don't want him around." He *thinks* that *our* company will agree. Even speaking before a friendly audience of his peers, he indicates that he is not confident that his opinion will decide the action taken. And he locates risk not in the factory, but in the worker, who "is a poor hazard" to the company—not the reverse.[19]

One option remained. Many physicians in lead factories steered a dangerous middle course between firing sick workers and retaining them by employing various strategies of prophylactic deleading. For decades, plant operators had tried to prevent lead absorption in their workers by providing free milk and administering various mineral dietary supplements. Once Aub's research confirmed the link between the calcium metabolism and lead, resourceful physicians turned to prevention by way of routine treatment. One physician at the Glidden plant began giving workers daily doses of ammonium chloride. Initially he prepared a liquid suspension, but after workers began to resist swallowing a teaspoonful of the nasty concoction, he found a way to administer it by pill. Workers' daily ammonium chloride at Dr. Koerber's NLC plant came coated in licorice.[20] Candy coating did nothing to eliminate the danger of administering toxic chemicals, and the practice of prophylactic deleading remained controversial. Debate, however, was largely confined to professional gatherings. Years later, when more effective deleading agents became available, the debate grew more heated—and public.[21]

In the years after World War I, American factories hired a veritable army of industrial physicians to monitor the health of their workers. These company doctors, proud of the services they rendered their patients yet zealous in their loyalty to their employers, wielded increasing power over factory conditions. Management often recognized the value of an active occupational hygiene program. L. G. Reichhard, an NLC officer, told physicians that he knew of a factory that had just spent "a hundred thousand dollars on sanitary equipment." But to the management of that factory, "it is worth a million dollars." He urged company doctors to be assertive: "Tell the management what you men know and they'll listen. Go tell them face to face what you are after and they will go all the way with you."[22]

Industrial physicians had to strike a delicate balance of loyalty to the company, responsibility to workers, and the prestige of their profession. One of their regular tasks was to testify in workmen's compensation cases, inevitably downplaying lead's contribution when doubt as to etiology existed. Here their loyalty to the company appeared clear, although they could rationalize, as did Carey McCord, that the system worked in most cases, since "unquestioned industrial lead poisoning cases seldom get into the courts." And when company physicians judged the success of their hygiene policies by the decline in claims for compensation, were they not also expressing pride in how much sickness those policies prevented?[23]

Carey McCord was one of the century's best-known company doctors. Born in Alabama in 1886, he earned his medical degree from the University of Michigan in 1912 and remained in Ann Arbor to work as a laboratory researcher in endocrinology. During World War I he served as an army surgeon, and his commanding officer convinced him to pursue clinical medicine, chiding him for his investigations of the physiology of tadpoles. "Working with men, not with tadpoles, is the real game," he had said.

After the war, McCord moved to Cincinnati, where he established a consulting service, the Industrial Health Conservancy Laboratories. After fifteen years of conducting investigations and serving as a consultant for manufacturers, government agencies, and medical societies, he became director of the Chrysler Corporation's medical department in 1935. Throughout his career he also contributed to his profession, teaching at the University of Cincinnati, Wayne State University, and the University of Michigan. He was always involved in professional activities, serving on committees of the American Association of Occupational Physicians and Surgeons and the American Public Health Association, as well as editing the journal *Industrial Medicine and Surgery,* for which he wrote his series on the history of lead poisoning in America.[24]

Two brief stories from McCord's autobiographical collection of "Vignettes of the Maladies of Workers" show the extremes to which industrial physicians went in service of all the recipients of their divided loyalties. During his time in Cincinnati, McCord conducted a routine course of blood tests in a battery factory, using the basophilic aggregation test (BAT).[25] The blood of one worker, Marshall Grant, showed over three times the basophilic aggregations usually present in mild cases of lead absorption, though he exhibited no signs of illness. McCord told the worker that "if this laboratory test is worth anything you ought to be in the hospital." McCord had more than an academic interest in the case, since he had developed the test. "'Go on, Doc. I never felt better in my life,'" Grant replied, and to prove it, he "picked up a fifty-pound tin of lead oxide and raised it above his head."

The next day, the plant foreman called McCord to report that Grant had "gone crazy," and had nearly fallen off the roof of his father's house. McCord and another young doctor drove far out into the Kentucky countryside to the elder Mr. Grant's home. Inside, they found a delirious Marshall. Three neighbors were "attempting to hold him on the bed," but Marshall, thrashing in violent convulsions, "from manifest and unusually severe lead encephalitis . . . was tossing them about with freedom." McCord's companion had brought an intravenous solution that could reduce the brain swelling, but the doctors quickly abandoned any hope of administering it by that route.[26]

Discouraged, they told Grant's family that without the facilities of a hospital there was little they could do, and that Marshall would probably die. But then they decided that perhaps the drug could be administered through other means, so they made a concentrated batch of the fluid and with the help of several of the neighbors, held Grant down and "kept up the fight until [they] had introduced some three-fourths of the quantity desired" through an enema tube. Exhausted, "with little hope that much would be accomplished," the doctors departed. The next morning, Sunday, the doctors again made the long drive, and "with that customary distress that attends any physician when a patient dies," asked about Marshall's condition. "'Oh, Marshall,'" his father beamed. "'He's at church now. He got up this morning and said he was all right. . . . That certainly was fine medicine you doctors gave him last night.'"

McCord told his "war stories" with a rich mixture of irony and gallows humor, suggesting that a steady stream of such tales inspired gales of laughter at industrial medicine conferences over the years. But such heroic acts were not always performed in the service of the patient. Another stock story had the doctor, by clever detective work or duplicity, undercutting the claimant in compensation cases. McCord's "The Doctor and the Rabbi" is a classic of the genre.[27] The insurers of a printing company hired him to determine whether lead poisoning had killed an employee whose dependents were suing. Since the printer had never been diagnosed with plumbism, had not complained of any symptoms, and had not worked directly with lead, McCord suggested that only laboratory studies of the decedent's bones could answer the question. Of course, this meant exhuming the body, "many months after the burial."

Ordinarily an arduous task, this autopsy was "attended by unprecedented difficulties" because the dead printer's family were orthodox Jews, and their rabbi forbade the removal of any tissues. However, McCord recalled, the family "had whispered to me that I should do whatever was necessary and not pay too much attention to the rabbi." So, after some negotiation, the rabbi agreed to observe the autopsy, which was to be performed in the cemetery chapel. Unfortunately, the grave diggers found that ground water had flooded the grave, and the body could not be removed. Doctor and rabbi proceeded to the graveside and McCord "descended into the grave," where, nearly overcome by the fumes of hydrogen sulfide gas, he examined the body.

Despite its state of extreme decomposition, McCord was able to determine that the printer had died of pneumonia, but knowing "that claims of plaintiffs' attorneys are not always reasonable, and that the plea would be made that both lead poisoning and pneumonia might have existed," he determined to remove some bones for analysis. But "the set-jawed rabbi watched every movement, rigidly determined that I should not remove any portion of the body." McCord played for time, spending over an hour in the cold, doing unnecessary examinations, calling for more bailing, sponging and cleansing. It worked. Suddenly, the rabbi fainted, "almost falling into the grave" with McCord. As the gravediggers carried the stricken rabbi to the chapel, McCord removed three bone samples. "These I thrust inside my boots, completely out of sight." After a moment the rabbi reappeared, "pale, wan, and wobbly," but resolved to resume his post. McCord announced he was finished and "reeled back to the chapel, gloating over the fact that necessary samples of bones were rubbing my ankles, secure in the belief that the rabbi was wholly satisfied that no violations had taken place." As he expected, the bones showed that "no more lead was present than would be expected in the average, unexposed adult."[28]

These two stories reveal a great deal: about Carey McCord, of course, but also about his generation of plant physicians, who sometimes worked under primitive conditions and for whom improvisation, experimentation, and estimation were the norm. More than anything, they underscore that industry-sponsored occupational hygiene was, for workers, a mixed blessing.

Industrial workers had good reason to question the degree to which the company doctor had their best interests at heart. But workers in smaller plants or in regions of the country where the new industrial culture of hygiene had not taken hold might have gladly accepted its harsher features in exchange for the protection it offered. Hygiene programs like those of the Lead Industry Association companies discussed here were by no means universal. With no outside body regulating plant conditions, many employers continued to ignore the economic and moral logic of protecting their workers.[29]

Anecdotal evidence suggests that region was a strong predictor of adequate medical protection on the job, although the very absence of such programs in nonindustrial states makes quantitative comparisons meaningless. For example, a Texas smelter established during the Depression to salvage lead from scrap employed no trained occupational hygienists. In 1932, a laborer from the plant was admitted to the City-County Hospital in Dallas with severe colic. The initial diagnosis resulted in his being scheduled for an emergency appendectomy, but "by some mischance" the surgery did not take place. The next morning, a young internist with experience in industrial hygiene and alert to occupational illness, questioned the diagnosis and ordered a lead test, which confirmed the true nature of the worker's pain. His interest piqued, the doctor visited the plant, where the concerned owner told him that "five or six of our boys have had similar trouble. I wonder if there might not be a problem here. I have asked a lot of people, but nobody seemed to be interested in it."[30]

In any region, small, unaffiliated plants were risky businesses. In 1925 a committee of the American Public Health Association found that 60 percent of America's industrial employees worked in plants that employed fewer than five hundred workers, and that less than 1 percent of these plants had "adequate medical departments," defined in part as at least part-time, on-site medical supervision in a clinic equipped with minimal supplies. The smaller the operation, the more likely that economies of scale made a medical department unprofitable. And as long as management could control the costs of sick workers—whether by rudimentary hygiene or by locating in a state with weak factory-inspection laws or with workmen's compensation that did not extend to industrial diseases—small plants could continue in business, taking a little piece of their workers for every ton of lead produced.[31]

The apparent correlation between company size and worker protection suggests why the painters' trades seemed almost immune to the improvements in occupational safety that workers in many other lead-using industries enjoyed. Painters, then as now, were often independently employed, free from the supervision that insurance companies required factories to provide. Painting has always involved contact with deadly chemicals. Until recently, the professional painter was expected to custom-blend bases and finishes for each job, and hence had to be familiar with a wide assortment of highly toxic chemicals, including benzene, naphtha, turpentine, and, of course, lead. The use of red lead oxide as a rust-preventive paint grew with the increased production of ships, bridges, and automobiles in the first decades of the century. The American house painter's preference for white lead paint persisted, and although some claimed in the 1920s that less lead was being used in house paint, production statistics indicate otherwise.[32] But even as safer pigments replaced lead in new coatings, painters' exposure to lead continued. Preparing to repaint a room with new nontoxic paints still required sanding away the old paint, usually with dry sandpaper, which produced clouds of white lead. Unfortunately, the closest most painters came to a respirator was a dust-choked bandanna worn over the face when the air got too thick. Many ate as they worked, most did not change clothes before lunch, and few jobs provided shower rooms or required clean coveralls. The health costs were staggering.

A sampling of statistics on the painting trades in the 1920s and 1930s reveals the price painters paid for their independence. A study in 1918 by New York City's Department of Health showed that 40 percent of the 402 painters surveyed had an "active case" of plumbism, and that sickness forced the average age of retirement for painters down to 50. Painters across the United States suffered a disproportionate share of fatal plumbism. Over half of the 421 Americans who died from lead poisoning in the years from 1925 to 1927 were painters. The number of fatally lead-poisoned painters was more than double the number of lead-poisoning deaths from all other occupational groups combined. In 1935, thirty-six painters died from lead poisoning—approximately 27 percent of the 130 lead-poisoning fatalities that year. Probably many of the thirty-six farmers who succumbed to plumbism that year were also poisoned while painting. In 1942 the number of lead-poisoning deaths fell 57 percent to fifty-six cases, but twenty-three were painters, representing a drop of only 36 percent for that occupation. These figures must be seen as very conservative because, as Alice Hamilton observed in 1913, "for many reasons mortality statistics are of little value in a study of lead poisoning. Rarely does a painter die of uncomplicated and typical lead poisoning."[33] Even this fails to account for the many doctors who would not have recognized an "uncomplicated and typical lead poisoning" if they saw one.

A number of unions represented American painters. Hamilton's 1913 survey of the painters' trades mentioned four, including those for sign painters, shipyard workers, and automobile and coach painters; she estimated membership at about one painter in four. In many ways, they resembled guilds more than industrial unions, focusing on strict apprentice-ship requirements, setting prices, and extending the employment season, which in northern cities might average less than two months.[34]

Although hours, wages, and expanding the number of union jobs domi-nated union activities, painters and their unions did make some sporadic attempts to promote safer working conditions. Hamilton noted a strike in 1913, in which the Brotherhood of Painters, Decorators and Paper Hangers (BPDP) demanded that painters using lead be provided with respirators, gloves, and adequate time (as well as hot water, soap, and towels) to wash paint off before eating or leaving the job. According to Hamilton, "this is said to be the first time the union has effectively dealt with health ques-tions."[35]

Ten years later, the New York chapter of the BPDP made a more con-certed effort by opening the Painter's Health Bureau, a union-funded clinic staffed by a physician, a dentist, a nurse, a roentgenologist (with x-ray ma-chine), and a lab technician. The proud union claimed that its clinic, which was open four evenings a week plus Saturday, was the first such union institution for the prevention of disease. Data from the first year showed the program's value: of 267 men enrolled, 59 percent showed lead anemia, and the staff uncovered four cases of clinical plumbism.[36]

The Painter's Health Bureau was established with the help of the Work-ers' Health Bureau of America, a small group of women who from 1921 to 1928 fought for improved working conditions through unionization and worker self-education. Founders Grace Burnham and Harriet Silverman sought to raise workers' consciousness of the dangers their workplaces pre-sented and of their status as an exploited class that should not count on management for protection. During the early 1920s the bureau benefited from organized labor's cooperation and funding. But that support gradu-ally faded over the course of the decade, as unions in general struggled merely to survive.[37]

The pages of *The Painter and Decorator*, the official magazine of the BPDP, suggest how little health issues impinged upon the union's day-to-day concerns and how untroubled was the union over the toxic paint its members employed. It would seem that the nation's largest painters' union remained an organization of "white-leaders." In a typical issue, lead manu-facturers bought over half the advertising space and a full-page ad for the National Lead Company (makers of Dutch Boy) would be prominently displayed on the inside front cover.[38]

The Painter and Decorator rarely published articles directly related to health. An exception was a series by one W. Schweisheimer, M.D., on the health of painters, which appeared in the late 1930s. Initially his articles focused on lead poisoning, its causes and prevention, a comparison of American and European statistics, and so on. He went so far as to discuss the long-term effects of plumbism on general mortality and the link between chronic lead poisoning and kidney disease—a correlation that was fairly well established, but one with costly implications for the lead industry. After this, his tone softened considerably. In an article on "Heart Trouble among Painters," he claimed that recent improvements in painters' cardiac health were largely due to reduced use of toxic materials including benzene, toluene, arsenic, and naphtha; "chromates" were also on his list, but notably these were not "lead chromates" or "chromates, especially lead." His avoidance of lead was more remarkable when he stated that "painters show frequently a tendency to constipation and the pressure which is necessary in such cases is bad for heart and blood vessels," without mentioning the most probable cause of that constipation.[39]

William Absalon, a commentator in *The Painter and Decorator*, warned his fellow painters in 1939 that they were suffering "from what the sociologists and technicians call 'cultural lag.' We have been notoriously backward," he chided, "about keeping in step with the other trades." The "culture" to which this critic from New England referred amounted to little more than technological innovation: painters wanted to keep mixing their own paints and making their own brushes.[40] He could have been talking about painters failing to keep step with the culture of occupational hygiene that prevailed in the lead industries. In part, their "backwardness" was due to their independence from the managerial control typical of the modern factory. But this independence cost dearly. When Progressive Era activism, both within and without government, dried up after World War I, its momentum continued to transform the industrial workplace. And while this transformation was far from complete and lead poisoning persisted even within the largest companies, painters by and large remained a part of the nineteenth-century's craft-oriented, producerist moral economy. The independent painters retained workers' control, together with the nineteenth-century's ethos of laying full responsibility for workers' sickness upon the worker and his fellow servants.

By the 1920s the public, having heard for over a decade of the hazards of producing lead, was increasingly susceptible to arguments that it, too, was at risk of lead poisoning. The lead industry had developed what most then agreed were adequate precautions for manufacturing lead-containing products. In the 1920s and 1930s, lead makers sought to employ occupational medicine's prestige, as well as its outlook, methodology, and—perhaps most important—its findings, to convince other scientists, the government, and the public that they, the producers, could also guarantee their product's safe consumption—in both the economic and metabolic senses of the word.

SEVEN

Introducing
Leaded Gasoline

The interwar years were critical for the lead industries. Modern technologies and a robust economy created tremendous opportunities for new lead-containing products. But public awareness of the potential toxicity of lead threatened the industries'

They died yelling.

DR. MCCANN, PHYSICIAN
ATTENDING POISONED
TETRAETHYL LEAD WORKERS,
DAYTON, OHIO, 1924

ability to tap those markets. This threat took two distinct forms. The troubled introduction of leaded gasoline in 1924 created an acute public relations crisis requiring a concerted, swift, and public response. A very different threat arose from the growing awareness of childhood lead poisoning, which is discussed in the next chapter. Although the two crises called for different responses, in both cases the lead-using industries employed their arsenal of medical experts, academic affiliations, and scientific know-how to convince any potential regulators that they better than anyone could protect the public's health. In doing so, they delayed for decades the public's demand for sharp reductions in environmental lead.

The use of lead paint and leaded gasoline directly caused most of the childhood lead poisoning that eventually burst into public consciousness in the 1960s. From the 1920s to the 1960s the number of childhood lead poisonings was high and rising, but the lead-using industries effectively silenced this "epidemic." This is not to say that the NLC or the LIA conspired to suppress evidence of childhood lead poisoning. Indeed, into the 1950s, top LIA executives did not even believe the epidemic existed. But in their efforts to promote "a useful metal," they set the terms by which lead poisoning would be defined. Their investments in basic research gave the lead industry a near monopoly in scientific knowledge. Much of their research was based on occupational standards and did not apply to public health. But the industrial hygienists' considerable success in the workplace endowed the researchers and their findings with credibility and the aura of scientific objectivity. Faith in industry's science produced an ill-informed public whose untroubled indifference allowed hundreds of thousands of children to die or be permanently disabled from exposure to these two consumer products.

For much of the twentieth century, the automobile was the lead industry's greatest friend. Lead consumption in the United States more than doubled between 1919 and 1929, largely because of automobile production.[1] Lead paint protected the chassis of millions of automobiles. Lead was a component of rubber tires for most of the century. Car batteries use both metallic lead and litharge. In 1993 battery manufacture accounted for 83 percent of all lead consumed in the United States (see Figure 7-1).[2] Manufacturing these products was always a potential hazard for workers, but the millions of tons of tetraethyl lead burned in gasoline between the 1920s and the 1980s threatened the general public, exposing the entire urban population to levels of atmospheric lead high enough to cause neurological damage in infants. The lead that coughed out of tailpipes raised almost every American's "body burden" of lead, making each additional exposure—through food, water, paint dust, and so on—all the more dangerous.[3] In 1965, the average American's blood-lead level met the current NIOSH definition of "elevated" and fell within the range that the CDC now deems of clinical concern in children. Most of that body lead was attributable directly to gasoline.[4]

But to the lead and automobile industries, tetraethyl lead was a technological marvel. They stressed that the fluid's ability to squeeze more energy out of gasoline made it an ecological boon. In the 1920s, an Ethyl Corporation spokesman went so far as to call tetraethyl lead a "gift of God." As late as 1985, Ethyl defended leaded gasoline, saying that its use was not associated with any health risks and that it "allowed high performance vehicles and saved billions of barrels of crude oil over 60 years. Furthermore, alkyl lead played a critical role in World War II, as it allowed us to produce high octane aviation fuel so important in our defense effort."[5] This public relations strategy assumed that there was no alternative to lead, that auto makers could not have developed high-efficiency engines much sooner had leaded gasoline not been available.[6]

In fact, the world came remarkably close to keeping the leaden tiger out of its tanks when, in late 1924, a series of industrial poisonings shook the nascent leaded-gas industry and nearly resulted in the federal government

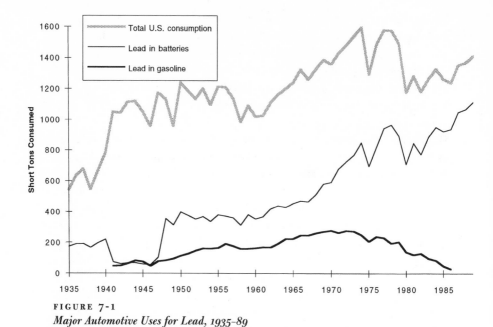

FIGURE 7-1
Major Automotive Uses for Lead, 1935–89
U.S. Department of the Interior, Bureau of Mines, "Lead," in *Minerals Yearbook* 1935–89
(Washington: GPO, 1936–90).

banning production. The mysterious death of one worker in a tetraethyl refinery began a crisis that quickly engulfed DuPont, General Motors, and Standard Oil in a potentially costly scandal.

Ernest Oelgert cracked. On 23 October 1924, a Thursday, he complained to his co-workers that someone was following him. On Friday he shouted there were "three coming at me at once" as he raced through the factory. He was caught and taken to a nearby hospital. He died in convulsions the next day. Others at Oelgert's workplace showed some of the same symptoms. On Monday, the *New York Times* reported Oelgert's death: "Odd Gas Kills One, Makes Four Insane," the headline shouted. By Thursday, all four demented workers—Oelgert's colleagues at the "loony gas" building—had joined Ernest at the morgue. On Sunday, Walter Dymock had risen from his sickbed and walked out a second-story window. Rushed to the hospital, he died on Monday in violent convulsions. William McSweeny went home sick on Monday, only to be taken off in a straitjacket later that day. He died Tuesday. Two other workers died later in the week—William Kresge on Wednesday, Herbert Fuson on Thursday. Postmortem exams showed that the poison had entered through their lungs, which showed signs of severe hemorrhage. The men's brains contained unusually high concentrations, as did other soft tissues, but unlike in conventional lead poisonings, there was very little lead in their bones.[7]

These five had been part of a forty-five-man workforce that mixed tetraethyl lead in Standard Oil's Bayway plant in Elizabeth, New Jersey. Crews worked around the clock in three shifts of fifteen men each. The men received eighty-five cents an hour—a full twenty cents more than laborers at the plant who did not work with tetraethyl lead. Their deaths initiated serious public debate over the hazards of Standard's new fuel additive. But they were not the first—nor would they be the last—to die from the new compound: prior to the Bayway accident, at least eight workers had been killed and hundreds sickened by the experimental manufacturing and initial distribution of tetraethyl lead.[8]

Thomas Midgley, a chemist at General Motors, had experimented with thousands of compounds, searching for an effective fuel additive to reduce the "knocking" that limited engine compression and fuel efficiency. If that "ceiling" could be raised, engineers would be free to design more powerful and fuel-efficient engines, and petroleum refiners could put more low-grade oils in their formulas. According to one industry legend, the search for such an additive had been underway since about 1912, when Cadillac owner Charles Kettering grew annoyed by his car pinging and backfiring with every hill. Kettering had invented the electric self-starter for automobiles, and in 1910 had formed the Dayton Engineering Laboratories Company (soon to be known as Delco) to market the starters and the batteries that ran them. He rose to become president of General Motors Research Corporation and Midgley's boss. Kettering himself said that he wanted to find an antiknock compound in order to silence critics (inspired, he suspected, by manufacturers of the hand-cranked magneto) who blamed engine knock on his Delco starter.[9]

According to historian William Kovarik, engine knock became a problem in the second decade of the twentieth century because gasoline refiners had responded to fast-growing demand for auto fuel by using inferior grades of crude.[10] By the early 1920s, researchers were experimenting with everything from iodine to benzol, anything that might raise compression ratios. Late in 1921, Midgley began experimenting with tetraethyl lead, a compound discovered in 1854 by a German chemist. Midgley's experiments showed that gasoline fortified with less than a teaspoon of tetraethyl lead per gallon burned without knocking, even when engine compression was doubled. The following summer General Motors contracted with Du-Pont, which at the time had a 38 percent interest in the auto manufacturer, to develop a process to produce tetraethyl lead on a large scale.[11] DuPont began limited production at its plant in Wilmington, Delaware, and began to plan construction of a larger facility. On 2 February 1923, motorists at a service station in Dayton were the first to buy the wine-red "ethyl fluid," mixed at the pump by the service station attendant.[12]

Before the Bayway disaster, accidental poisonings occurred in every plant that produced tetraethyl lead. At least fifty men contracted serious lead poisoning at GM's experimental facility in Dayton, and at least two died. Several more workers died in DuPont's Wilmington plant. DuPont's facility in Deepwater, New Jersey, designed to produce 1300 pounds per day, began operation before proper ventilation was installed and before technicians had established a safe process. As a result, hundreds of workers were poisoned, and at least six died before the Bayway disaster.[13]

Two factors impelled the ongoing accidents. The first was simply the new fluid's toxicity. Tetraethyl lead was far more poisonous than traditional lead oxides. Tanquerel des Planches, the nineteenth-century pioneer in industrial medicine, proved that the body takes most forms of lead into the blood only through the digestive or respiratory systems. Handling the most toxic lead oxides is not very dangerous as long as dust and fumes do not enter the mouth or nose.[14] But tetraethyl lead, an organic compound, is fat soluble, readily absorbed by the skin, and extremely damaging to the nervous system. The compound's toxicity was no secret in 1923. The United States War Department had tested it as a nerve gas, but found it less effective than mustard gas for combat purposes. Just the same, the Chemical Warfare Service's investigation estimated that a mere 21 cubic centimeters—about 5 teaspoons—applied to healthy skin could be fatal.[15]

The second factor that invited danger at tetraethyl lead plants was the tremendous payoff that would result from putting an effective antiknock additive on the gasoline market. Tetraethyl lead was clearly superior to any of the other additives then being considered, making the possibility of cornering the antiknock market very real. And lucrative. General Motors projected that treating every gallon of gas Americans burned would require 60 million tons of tetraethyl lead annually.

The very small quantities of ethyl fluid required to treat gasoline permitted test marketing long before full-scale production facilities were built. This fact, amplified by the obvious advantages of being first on the market, strained the informal cooperative research and marketing agreement GM and Standard Oil had worked out when both companies had been desperately seeking antiknock solutions.[16] Initially, DuPont was to produce the fluid which Standard Oil would distribute according to GM's price schedule. Chafing at these restrictions, Standard hired its own chemist, who developed a method of producing tetraethyl lead at one-fifth the cost of the DuPont process.[17] In mid-1924, Standard began work on a facility within its Bayway plant to mass-produce tetraethyl lead with this new process. General Motors was eager to expand distribution beyond what DuPont's experimental facility at Wilmington was able to support, and even more eager to acquire Standard's formula. Accordingly, on 18 August 1924, GM and Standard established the Ethyl Gasoline Corporation as equal partners. Kettering was named president, with Thomas Midgley and Standard's Frank Howard as vice-presidents.

Even under the safest conditions, tetraethyl lead was potentially deadly, but in the rush to get production up to speed, the companies frequently disregarded precautions. It was clear by early 1924 that the toxicity of tetraethyl lead raised serious questions about both occupational and public health risks. In April, after the deaths of two workers at the Dayton mixing plant and another at the DuPont plant, GM established a medical committee, made up of Drs. Thompson and Smith, medical directors of Standard Oil and DuPont, respectively, and Robert Kehoe, a young physiologist from the University of Cincinnati whom GM had recently hired.[18]

The committee concentrated on three tasks: to collect data on conditions in the plants and on the effects of tetraethyl lead; to make specific recommendations for improving plant safety; and to assess the public health risks and assist management in controlling adverse publicity. The physicians made their first plant inspection in May 1924, at the Dayton facility. In August they produced a report that gave detailed recommendations for ventilation, protective clothing, and dealing with publicity. From their internal reports and the press accounts that appeared after the Bayway accident in October, a disturbing picture of conditions inside the early tetraethyl lead plants appears.

Initially, workers and operators alike seemed ignorant of the dangers that the new compound posed. At DuPont's first commercial plant, workers frequently carried tetraethyl lead around in open buckets and were quite casual about getting the fluid on their fingers. Before manufacturers built large plants specifically designed for tetraethyl lead production, company engineers often crowded refining equipment into buildings serving several functions, some completely unrelated to tetraethyl lead manufacture. They made little attempt to segregate the most hazardous processes from less dangerous ones. Every worker in the building, from chemist to carpenter, faced the same risks. Especially dangerous were the "blowouts" that plagued all the plants. Uncontrolled pressure changes in the processing tanks blew open safety valves, releasing clouds of tetraethyl lead vapors that forced workers to evacuate.[19]

Workers at the Elizabeth tetraethyl lead plant called their workplace the "loony gas building." DuPont's Deepwater plant carried the moniker "the House of Butterflies" for the creatures that poisoned workers imagined fluttering around them as they worked.[20] The neurologic damage initially caused insomnia and restlessness. Unlike in traditional plumbism, there was no localized paralysis; instead, the victim staggered like "a drunken man." Eventually the patient became "violently maniacal, shouting loudly, leaping out of bed." One of the workers at Dayton smashed a window, cutting his hand badly. Another "saw the wallpaper converted into swarms of moving flies and thought pictures of his family on the walls were alive and moving about." The horror of the terminal symptoms profoundly affected the attending doctors, one of whom stated that he "never in his life had seen anyone die in such agony."[21]

Unfortunately, once a worker was poisoned, there was very little the doctors could do. As in other lead-using industries, reducing exposure was the best hope for saving workers' lives. In August 1924, before the disaster at Bayway brought the issue to public attention, the medical committee issued strict but practical rules that could minimize the risks. These differed little from those Alice Hamilton had recommended for potteries and paint factories a decade earlier: isolate the most hazardous processes; reduce the amount of dust; install adequate ventilation; provide safety equipment and clothing and insist on their use.[22]

The Ethyl Corporation would eventually embrace this medical advice, but prior to the Bayway disaster, the tetraethyl lead manufacturers seemed more concerned with using medical experts to reassure the public than to protect workers. Although this sudden show of concern for public health was strictly defensive, the companies' recognition of the potential need to weigh their product's impact on the environment was rarely expressed before the 1960s.

In the spring of 1922, Thomas Midgley consulted Yandell Henderson of Yale's Applied Physiology Lab. Henderson had worked on poison gas in World War I for the Chemical Warfare Service and the Bureau of Mines, and after the war he studied carbon monoxide poisoning for the bureau. Midgley sought Henderson's opinion on the potential impact of tetraethyl lead on public health. Henderson replied that he believed "widespread lead poisoning was almost certain to result," and agreed to undertake a study, "provided we could do it freely, without any dictation, and simply to find the facts." Midgley did not follow up.[23]

During 1922, growing concern among experts such as Henderson and increasing interest at the Public Health Service prompted an investigation of the issue of exposure to tetraethyl lead. Public distrust of industry required that a neutral party conduct any such study. Economic necessity dictated, however, that the investigation exonerate tetraethyl lead. Early in 1923, at the same time General Motors was rushing headlong into marketing the new fuel additive, the corporation financed a Bureau of Mines investigation. By providing the funding and exerting editorial control over the bureau's report, GM all but assured a favorable outcome, complete with government sanction—a green light for continued production.[24]

Tetraethyl lead posed three distinct risks. First, in its pure form, it was unquestionably deadly. But only those who refined and distributed it faced this danger. Midgley took direct action on this. He hired Robert Kehoe, moving him from Cincinnati and setting him up with his own lab in Dayton, where after July the young physiologist conducted animal tests and on-site surveys.[25] The second risk, public exposure to tetraethyl lead mixed in gasoline—through spills, misuse of gasoline for cleaning, or other accidents—received little public attention.[26] Today it is obvious that the third hazard, lead oxides in automobile exhaust, was by far the most injurious to public health. Aware of the economic danger that might arise over "ignorant and alarmist assertions" about public risk, General Motors, Standard Oil, and DuPont specified that the Bureau of Mines survey include a study of tetraethyl's combustion byproducts.[27] The bureau's data, based on exposing rabbits to exhaust, suggested that there would be little effect on the public's health.[28]

The investigation came under harsh criticism while it was under way and when it was released—and by historians ever since. Most of that criticism focused on the sponsors' obvious self-interest, the assumption being that an "objective" study would have yielded significantly different results, that rabbits exposed to automobile exhaust in an objective study would have died. But the widely accepted definitions of lead poisoning in the 1920s made any other outcome unlikely. The fact that Midgley made almost no effort to assess the risks prior to 1924 indicates either reckless abandon or confidence in the existing epidemiology of lead poisoning. Before the investigation, Midgley had a fairly good notion about the amount and nature of the lead that would appear in automobile exhausts "in the form of a finely divided dust comparable to cigar smoke." This dust was litharge, basically the same lead oxide to which battery workers were exposed. Based on his optimistic interpretation of the best information available from other lead-using industries, no measurable health effects should have accompanied even the highest probable public exposures.[29]

Nagging uncertainties about public health, trepidation over possible controversy, and the fear of potential government regulations could not outweigh the manufacturers' faith in the medical status quo. More important, they could not allow the public's irrational fears to slow the race to exploit "this highly useful product." Ethyl fluid was proving very popular, and marketing had expanded deeper into the Midwest and into the South. By the fall of 1924, motorists from Minneapolis to Jacksonville could order a shot of ethyl when they filled their tanks. Producers of tetraethyl lead sought to keep up with the growing demand, and at Standard's "loony gas building," production went on triple shifts. This sudden haste may not have precipitated the Bayway disaster, but the deaths at Standard Oil's experimental plant brought the furious rate of production to a halt. On Saturday, October 25, the same day that Ernest Oelgert died, Standard Oil's medical committee "gave instructions that all work there be suspended until such time as the plant may be made as nearly safe as possible for the health and lives of the workmen." Soon afterward, "a large gang of men . . . clad in rubber from head to foot" washed down the building and dismantled the tetraethyl reactor.[30]

When the New York papers reported the first death and four hospitalizations for insanity, Standard's spokesman said that "the men probably went insane because they worked too hard." But when it was revealed that of forty-nine workers in the tetraethyl plant, five had died and thirty-five had been hospitalized, the company took a defensive position by blaming its employees. Gas masks and protective clothing were available, the company insisted, but it was difficult to get the workers to use them.[31]

Gradually, reporters uncovered the earlier accidents in tetraethyl plants. The *Nation* reported the two deaths at General Motors' research facility in Dayton, Ohio; DuPont admitted that "some" had died in the Wilmington plant. But, DuPont's spokesman remarked, any new technology presented possible dangers. He admitted there had been some trouble, but things were improving: "In the past several months, under full production, only slight difficulties have been encountered." But workers at DuPont's tetraethyl plant in Deepwater told the *New York Times* that nine had died there in the two years it had produced tetraethyl lead. No word of this disaster had reached the public before because DuPont maintained such tight control on local hospitals and newspapers. DuPont denied the workers' claims: "only" three workers had died, and conditions were becoming safer all the time.[32]

In the six months after the Bayway accident, the *New York Times* found hospital records of over three hundred poisonings at DuPont's Deepwater facility and concluded that eight tetraethyl lead workers had died. Early in 1925, the Deepwater plant began producing tetraethyl lead by the process Standard Oil had developed, but DuPont's chemists were no better able to control this more volatile process, and another six workers died. DuPont does seem to have controlled publicity better than Standard, but in the wake of this accident the state Labor Commission quietly closed the plant.[33]

Within a week of the deaths at the Bayway plant, the Bureau of Mines released its preliminary findings, even though Charles Kettering had negotiated a promise of no preliminary press releases because "the newspapers are apt to give scare headlines and false impressions." Now it was hoped that the study's soothing data would help quell public fears about this new potential source of lead poisoning. The headline in the *Times* read "No Peril to Public Seen in Ethyl Gas . . . More Deaths Unlikely."[34]

The Ethyl Corporation assured the public that "ethylized gasoline has been on sale for one year and nine months in about 20,000 filling stations covering one-third of the territory of the United States" and that "about 200,000,000 gallons have been used by more than 1,000,000 motorists with complete safety and satisfaction."[35] But Ethyl's promise of complete safety came to appear an idle boast. The Bureau of Mines report backfired. In late 1924 and into 1925, the growing reports of fatalities among tetraethyl lead workers outweighed the bureau's sunny pronouncements. The scientific community smelled a cover-up.[36]

Yandell Henderson, the Yale physiologist, led the public outcry against tetraethyl lead. Some of his predictions seem remarkably prescient. For the immediate future, he warned, "there will be vast numbers of people in all our cities who throughout their lives will have a continual low grade of lead poisoning." He also accurately foresaw the long-range effects, which "will not appear to-morrow, or next year. The important effects will appear ten, or even twenty-five years from now." On the other hand, some of his forecasts were ridiculously overstated. He warned that residents of certain areas in New York City faced imminent mortal danger from lead in car exhausts. He gave this warning to people who lived in areas of the city "in which dust does accumulate": "If this is true, then when you go out in front of your house or shop when the dust is stirred up, as New York dust always is, and snuff and snuff until you have inhaled from two to three milligrams, and do this every day for a week, then you will be a goner."[37]

Although in the long run such drastic pronouncements probably hurt the cause of tetraethyl's opponents, public fear was great enough to spur New York City, and then the state of New York, to ban the sale of leaded gasoline. Concerns voiced by the medical community, social activist organizations such as the Workers' Health Bureau, and the general public prompted Surgeon General Hugh Cumming, in April 1925, to call for a conference on tetraethyl lead. On May 5, the Ethyl Corporation suspended sales, pending the outcome of the conference, which began May 20 in Washington, D.C. Approximately sixty participants from industry, labor, and the scientific and medical communities appeared to air their views.[38]

Historians David Rosner and Gerald Markowitz assert that Ethyl's case for leaded gasoline rested on three key elements. First was progress in technology—America's industrial growth demanded tetraethyl lead. As the *Monthly Labor Review* reported, tetraethyl lead "was recognized as one of the greatest accomplishments of the year in chemical engineering." A chemical that doubled the mileage from a gallon of gasoline was not to be discarded. The report concluded that leaded gas "has come to stay and therefore it is only a question of furnishing adequate protection to workers coming in contact with it."[39] But complete protection was impossible, which suggested Ethyl's second argument: every innovation, every discovery, every step forward, involved some risk. Ethyl's third claim was the old saw that careless workers brought most poisonings on themselves. As an industry medical director commented in 1933, "the individual whose appearance indicates chronic slovenliness is always a difficult problem and may, in spite of [the company physician's] best efforts, succeed in getting himself into serious difficulties." This was the same doctor who warned against "wise guys" who flaunted rules, among whom must be counted the inventor of leaded gasoline himself. At a 1924 press conference, Thomas Midgley washed his hands in undiluted ethyl fluid and declared (while quickly drying them) that he could do so every day without risk—this despite the fact that in 1923 he had taken an extended leave of absence in Florida for what he called "a short vacation in order to shake off the toxic effect of the lead compounds."[40]

Ethyl's opponents at the Washington conference stressed the slow, cumulative nature of lead exposure, bemoaned the dearth of hard evidence regarding emissions from leaded gas, and demanded federal action to protect the public's health. In their view, the lack of statistics should not be used as an excuse for government inaction. Instead, they recommended a halt in production until further studies were done. It was not the federal government's duty to prove that tetraethyl lead was harmful; it was the Ethyl Corporation's duty to prove that it was safe, before distribution could be resumed. In defense of the workers, they reminded the Surgeon General that lead poisoning came from dust and vapors and not from poor worker hygiene. Alice Hamilton suggested that the convention's focus on occupational hazards was too narrow, that leaded gasoline was a potential menace to the public's health: "You may control conditions within a factory," she declared, "but how are you going to control the whole country?"[41]

The Surgeon General appointed a committee of seven scientists and physicians, including David Edsall of Harvard and W. H. Howell of Johns Hopkins, to gather hard evidence for or against leaded gasoline. The panel was to announce its findings by January 1926. With just seven months they were able to conduct only a very limited study, involving 252 gasoline station employees and chauffeurs in Ohio. Blood and stool samples from drivers and attendants who used leaded gas were compared to samples from drivers who did not. The data showed that all gas station employees had elevated lead absorption, regardless of whether they pumped leaded gasoline. The correlation between leaded gasoline and increased lead absorption was therefore not strong enough to warrant a prohibition of tetraethyl lead production and distribution. The panel's report did not suggest there was *no* danger, however, warning that "longer experience may show that even such slight storage of lead as was observed in these studies may lead eventually in susceptible individuals to recognizable lead poisoning or to chronic degenerative diseases of a less obvious character."[42]

The Washington conference presented great opportunities for the lead industry—and for the health of the nation. America had the rare chance to pause at a critical juncture, examine its options, and choose a path based on careful study. The government had the chance to make industry find a safer way to build more powerful engines and fuel. In deliberately letting this opportunity slip by, the government put millions of its citizens in the same category as Ernest Oelgert and his co-workers. "They were human material," wrote Mary Ross for the *Nation*, "bought in the labor market at eighty-five cents an hour, and scrapped in the feverish rush to try out and market a new product which promised tremendous financial returns."[43]

The Ethyl Corporation, on the other hand, seized two golden opportunities offered by the conference. First, it was allowed to continue production of tetraethyl lead; and second, it obtained a virtual monopoly on lead-poisoning research. Ethyl produced 6.6 million tons of tetraethyl lead over the next sixty years. As large as this figure sounds, sales never accounted for more than 21 percent of the lead consumed in any year but they more than compensated for the falling demand for white lead (see Figure 2-1).[44] Far greater, but unmeasurable, was tetraethyl's value to the petroleum and automotive industries, represented in more than a symbolic sense by Ethyl's two owners, General Motors and Standard Oil.

"ETHYL IS BACK," read signs in filling stations shortly after the Ethyl Gasoline Corporation resumed sales of its only product on 1 June 1926. The business was now three million dollars in debt but had spent the downtime preparing for an impressive comeback. DuPont had improved the ethyl chlorine process that had been so disastrous at Bayway and Deepwater, and full production was underway at its plants. Distribution had been made safer as well. No longer would gasoline be "ethylized" at the pump by the station attendant; mixing would be done by the tankful at the refinery.[45] The advertisements for Ethyl that appeared in the late 1920s and into the 1930s stressed fuel economy, not product safety. Significantly, many of these ads mentioned in small print that "the active ingredient in Ethyl is lead." But ethyl's active ingredient was immaterial. The real products being offered were fuel efficiency, pep, and power.[46]

Gasoline distributors had to balance doubts about whether the public would accept tetraethyl lead with the certainty that customers would demand some potent antiknock agent. And no one had a cheaper or more effective antiknock agent than Ethyl. The Bayway disaster had convinced Standard Oil to leave the manufacture of tetraethyl lead to DuPont. Now, Standard's executives were uncertain whether to resume distribution. "ESSO," the company's premium gasoline, had been formulated to replace "Standard Ethyl Gasoline." It consisted of very expensive grades of oil fortified with costly benzol. When Ethyl went back on the market, Standard Oil shied away from putting ethyl fluid back in its premium grade, even though it owned half of the Ethyl Gasoline Corporation. By the end of 1927, however, the economics of the situation prevailed, and ESSO was ethylized, although the word *Ethyl* appeared only on gas pumps, not in their advertisements.[47]

The second opportunity presented by the Washington conference and its subsequent investigation was less tangible than the economic benefits, but of equal significance. Although the Surgeon General's panel gave the go-ahead to resume production of tetraethyl lead, it also recommended further study. Public health officials had greeted the Bureau of Mines report with great skepticism. The outcome of further government investigations was uncertain, especially given the combative atmosphere that prevailed after the Surgeon General's conference. Such considerations probably helped fix in Kettering's mind the importance of establishing a credible, industry-supported, and lead-friendly research body. Ethyl had used the hiatus to shore up its medical and research program. In the absence of further government-sponsored surveys, Ethyl and—by virtue of their interrelated interests—the lead industry as a whole seized the opportunity to take a leadership role in setting the research agenda. For the next forty years, industry-owned or -financed centers conducted the most influential studies, usually with pro-lead results.[48]

This decision marks a transition with broad significance for industrial toxicology and, because of that discipline's direct link to modern environmental science, for environmental and social history. From 1910 to 1920, the most noted figure in occupational medicine was Alice Hamilton, an activist dedicated to government regulation and an assertive advocate for worker protection. But after the tetraethyl lead scandal it was Kehoe—an industry insider—whose work would be most cited, whose opinions were most sought, and whose data influenced the work of industrial hygienists and public health workers as well as that of general practitioners and pediatricians.

Robert Kehoe was born in 1893, one of five children in the family of Jeremiah and Jessie Kehoe of Georgetown, Ohio.[49] Robert studied at Ohio State University and finished medical school in 1920 at the University of Cincinnati, where he would spend most of his career. Shortly after receiving his M.D., he married Lucille Marshall. In 1923, when Charles Kettering and Thomas Midgley realized they needed a medical specialist, Kehoe was working in the university's physiology lab, studying the effects of toxic materials—including lead—upon proteins. Midgley had met Kehoe at a Chemical Society meeting, where Kehoe had presented a paper on lead poisoning, and the head of the physiology lab at the university was an old friend of Kettering's. Kehoe was offered the job of investigating the strange new form of lead poisoning that Midgley's invention was causing, and although he was still teaching three days a week in Cincinnati, he agreed to move his family to Dayton almost at once and commuted for the rest of the semester.

Early in 1925, Kehoe was named medical director of the Ethyl Gasoline Corporation and special consultant to DuPont. Ethyl and DuPont granted tremendous administrative powers to the young doctor, probably the result of Kehoe's personal drive and the companies' desperation after the embarrassing and costly plant disasters. Kehoe designed strict controls and procedures concerning every process from distillation to the cleaning of storage tanks. These rules appeared in every contract Ethyl made with distributors, and Kehoe jealously monitored their implementation. He collected reports on every documented case of lead poisoning associated with tetraethyl lead, both domestic and foreign. So closely did he watch for unusual activities that once, during World War II, he chased the paper trail of a stray 55-gallon drum of pure tetraethyl lead across the country and back. The trail stopped at the desk of a general in Washington who told him that "the material was in good hands and was being handled per his instructions." Though Kehoe's appeals to determine the situation more exactly went as far as the office of General George Marshall, it was not until long after the war that Kehoe learned the drum had been sent to a secret Manhattan Project facility in Oak Ridge, Tennessee.[50]

By limiting the scope of most lead research to occupational hazards, industrial medicine became a tool for promoting lead's image. Robert Kehoe was the physical and intellectual link between the lead industry and the medical sciences. He served as Ethyl's medical director until 1958, but was also a professor of physiology. With funding from the Ethyl Gasoline Corporation, DuPont, and the Frigidaire Corporation, Kehoe established the Kettering Laboratory of Applied Physiology at the University of Cincinnati. The laboratory soon became the center of the lead industry's research programs and the strongest institutional link between lead-using industries and the academic research community. Kehoe was its first director and perennial figurehead.[51] He wrote dozens of journal articles from the 1930s to the 1960s; citations to his work pepper the bibliographies of almost every medical report on lead during this time.

Kehoe and the Kettering Laboratory forged and defended a body of work that acknowledged some of lead's dangers while maintaining its essential harmlessness. Their interpretation of data rested on four principles. First, lead absorption is natural, and every human body contains some lead. Second, the body has mechanisms to cope with these "natural" levels of exposure. Third, below the "threshold" of a certain blood-lead level, no ill effects will occur. And fourth, the general population's exposure to lead from pollution put it so far below this threshold that environmental hazards should be of little concern, and resources should therefore be directed toward protecting lead workers.

The empirical underpinnings for these four theses came from Kehoe's "balance studies," a series of experiments conducted in the Kettering Lab in the 1930s to examine how the body metabolizes lead.[52] According to his conclusions, the body regulates absorption and excretion of lead well enough to protect itself from all but the most toxic levels of exposure. According to the balance model, if a worker faces greater exposure than the body is accustomed to, blood-lead levels will shoot up until the body's excretion rate rises to the challenge. But over time, the body "learns" to excrete almost all the lead taken in, achieving a new "balance." As long as the resulting blood level does not exceed the threshold of toxicity, the worker is safe. If the worker's normal exposure to lead declines, the body will gradually cleanse itself.

But what is the "threshold"? What magic number defines for all workers the line of demarcation between safety and danger? Kehoe used 80 μg/dL because he had not seen workers with clinical signs of lead poisoning whose blood lead was lower than this. Kehoe cautioned against using the threshold as "a magic number," but by this he meant that he had seen plenty of workers with *higher* blood leads who showed no symptoms. Plant physicians need not automatically assume, therefore, that a worker with blood-lead levels above the threshold was poisoned.[53]

The number chosen as *the* threshold is not nearly as important as the fact that the medical community believed one existed. The influence of "balance" and "threshold" concepts in industrial hygiene is difficult to exaggerate. If it is not the level of exposure as much as the individual's metabolic response that determines sickness, then the burden of keeping blood-lead levels low can be placed on the worker. The manufacturer may then safely shift the emphasis from maintaining a clean workplace to monitoring the workers and replacing those whose metabolisms cannot "get the lead out."[54] Such workers faced an endless cycle. They worked, had their blood tested at regular intervals, and when their blood lead crossed the threshold, they were removed from hazardous exposure until it returned to a "safe" level—at which point they went back to the work that poisoned them. Workers in lead plants in the 1990s still went through the same rounds of testing, removal for deleading, and reinstatement.[55]

Throughout his career, Kehoe sounded warnings about the potential public health threats involved with lead products, although he raised this voice only in the confines of the lead industries. This politic criticism was further constrained by Kehoe's fundamental—if understandable—misinterpretation of data. Lead appeared in the blood of every subject Kehoe examined, regardless of the individual's history of exposure; he therefore assumed that some degree of lead absorption was "natural." This assumption would stand until 1965, when geologist Clair Patterson linked increases in human lead absorption to the rise of industrialization. A second fallacy, following from the first, was the notion of a threshold for harmlessness. If a hazard is ubiquitous, it is normal—or, by extension, natural. Without adequate baseline data, depressed mental function in the general population cannot be determined, nor can lead's role in kidney and blood diseases.[56]

The threshold concept, developed to define when a worker should be removed from a poisonous environment, had a harmful impact well beyond occupational medicine.[57] Many doctors adopted the 80 μg/dL level as a constant for all patients, including 2-year-olds. Kehoe was less absolute than the doctors who cited his authority. In a presentation before the American Medical Association in 1933 he emphasized that neurological symptoms arose far quicker in children than in adults, and perhaps at lower levels of exposure, making prevention through environmental control paramount.[58] To be sure, most of his public pronouncements clung to the universal safety of the threshold, and the lead industry relied on the concept to market lead-based paint and leaded gasoline years after it was known that exposure to both raised everyone's blood lead. Kehoe testified in a 1966 Senate hearing that he still had found no harmful effects to the general public from the burning of leaded gasoline, that there was no danger from such "low" levels of exposure. Doctors and public officials listened attentively to the sanguine industrial hygienist, acting—or failing to act—on issues affecting children in accordance with their belief in the magic boundaries. Well into the 1970s many pediatricians continued to apply the industry standard for lead workers to the diagnosis of infants and children.[59]

General Motors developed tetraethyl lead to satisfy the American public's seemingly limitless passion for automotive power. In the process of containing the economic fallout from the Bayway disaster, the Ethyl Corporation produced a very different kind of power. Robert Kehoe's aggressive research programs convinced the government and the public that Ethyl had controlled the occupational hazards related to manufacturing and distributing tetraethyl lead. Furthermore, Kettering Lab research produced "scientific" proof that nature had equipped the human body with mechanisms to cope with the levels of environmental lead that the use of leaded gasoline might produce. A crucial lead-using industry, with strong ties to the nation's largest automotive and chemical concerns, acquired the power to define lead poisoning.

But while consumers confidently burned ethylized gasoline and relished the enhanced automotive performance that resulted from the newest lead product, more and more pediatricians were on the lookout for signs of childhood lead poisoning stemming from one of the oldest. Just as Ethyl had directed its interrogators' focus from public to occupational hazards, the lead-paint industry would focus on those aspects of the child's "lead world" it could control, while downplaying or ignoring the risks from children's general environment. The growing association of childhood lead poisoning with poverty made this side-stepping much easier. But sooner or later the lead industry would have to confront the insidious possibility that lead paint was killing children in their cribs.

EIGHT

Defining Childhood Lead Poisoning as a Disease of Poverty

This cheerful verse appeared in "The Dutch Boy's Lead Party," a booklet sent by the thousands to paint stores in 1923. A few years later, the National Lead Company distributed "The Dutch Boy's Hobby: A Paint Book for Girls and Boys." The famous trademark character appeared on the cover, brandishing a ladder and a loaded paint brush, mounted upon a strange pony

A little Toy Lead Soldier
Once to the Dutch Boy said,
"We have some fine relations
Who all contain some lead.
"Why don't you give a party
So folks can meet and see
The other happy members
Of the great Lead family?"

whose body was a lead ingot and whose head was a bucket of paint. The coloring book, which told "in nursery rhyme of the Dutch Boy's adventures," came with "a sheet of real water colors."[1] The NLC intended these books to do more than please children. As one trade advertisement advised paint dealers, "This '*Paint Book*' sells two generations at once." Each book contained "a little pamphlet intended for [the parents] which may be readily detached. The little pamphlet gives suggestions on decorating. It starts the reader thinking 'paint thoughts.'" The promotional literature boasted of the psychological game they were playing. A similar announcement explained that "there is a certain psychology in having the two books bound together. The bright appealing colors of the children's section unconsciously starts the parent to thinking about color and the improvements a judicious use of it might bring about in his home. This is what salesmen call a buying frame of mind."

The NLC's blithe juxtaposition of children and lead in these promotional materials suggests just how little concern pediatric lead poisoning evoked in the 1920s. Such deliberate promotion of lead for children would be unheard of twenty years later, when the Dutch Boy would seem far more

FIGURE 8-1
Illustration from a National Lead Company "Paint Book"
Dutch Boy Painter 16 (Sept. 1923): 125.

open-minded about what minerals went into the best paints and the paint industry would establish a voluntary restriction on the amount of lead in paints sold for interior use. These changes had nothing to do with protecting workers, the chief motivation behind European restrictions on white lead. Instead, they were responses to a fundamental change in the perception of lead poisoning, which by the 1950s would be as closely associated with children as with painters and lead workers.

This transformation is readily visible in census mortality records. Although federal reports of lead-poisoning deaths are a poor index of actual lead poisoning, they do reveal a dramatic shift in the public health community's awareness of the disease. What was formerly a hazard of "the dangerous trades" would come to be perceived as essentially a childhood malady. Figure 8-2 shows that until the 1930s, children accounted for fewer than 10 percent of reported deaths from lead poisoning. This percentage doubled during the Great Depression, and by the postwar years, children under 5 accounted for about 30 percent of reported deaths. Moreover, both the figure and Table A-2 of the Appendix reveal that as lead poisoning's new definition as a disease of childhood took root, it had an increasingly racial cast. These data reflect an underlying epidemiological and historical reality: lead poisoning has affected a disproportionate number of black children. But before the 1950s, race played a smaller part than economics in setting the new definition. The public and professional perceptions of the day defined childhood lead poisoning as a disease of poverty, another troubling problem among the myriad others that blighted the nation's growing population of urban poor. From the perspective of the public health community of the time, lead poisoning was a problem as intractable as poverty itself.

From today's perspective, the growing awareness of pediatric plumbism in the 1920s and 1930s might be seen as a victory for public health and the first step toward the lead industry's inevitable day of reckoning. This assumes, however, that regulations to reduce the public's exposure to lead could not have been put in place before the activist 1960s. But awareness of childhood lead poisoning grew at a time when the Progressive Era campaigns against occupational lead poisoning were alive in public memory, when European nations were enforcing sharp restrictions on lead paint for health reasons, and when the controversies over leaded gasoline made lead a volatile political issue concerning the health of the general public. In this context, the 1960s were a long time to wait for "the inevitable."

If the tetraethyl lead scandals had any direct impact on public awareness of pediatric lead poisoning, little evidence of it appears in the popular press. Curiously, although the late 1920s saw a marked increase in the attention pediatric researchers paid the subject of childhood lead poisoning, their published reports never mentioned leaded gasoline. Nevertheless, it seems likely that the tetraethyl scare added to the increasing tendency to associate any use of lead with the threat of possible poisoning.[2] As more physicians became aware of the potential for pediatric cases, the more cases they found, producing a rising spiral that underscored the dark implications of John Ruddock's statement that "the child lives in a lead world." Containing the potential panic became a perennial issue for the newly formed Lead Industries Association and its secretary-treasurer, Felix Wormser. Between the late 1920s and the 1940s, pediatricians, city health departments, and the lead industries forged a definition of childhood lead poisoning that lasted into the 1960s, a definition that minimized lead's apparent threat to the general public health by limiting it to the poor.

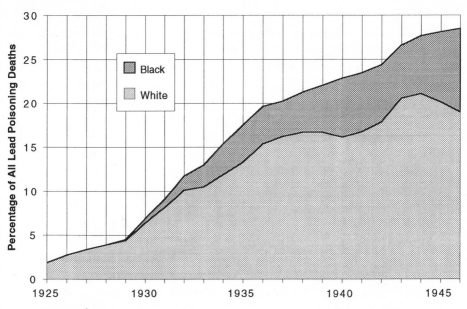

FIGURE 8-2

Reported Lead-Poisoning Deaths among U.S. Children under 5 Years of Age,
1925-46

Figure represents a moving five-year average. Data for 1925-44 drawn from Bureau of the Census,
Mortality Statistics 1923-44 (Washington: GPO, 1926-46); for 1945-48, U.S. Public Health Service,
Federal Security Agency, *Vital Statistics of the United States 1945-48* (Washington: GPO, 1947-50).

* * *

Late in the summer of 1930, Louis Dublin, statistician for the Metropolitan
Life Insurance Company, was perplexed by the sudden appearance of
childhood lead poisoning among his policyholders. He found it remarkable that "several" children died of lead poisoning that summer, because
prior to these deaths only one child "among our more than nineteen million Industrial policyholders" had died from lead, and that case had been
several years earlier. All other lead-poisoning deaths among Metropolitan's
customers had been chronic occupational cases. In September, Dublin
sent a survey to seventy-five pediatricians around the country, asking them
to describe their understanding of and professional experience with childhood lead poisoning. Almost all of the thirty-three pediatricians who responded expressed some familiarity with its symptoms and causes, and
many wrote "that they had seen cases of lead poisoning in children and
mentioned cribs, toys, woodwork, furniture, lead nipple shields and breast
ointments with a lead base as the sources."[3]

Dublin published the results of this survey in the Metropolitan's "Statistical Bulletin," where apparently it received "a great deal of publicity" and "strong remonstrance by the Lead Industries Association." Three years later, Ella Oppenheimer of the U.S. Children's Bureau was preparing "a popular folder on the Prevention of Lead Poisoning in Children," and wrote Dublin for details of the survey. Dublin offered to show her the complete files, but since Metropolitan wanted "to avoid any controversy with the lead people," the Children's Bureau was cautioned not to mention the insurance company "either directly or by inference" in any press releases or publications. Oppenheimer assured Dublin: "We shall be very careful, indeed, not to mention the Metropolitan bulletin in any way."[4]

This exchange, although limited to three very brief letters, speaks volumes about the relations among the federal government, the insurance industry, and the lead industry, as well as the dearth of information available in the 1930s about childhood lead poisoning. Oppenheimer, a physician and high-ranking bureaucrat in the Department of Labor, relied on an insurance company for basic information. Although the company's previous experience reinforced the prevailing belief that lead poisoning was almost exclusively an occupational problem, it had undertaken a fact-finding project to test that assumption. For obvious economic reasons, the insurance company did not want to ruffle the lead industry's feathers; the federal government, for less than obvious reasons, acquiesced in that appeasement. Oppenheimer saw little reason to cause "any controversy with the lead people." She was not publishing an exposé of the lead industry, just "a popular folder"—which was about as far as any federal agency involved itself in the issue of childhood lead poisoning until the 1960s.

As was discussed in Chapter 2, the sudden increase in reported cases of childhood lead poisoning stemmed from growing sensitivity and interest in the disease on the part of a number of prominent pediatricians in the mid-1920s. Occasional outbreaks of occupational lead poisoning and noteworthy workmen's compensation cases probably kept it in the back of many physicians' minds. The public health scares at the time of the introduction of tetraethyl lead heightened that awareness. The development of better diagnostic tools also increased the probability of identifying cases in which environmental diseases mimicked infectious disease. Increasingly, lead poisoning was found to be one such "aping" disease.[5]

By the 1930s, pediatric researchers had established a fairly accurate picture of the nature and causes of acute childhood lead poisoning. They became aware of important differences between childhood and adult plumbism. Where onset in adults was almost always signaled by colic and constipation, the lead-poisoned child might exhibit these symptoms—or diarrhea and vomiting. Severe plumbism in adults tended to affect the peripheral nervous system, producing "wrist drop" or "foot drop"; involvement of the central nervous system was comparatively rare. Lead-poisoned children also showed peripheral nerve damage, but they seemed far more prone to central nervous system involvement. Those children who recovered were more likely than adults to suffer permanent brain damage.[6]

Physicians also learned to distinguish age-specific differences within the range of pediatric lead poisoning. They quickly discovered that toddlers were at far greater risk than infants and school-age children. The only likely source of lead for nursing infants was from their mothers' bodies: from lead-containing cosmetics or medications, or from breast milk tainted by prolonged contact with solid lead nipple shields. Older children seemed to contract lead poisoning only from severe environmental pollution or occupational sources. It became clear that children from ages 1 to 3 were most at risk, and they almost always got sick by eating paint—from their cribs, their toys, or any painted object within their reach.[7]

In searching for the sources of this form of lead poisoning, child health specialists focused on objects that teething children were expected to encounter, such as toys and cribs or other baby furniture. The Children's Bureau sought to warn the public of these everyday dangers. In September 1931, Ella Oppenheimer told listeners to a national radio broadcast that lead-painted toys posed a serious threat to health. The bureau's pamphlet "Infant Care" recommended that concerned parents buy cribs and toys that contained only nontoxic paints.[8]

Parents and toy makers alike wrote the Children's Bureau to see how this could be done. Mrs. Michael Sacharoff of Beverly, Massachusetts, was shopping for a crib, and was "considering a model offered by Sears Roebuck . . . finished with a washable enamel paint." Sears apparently could not assure her that this enamel was nontoxic, and Sacharoff asked the Children's Bureau to recommend a leadless paint. Ella Oppenheimer regretted that she could not provide the names of leadless paints or of toy firms that used them. Instead, she suggested that Mrs. Sacharoff write the Toy Manufacturers of the U.S.A., who, she assured the worried mother, would know which pigments were nontoxic, since they were "doing everything they can to make toys safe."[9]

Consumer concern prompted retailers to ask for safer toys. From around the late 1920s Macy's department store demanded that toy suppliers guarantee their products were painted with nontoxic pigments, although it is uncertain how carefully the store monitored this policy or how often it was invoked only to force manufacturers to take back unsold merchandise.[10] Several toy manufacturers responding to a 1935 survey by Ella Oppenheimer spoke of their problems in finding substitutes for lead, arsenic, mercury, and other poisonous pigments. But most claimed that despite the Depression's "extreme pressure on every hand for lowered production costs," their pigments were safe and nearly lead free.[11]

Questions about lead paints and toy safety, however, never generated the kind of furor attending the tetraethyl lead accidents. But Felix Wormser, secretary-treasurer of the LIA, was quick to head off scandals before they materialized.[12] In November 1930, before Ella Oppenheimer took to the airwaves and childhood lead poisoning was still an issue only in medical journals, Wormser conducted an informal survey of crib and children's bed and furniture manufacturers, asking them to "kindly let us know if it is your practice to use any white lead in painting this type of furniture." It is unclear how many letters he sent, but he released the responses from twelve companies. Only one acknowledged using lead paint, and it gave the amount as "very little." On the other hand, only one company flatly asserted that it did not use "any lead paint." Most simply answered Wormser's question, which specifically targeted *white* lead—which to anyone familiar with paints in the 1930s meant pure carbonate of lead in oil. Four of the respondents pointed out that for their purposes enamel or lacquer made a better surface, and it is probable that the majority of the survey respondents used enamels.[13] These manufacturers may or may not have known how much lead was in the pigments in their enamels, but their answers probably reflected their faith that their paints were nontoxic, and in any case were not "white lead."

This faith had adherents in high places. Several years after Wormser's survey, Alice Hamilton wrote to Martha Eliot, assistant chief of the Children's Bureau, on behalf of a friend, "a young married woman . . . whose husband is in his third year at Harvard Medical School," who had asked Hamilton about paints for baby furniture. Hamilton remarked that "it has always been my impression—not very substantially founded—that furniture and toys were painted with enamel paint, lead-free." She excused her ignorance of pediatric lead poisoning by pleading that such cases "are not industrial in origin." A quick call to the Bureau of Standards corrected her, revealing that enamel paints often did contain lead. "This would mean that tests would have to be made of furniture paints and toy paints." She concluded, "Do not you think that it is an important question, and that it lies within the field of the Children's Bureau?"[14]

Given the prevailing misunderstandings of lead epidemiology and paint chemistry, Wormser's and Oppenheimer's surveys were reassuring, since they found that new cribs and "better" toys seldom contained lead paint. Charles McKhann, then the foremost authority on childhood lead poisoning, cited Wormser's survey approvingly. Statistician Frederick Hoffman suggested that Wormser's survey proved that the diagnoses in many reported nonindustrial lead-poisoning cases were "superficial and possibly erroneously arrived at."[15]

Neither Wormser's nor Oppenheimer's surveys addressed the issue of children whose parents bought cheaper toys or repainted old cribs rather than buying new ones. The vague replies from some manufacturers suggest why the problem continued to be ignored. One maker responded to Wormser, "Wish to advise you that very little of this lead paint is used in painting our material." The manufacturer of what Oppenheimer called "cheaper blocks" asserted that his blocks were "only the highest grade merchandise" sold "at popular prices." But he would not guarantee that his blocks were lead free—only that "all of the materials used, not only paints, are definitely harmless to the user, and, of course, we refer to the children."[16] It would seem that both Wormser and Oppenheimer found what they expected: many manufacturers used lead-free pigments (or believed they did), and the lead in products made by the rest was "definitely harmless."

Children's Bureau records show no evidence that Alice Hamilton's advice to investigate further was followed.[17] It would seem that the bureau's interest in childhood lead poisoning was limited to encouraging safer toys and baby furniture. With the assurances of the toy manufacturers and Felix Wormser's word that the LIA was policing its own, Oppenheimer resumed writing "popular folders" on the problem. A generation later, Wormser was still telling an audience of physicians at an American Medical Association meeting how his survey had "cast grave doubts" on the common belief "that the eating of paint on cribs and toys has been a source of lead poisoning among children."[18]

With toys and cribs declared safe, lead-poisoning awareness shifted away from the "normal" world of teething children. If lead paint in the home did not threaten the average child, there was no threat to the general public. Reports of childhood lead poisoning during the 1930s increasingly associated it with the poor, the culturally deprived, or the "foreign." Perhaps nothing did more to confirm childhood lead poisoning's status as an illness of poverty than the outbreaks of "Depression disease" that recurred almost every winter during the mid- to late 1930s: lead poisoning caused by burning discarded automotive battery cases for heating fuel.

This unusual source of lead poisoning was first reported in 1932 by

Baltimore physicians. By then, according to historian Elizabeth Fee, doctors at Johns Hopkins were alert to childhood lead poisoning, and Baltimore's new commissioner of health, Huntington Williams, had already begun his famous campaigns against the disease. But when Baltimore-area hospitals reported forty cases of lead poisoning in three months, "all in Negro families," there seemed to be something more than crib-chewing going on.[19]

Miriam E. Brailey, an intern at the Harriet Lane Home, told of the first case she encountered, "a colored girl of seven years" whose family brought her, unconscious, to the hospital in June. "A little prompting" from Brailey produced a "fairly convincing though rather vague picture" of a family wracked by tuberculosis. The girl's father had died from TB the year before, and her mother was now disabled with the disease. A TB test confirmed that the child "was strongly positive," but her lumbar puncture was more consistent with lead poisoning. Thus alerted, Brailey found the telltale blood stippling and even a lead-line along the girl's molars. But where had a 7-year-old contracted lead poisoning?

Brailey visited the girl's home, "in a squalid Negro quarter," and found the mother, sick not with pulmonary disease but "confused mentally" and somewhat disabled physically. The mother referred most questions to "a large Negro, Melrose Easter, whose eyes were bloodshot and whose breath was strong with whisky." According to him, the mother suffered from convulsions and seizures. The young doctor and her two informants examined the home for loose paint, chewed plaster, or other possible sources of lead poisoning. Nothing. "After a fruitless search Melrose bethought himself and brought out the suggestion which proved to be the clue in the investigation." The mother had been "not herself since midwinter." Could they have gotten sick from burning pieces of batteries? "The smell was bad," recalled Easter, "even made the food taste." He produced a large piece of the wood, and analysis confirmed that it contained massive quantities of lead salts.[20]

Local junkyards had been supplying the battery cases to poor persons "free or at small cost" after salvaging the lead plates from them. The Baltimore Health Department alerted junk dealers to the problem and according to Huntington Williams's account, these cooperative measures "apparently resulted in eliminating this unusual health hazard in the city of Baltimore." Poor families in other cities would have to wait. Over the next few years, epidemics of "Depression disease" were reported in Philadelphia, Long Island, Detroit, Chicago, and Lexington, Kentucky. As late as 1947, a doctor reported seventeen children poisoned by battery cases in

Staunton, Virginia. The problem was not limited to the United States. In December 1954, in the village of Canklow in Yorkshire, England, two children died and more than two dozen were hospitalized when their families burned the wares of "a junk dealer who sold old auto battery cases for only a shilling a sackful."[21]

If the LIA sought to divert attention from children chewing on letter blocks and to define childhood lead poisoning as a tragic but inescapable consequence of poverty and ignorance, it could not have asked for a more helpful development. Vivid narratives such as Brailey's cast the victims as stock figures in a paternalist melodrama, and the readily available solution provided the requisite self-congratulatory happy ending. Felix Wormser made no attempt to play down stories about battery burnings. In 1946 he widely publicized the results of his challenge to a journal that reported fifty-five Chicago children had died from lead paint. According to Wormser, "there was no lead poisoning of the kind described, nor were lead toys or lead painted cribs involved." His investigation revealed "that over the same period five children had taken some old lead battery boxes, used them as fuel and breathed the fumes, with fatal results." Wormser crowed, "A year later we received a retraction."[22]

A problem like battery burning was tailor-made for proving that lead, while potentially dangerous, was a public health problem only for those whom poverty forced to take drastic measures or those who, through ignorance of the dangers, used lead products in ways for which they were not intended. In the Prohibition era, moonshiners who distilled liquor with leaden equipment produced another "deviant" source of lead poisoning. Years later, before a panel convened to discuss potential public health threats from lead, Robert Kehoe laid out a number of "situations in which lead poisoning occurs . . . through innocence, ignorance, or irresponsibility." These included ill-attended infants who chewed lead paint, homeowners who installed lead water pipes for untreated spring water, "the frugal collector of maple sap" who painted his buckets with white lead; or "the art-conscious but technically incompetent potter" who produced and sold poorly glazed dishes "with which to beguile aesthetic housewives into feeding their families dangerously."[23]

Doctors in the late 1920s and 1930s noted that, along with extreme poverty and extraordinary ignorance, "culture" was a factor in cases of lead poisoning. In the mid-1920s, Japanese pediatricians discovered that lead-based face powder was the cause of a form of meningitis that had been killing nursing infants there for over two hundred years. This startling discovery received considerable attention in the American medical press, which probably alerted pediatricians to the danger of nursing women applying lead-oxide ointments or using lead nipple shields. But the foreign source of the report helped to heighten the sense that childhood lead poisoning was "exotic." Well into the late 1930s the most famous cases were those reported in Australia by Lockhart Gibson in 1904. American discussions of Gibson's work emphasized the climate and unusual architecture of tropical Queensland, ignoring the wider implications for children living in any lead-painted environment.[24]

Race was not yet seen to be a "causative" factor—medical literature of the 1930s associated lead poisoning far more with poverty, carelessness, or "culture." A 1939 study of battery-case burnings in Chicago found that the number of black victims was almost twice that of white. But, the authors observed, "we do not believe race plays any role in this condition except that more negro people, because of their poor economic status, use battery casings for heating their homes than do white people." In 1933, Frederick Hoffman observed of four children whose deaths were attributed to eating paint on furniture, "it is significant . . . that all these children were of foreign parentage, two having been born in Russia and two in Italy, no doubt suggestive of the ignorance of the parents in watching the children's habits."[25]

By the 1930s, the likely victims and probable causes of lead poisoning seemed clear. The highly publicized programs of industrial hygiene begun in the Progressive Era defined lead poisoning as an occupational risk. The lead industry's management of the tetraethyl lead scandal reassured a skittish public that leaded gasoline posed no general health threat and secured the industry's role as arbiter of that risk. And the initial epidemiology of pediatric lead poisoning defined the causes as poverty, ignorance, or culture. The plight of one Indiana family in the early Depression demonstrates the tragic consequences of such definitions and attitudes.

In April 1933, Anne White Mathews of Gary, Indiana, wrote to Frances Perkins, "I wish to congratulate you on your appointment as Secretary of Labor. I believe a woman has higher ideas and ideals than a man, and, also a greater concern for human welfare." She went on: "I'd like to tell you about my baby. She will be three years old, July 29. She cannot walk, talk, sit up, or even hold up her own head." Anne Mathews's ninth child, Martha, had been born with spina bifida—caused, her doctor believed, by her father's exposure to lead at his job. "'Industry's child' the doctor calls her." At the time Martha was conceived, Frank Mathews was a chemist at the Grasselli Chemical Company, a DuPont-owned insecticide manufacturer, working upward of twelve hours a day making arsenate of lead and other lead compounds. Mathews recalled that Martha "had no life while I carried her, like carrying, I used to say, 'a lump of lead.'"[26]

When Martha was a year and a half, Frank Mathews developed a severe case of lead poisoning, which kept him home for three months—an absence that cost him his job. Mrs. Mathews went to her husband's employer and extracted a verbal promise of reemployment for her husband in an uncontaminated area of the plant and compensation for Martha's condition. Mathews was clear in her purpose in writing: "Please Miss Perkins, I am not a lobbyist, I am a Mother, who has seen her own, her husbands [sic] and her child's life practically wrecked by industry . . . I want drastic legislation—protecting the lead worker and his family."[27]

Several months went by with no response. In November, Mathews wrote to President Roosevelt, telling him about her husband and child and reporting on the outcome of her promise from Grasselli: "Needless to say, a job was all he has received. So you have the spectacle of a company, employing a man, because it is afraid to let him go."[28] Both this and the earlier letter to Frances Perkins were forwarded to Clara Beyer at the Children's Bureau, who enlisted the help of Mrs. J. W. Moore, chair of the legislation department of the Indiana Federation of Clubs. "There is no way in which we can do anything for her from Washington," Beyer explained to Moore, although, she added, "we are planning a little pamphlet" on the subject. Perhaps Moore could enlist "some private individual or organization" to tackle the problem at the state level.[29]

Over the next eighteen months Moore, Beyer, and other officials in Indiana wrestled with the immediate problem of what to do for Martha, as well as the long-range problem of how to protect other Frank Mathewses, but were left shaking their heads at the tragedy, powerless in the absence of clear jurisdiction or compensation laws. Undaunted, Anne Mathews found a clinic in Chicago that could help Martha, and started looking for a lawyer who would be willing to bring suit against Frank's employer in order to secure compensation for Martha's ongoing medical expenses.[30]

The cases of Frank and Martha Mathews tell us a good deal about the status of lead poisoning in the early 1930s. Frank Mathews acquired his occupational plumbism twenty years after Alice Hamilton's highly celebrated exposés of lead-using industries in the United States. Yet the plant that employed Mathews provided no protective devices, and he and his co-workers typically worked fourteen-hour days. Apparently the managers of Frank Mathews's plant—located in East Chicago, Indiana, right in Hamilton's back yard—had not read her reports that told of the economic benefits accruing from simple improvements in plant hygiene.[31]

It can never be determined whether lead had anything to do with Martha's condition. Medical researchers have not established a causal connection between parents' occupational lead exposure and spina bifida. The research basis for a link between occupational lead poisoning and genetic damage (as opposed to transplacental lead poisoning) is weak, so there is some question about whether in cases like the Martha's, the mother's "second-hand" exposure to her husband's lead might be the culprit. One area often ignored in the pioneering nineteenth-century French and British research is the impact of the father's lead exposure upon the mother. How much exposure did Anne have from laundering Frank's clothes, from the dust he dragged home in his hair, shoes, and so on?[32]

Regardless of whether lead can be implicated in Martha's illness, her case has significance beyond its obvious pathos. It points up the cyclical nature of lead-poisoning research and the wasted energies and loss of human potential these cycles of concern and neglect have brought. Martha was "out of sync" with the researchers' interests. Hereditary or congenital lead poisoning was of great concern to Victorian and Progressive Era doctors and legal reformers interested in gender-specific "protective" legislation. But by the 1930s, professional interest in lead poisoning had shifted from nature to nurture. Pediatric lead poisoning was something that happened to ghetto children who inhaled fumes from burning battery cases or ate crumbling lead paint from walls and windowsills. Few researchers were studying the cases the Mathews family doctor referred to as "industry's children."

Decades later, the scientific and legal controversies surrounding fetal protection would direct the focus of new research back to earlier generations' interest in congenital lead poisoning.[33] Martha was born during the years between the first and second waves of interest. Had the first wave continued to swell instead of cresting and falling after 1918, Frank Mathews's employers might have seen the efficacy of installing simple protective devices in their plant.

Anne Mathews's role in the story seems similarly anachronistic. She does not quite fit the prevailing model of politically active women of her time. No evidence links her to any labor or women's association. She was busy raising ten children during the hardest years of the Depression, struggling to make the payments on her sewing machine and to provide feed for the family's milk cow. Yet this "ordinary" housewife was a political activist in her own way. Her educational background remains a mystery, but although she apologized for her handwriting, she wrote eloquently and fervently, maintaining correspondence with DuPont, the Children's Bureau, the Indiana Department of Commerce and Industry, doctors, and insurers. She educated herself about the lead industry (DuPont in particular), the known causes of her daughter's condition, and federal and state compensation laws. And she negotiated with agents of her husband's employer in order to save his job and protect her family.

To some extent, Anne Mathews foreshadows the activism of mothers and fathers in the late 1960s. These parents—often poor and politically powerless—would fight landlords, paint companies, and governments to protect their children from lead in their homes and in the general environment. Like most lead-poisoning activists of her generation, Anne Mathews did not know the extent of pediatric lead poisoning that could arise from a child's immediate surroundings. She fought to increase awareness about the consequences of occupational lead poisoning in order to make the workplace safer. Lead was a pediatric issue only when a lead worker's children were affected. This narrowed vista was more the result of the lead industry's successes in reducing occupational lead poisoning during the first quarter of the century than of deliberate dissembling on the part of paint companies. Over the next two decades, improved diagnostics and treatments developed for the workplace would encourage further research into childhood lead poisoning. Thus the lead industry inadvertently played a critical role in giving voice to the epidemic it had created, defined, and silenced.

A handful of doctors who worked closely with childhood lead poisoning refuted its image as the product of its victims' deficiency or their improbable misadventure. Researchers such as Lockhart Gibson, Kenneth Blackfan, John Ruddock, and Charles McKhann maintained that pediatric plumbism was the logical result of the normal hand-to-mouth activities of children who happened to live in lead-contaminated environments.[34] But those who correctly described as an everyday menace the lead paint used on walls in the homes of rich and poor alike could barely be heard above the din of dramatic stories of battery-case burnings and cases that arose "from unusual sources." An *Archives of Pediatrics* article in 1939 concluded that "the most important diagnostic points in lead intoxication are: History of burning of battery casings, the use of nipple shields and of lead intoxication in other members of the family." And a 1940 survey of the sources of lead poisoning concluded that "lead in toys and in paint used on toys and household articles was also responsible, formerly, for many cases of poisoning in small children, and in spite of a vigorous campaign waged against this practice, poisoning from these sources still occurs."[35] A generation later, pediatricians, and the paint industry itself, would come to realize that the lead paint on the walls and ceilings of almost every American child's home posed a far greater threat than cheap blocks or ignorant junkyard merchants.

Lead poisoning's confirmed status as a disease of poverty quelled most of the urgent calls for aggressive remediation. Indeed, to many the mere fact that the "Depression disease" had been recognized was sufficient. Proper diagnosis and prompt treatment would prevent the needless deaths of unfortunate poor children; increased awareness would lead to earlier interventions and more complete cures. But a study published in 1943 dashed this optimism, revealing that society's cost for each case of lead poisoning was much higher than imagined.

In the late 1930s, Randolph Byers, a pediatric neurologist, and Elizabeth Lord, a psychologist, both worked in the Neurologic Service of Boston's Children's Hospital. Independently, each came to suspect that lead poisoning had affected many of the school-aged students referred to the hospital because of learning disabilities or "behavioral difficulties." Scanning admission records for the previous decade, the researchers identified 128 lead-poisoning cases.[36] Over the next several years, Lord conducted periodic psychological evaluations of the twenty children who still lived in the Boston area and tracked their progress in school. With small grants from the Earhart Foundation and the Commonwealth Fund, Byers and Lord began writing their report. Suddenly, early in 1943, Lord died from leukemia. But with the help of one of her students, Byers interpreted Lord's notes and completed the paper, which appeared in the November issue of the *American Journal of Diseases of Children.*

Their findings undercut the premise that noncerebral lead poisoning left no after-effects. Although none of the study's patients had shown "gross evidence of cerebral damage at the time of discharge," only one child was doing well in school. Despite the fact that most had IQs in the normal range, they showed attention deficits, learning disabilities, and erratic behavior. More than a few also showed clear symptoms of neurologic damage many years after having been sent home "cured."[37]

The media response to the paper was swift but in its breathlessness missed the study's implications for lead poisoning as a public health issue. In December, *Time* reported the story in an article entitled "Paint Eaters." It focused on the more sensational aspects of some of the studied children's behavior, warning parents that "if your child is slow with building blocks, but quick on tantrums, he may be a lead eater."[38]

Lead-industry insiders did not miss the potential ramifications of the study. Felix Wormser found the Byers and Lord study "a most alarming revelation" and worried that "if what this article describes is correct, then we have indeed a most serious public health hazard." Of course, Wormser had little doubt that the study was incorrect, and wrote to Robert Kehoe, hoping to enlist his help as chairman of the American Public Health Association's lead-poisoning committee: "To any one who has studied lead poisoning intensively . . . and who wishes to sift fact from fancy, the assertions made appear to be far from proven." The flaws in Byers and Lord's study had to be exposed, Wormser insisted, before "other doctors . . . accept as authoritative this paper . . . and probably build upon it still more fantastic assertions."[39]

Kehoe's response could not have comforted Wormser. He apologized, but argued that the study's evidence "if not entirely adequate, is worthy of very serious consideration." The veteran lead researcher was inclined to believe Byers and Lord's findings, since he had "seen cases of serious mental retardation in children that have recovered from lead poisoning." Setting aside any personal predisposition, Kehoe believed that the study stood up to skeptical evaluation of its evidence.[40]

Most of the children in the study ate paint from their cribs. Wormser made much of the fact that the initial medical teams failed to test these paints for lead content. Were Byers and Lord unaware of the fact that years earlier, the LIA had worked tirelessly to prove that crib and toy manufacturers employed almost no lead paint? No matter, argued Kehoe. Parents repaint cribs, and a child who chews paint from a crib also eats paint around the house. "'Pica' is at the bottom of most of these cases," Kehoe insisted, "and unfortunately the environment of small children is not sufficiently free of lead for their safety." Kehoe also dismissed Wormser's skepticism regarding what he thought were dubious clinical measures taken during the children's initial hospitalization. Wormser was especially critical of x-ray evidence showing lead-lines along the margins of growing bone. But again, Kehoe admonished, "I'm afraid you are not on good ground. . . . The work . . . on this point is much too extensive and good to be dismissed so easily."[41]

Whether daunted by Kehoe's response or reminded of the beneficial relationship the lead industries had developed with Joseph Aub at Harvard, Wormser took a different approach with Dr. Byers: instead of confronting and refuting his science, he sought his cooperation, if not his silence. According to Byers, the Lead Industries Association threatened to sue him for a million dollars, but this threat could have been nothing more than a stick to complement the carrot they planned to extend. Byers and his boss, Dr. Bronson Crothers, were invited to lunch at the Boston Ritz, with Aub as host. Wormser represented the LIA, with support from Aub and Manfred Bowditch of the Massachusetts Division of Industrial Hygiene. As Wormser reported the meeting to Kehoe, there followed a frank discussion of the study's weaknesses and the urgent need for further, more rigorous, study.[42] Byers recalled the meeting as a less academically oriented affair: "When we were through, the lead people gave me a grant which they continued for eight or ten years to continue and expand my studies. This fortunate result may have been abetted by the four or five martinis the lead executives drank, but more likely it resulted from the quiet and skillful backup given my work by Dr. Crothers."[43]

The simple epidemiological and statistical tools employed by Byers and Lord would never cut muster today. In addition to weaknesses in statistical rigor—sample size, failure to test for confounding variables, and so on—most of Wormser's objections would have reduced the study's credibility. The Byers and Lord study lacked the sophistication of Robert Kehoe's work done under the aegis of the Ethyl Corporation. Recent scholars who maintain that the lead industry sought to silence dissent by defining the terms of scientific discourse would argue that Byers was "bought off" to maintain control of the production and interpretation of data. The work that Byers published with LIA support *was* far more sophisticated in method, circumscribed in scope, and circumspect in its findings—in short, more "scientific."[44] But if the lead industry had in fact maintained control of lead-poisoning research, Byers and Lord's paper would never have been published. If Byers had worked at Massachusetts General instead of Children's Hospital, Joseph Aub might have bridled his enthusiasm, or at least discouraged publication. It would seem that the few blocks separating the Harvard School of Medicine and the Boston Children's Hospital exceeded the compass of the lead industry's supposed hegemonic tentacles. Instead, Felix Wormser had to come up to Boston to do damage control in the time-honored way he had dealt with Harvard's researchers since the 1920s: by buying their cooperation with research funds. Too late, it would seem; the damage had been done.

Most serious studies of lead poisoning after 1943 acknowledged the fundamental impact of Byers and Lord's study, and Byers's subsequent articles appeared in a transformed epidemiological world. Still, Byers and Lord did not break some imagined stranglehold industry maintained on research; rather, their research revealed the more simple truth that only those interested in lead poisoning study lead poisoning. Commenting on the erratic swings in cases of lead poisoning admitted annually to Children's Hospital, Byers and Lord posited that "such cycles may . . . tend to occur in relation to the interest of the staff in the subject."[45] The postwar years would see a tremendous rise in such interest.

NINE

Urban Physicians Discover the Silent Epidemic

After World War II, the world bristled with new and terrifying dangers. In an age imperiled by nuclear weapons, invisible radiation, and virulent ideological pathogens, lead poisoning would seem an outdated and trivial problem. But while the

These young Baltimore paint eaters are a real headache. . . .

MANFRED BOWDITCH, LEAD INDUSTRIES ASSOCIATION, 1949

market for lead changed as much as it expanded during the years after the war, those years saw also a transformation in the definition and treatment of lead poisoning. Postwar America's fears of secret poisons and domestic infiltration, as well as its new-found devotion to domesticity and scientific baby care, enhanced the probability that the epidemic of lead poisoning silently maiming and killing its children would finally be discovered.

Although the general public would remain uninformed of the dimensions of the "silent epidemic" in the nation's cities until the mid-1960s, the public health community's awareness of lead-paint poisoning mounted through the 1940s and 1950s. Several factors promoted this development. The data collected in preliminary case-finding efforts revealed that every death from lead poisoning bespoke thousands of lead-exposed children. The new epidemiology was especially troubling in light of growing evidence that permanent psychological and neurological damage followed "milder" cases of lead poisoning. New treatments developed in the years just after World War II also transformed the disease, with mixed implications. The ability to treat serious lead poisoning encouraged doctors to seek out children to cure, giving a fillip to case-finding programs. But the presence of a "cure" also lulled many into a false sense of security.

The increased interest in pediatric lead poisoning led to two important developments. Medical and industrial associations defined a standard for the content of lead in interior household paints, and a nationwide system of poison-control centers was established. The one would reduce future exposures; the other would enhance the probability that present poisonings would receive effective treatments.

In the early postwar years, the lead-using industries continued to sponsor primary research and participate in case-finding programs, materials testing, and consumer safety projects. However, by the mid-1960s they could no longer direct research and publicity to the extent they had in the 1930s. As the base of health professionals and agencies interested in lead poisoning increased along with funds available through federal research grants, the industries' participation would become superfluous, their researchers' findings distrusted. Thus the first two postwar decades mark a transitional period in the history of lead poisoning, both for the lead industry and for the medical and public health professionals who worked the urban lead mines.

"Let's all help Pass the Ammunition," shouted the little Dutch Boy from the pages of a women's magazine in 1944.[1] Throughout World War II, the lead industries gloried in their metal's ability to "carry its weight" in service of the nation. Besides the obvious demand for lead bullets, the need for other lead products soared: steel ships required tons of red lead to prevent corrosion; aircraft designers deemed tetraethyl lead indispensable for formulating high-octane fuel; motor bearings in ships, planes, and tanks employed lead as a hardener. In addition, wartime shortages of steel, aluminum, tin, and other metals prompted hastily improvised leaden substitutes.[2] As a result, consumption of lead rose by almost 70 percent between 1939 and 1944.[3]

Lead played its part on the home front, too. Despite early fears that shortages would bring restrictions, lead mines in the United States, Mexico, and Canada instead increased production, and from 1942, few limits were placed on domestic use of lead pigments. The NLC told Americans that the federal Housing Administration wanted them to maintain the finish on their homes: "To paint where paint is needed is the patriotic duty of every citizen responsible for the upkeep of property—whether it be a home, a farm building, factory, bridge, storage tank or any similarly irreplaceable structures." And Dutch Boy was there to help, "for pure white lead and pure red lead paints have protected America's home front during every war since the Revolution!"[4]

Looking ahead to the coming peace, Ernest Trigg, president of the National Paint, Varnish and Lacquer Association, predicted bright days for

paints. Thirty billion dollars worth of building and equipment maintenance had been deferred during the war, and paint and varnish companies
would share in the fix-up. But it was in new consumer products, especially
automobiles, where the "rapidly expanding future" lay. "The demand for
new cars . . . will be by far the most overwhelming in its history . . . the
demand for automobile finishes will be correspondingly enormous."[5] Still
greater would be the demand for housing, though even Trigg underestimated the extent. Estimates that one million housing units would be required annually for the next decade were overly optimistic, he wrote. In
fact, there were about 13.4 million housing starts between 1945 and 1954,
and the number never fell below 1.2 million per year from 1955 to 1970.[6]
Hundreds of tons of lead went into these new homes—in their plumbing,
in solder in their electrical systems and appliances, and on every paintable
surface, inside and out. Every home in the Levittowns of New York, Pennsylvania, and New Jersey was fitted with a lead waste-drainage system;
many also contained lead supply pipes. And leaded gasoline powered the
cars that shuttled the new suburbanites to their jobs, to shopping centers
and churches, and back home.[7]

In unprecedented numbers, the owners of these new homes took on the
task of painting their suburban castles themselves, creating a market for
easy-to-use paints. The change was especially profound for women. In the
1910s, one paint advertisement stated flatly, "Women do not paint," though
they might be shown conferring with painters on the choice of colors. In
the 1940s and '50s, a number of factors—including development of water-
and latex-based paints and improved paint rollers, painters who demanded
higher wages, changing expectations for the role of wives as homemakers,
and a revolution in working-class home ownership—put paintbrushes in
the hands of millions of homeowners, both men and women.[8]

Nuclear technologies added new uses as well. The LIA's Felix Wormser
prophesied, "Just now at the dawn of the atomic age it appears that lead
will be in especially strong demand to shield us from harmful rays." Lead
was promoted as a shield for much more than buildings. A 1964 issue of
Lead, the LIA's promotional magazine, advised the "sensible girl who
works with dental X-ray, in a radiological lab, or near isotopes" to "look
your loveliest in lead"—in a lead-lined girdle. This "news" item, complete
with a department store catalog-style photograph of a woman modeling
the garment, claimed that the girdle provided "unobtrusive protection
against gene damage."[9]

Although the nation eagerly purchased ever greater quantities of leaded
products, concerns over lead's toxicity continued to mount. Shortly after

the end of World War II, William Wilentz, medical director at the NLC's plant in Perth Amboy, New Jersey, wrote an article for *Industrial Medicine* to counter "the unfair and insidious propaganda" that had put the lead industries "on the defensive far too long and unnecessarily." Uninformed propagandists had instilled in the public a condition that Wilentz named "Plumbophobia." Wilentz's thesis was not a simple-minded declaration that "we have nothing to fear but Plumbophobia itself." Most of his article deals with the nuts and bolts of preventive hygiene. As to cure, Wilentz believed that most cases of occupational lead poisoning could be managed through the well-worn method of encouraging the worker's body to lay its burden of lead down in the bone. In obeisance to Kehoe's precepts, de-leading was "not advocated."[10] For Wilentz and other lead-industry insiders, the years since the war must have seemed a long descent into panic and distrust of their "useful metal," as physicians, activists, and politicians converted plumbophobia into a force for community action and legislative and institutional solutions.

When Wilentz's article appeared in 1946, a growing number of public health departments and pediatricians were monitoring pediatric lead poisoning. This increased interest in the postwar years was due partly to the Byers and Lord study and the ongoing campaigns in Baltimore and other cities, but also to new treatments that had transformed lead poisoning from an insoluble problem into a tragedy of neglect. Just as germ theory transformed fundamental principles and practices of public health a century earlier, effective treatments for lead poisoning worked a similar transformation in the way pediatricians and clinicians conceived of childhood plumbism.[11]

In World War I, the U.S. War Department considered using tetraethyl lead as a nerve gas, so it is fitting that a drug concocted during World War II as an antidote to nerve gas became the first effective treatment for severe lead poisoning.[12] In the early 1940s, a number of researchers in the United States began experimenting with a secret formula, British anti-Lewisite (BAL), a chemical chelator that binds with many heavy metal salts to form less toxic compounds that the body can safely excrete.[13] After experimenting briefly with animals, doctors began testing BAL compounds on human subjects. A physician with the National Cancer Institute reported that the first human studies involved "some prisoners who volunteered" and a number of patients in a syphilis treatment program.[14]

Soon lead-industry researchers were dabbling with it. Late in 1946, researchers at Robert Kehoe's Kettering Laboratory of Applied Physiology began small human studies. They found that injecting BAL immediately reduced the amount of lead in the bloodstream and dramatically increased the level in urine. But the initial readings returned within a day or so, so that "no quantitatively important proportion of the absorbed lead of a poisoned or endangered individual has been removed." Kehoe therefore remained consistent in his opposition to deleading.[15]

Other researchers were more optimistic. James Tefler, a U.S. Public Health Service (PHS) surgeon stationed in San Francisco, reported successfully treating a severe case of lead poisoning with BAL.[16] The 40-year-old boatswain in the Navy, identified as "R.C.A.," customarily worked as a painter on oceangoing ships and was a "white-leader" who "always made a special effort to get 'good lead paint.'" After several years in the confined quarters of his workshop he developed chronic constipation, which suddenly flared into severe abdominal cramping and diarrhea. On admission to the hospital, his doctors prescribed the traditional calcium treatment. After ten days he began a three-day series of twelve BAL injections. Two weeks later, doctors administered a five-day course entailing nineteen injections of a slightly more potent concentration. Although his blood and urine still showed signs of lead, he was released.[17]

The early studies of BAL were not conducted with a view to altering radically the treatment of occupational lead poisoning—most such cases were of the less severe chronic type, thought to abate with diet and temporary removal from lead exposure.[18] Pediatric researchers, however, were very interested in the potential cure. In the same years that occupational medicine was rejecting BAL, lead poisoning killed fifteen children in the Baltimore area, nine despite the best efforts of hospital staffs that were by then more familiar with pediatric lead poisoning than any in the world. Once lead poisoning reached the level of brain involvement, the results of treatment were "uniformly poor."[19] And those who survived encephalopathy faced a life of severe mental retardation, recurring convulsions, and the risk of permanent blindness. The promise of BAL was worth the associated dangers of reported toxicity or limited effects.

An initial study, conducted by doctors in four Baltimore-area hospitals, compared thirty-one children admitted with encephalopathy from 1943 to 1947 with sixteen admitted between 1948 and 1950. Children in the later period were treated with BAL. The results in severe cases were dramatic. Eighteen of the thirty-one in the control group showed severe encephalopathy; of these, twelve (67%) died. In the BAL group, nine cases were deemed "severe," and of these, two (22%) died. A follow-up study three years later showed that with BAL, the case-mortality rate had fallen to 6.25% (see Table 2).[20]

TABLE 2   *Mortality Rates before and after the Introduction of British anti-Lewisite in the Treatment of Acute Lead Encephalopathy, Baltimore, 1931–51*

	Preintroduction (1931–48)	Postintroduction (1948–51)	Total
Total number of patients	38	16	54
Total number improved	28	15	43
Total deaths	10	1	11
Mortality rate	26.31%*	6.25%	24.6%

SOURCE: Garrett Deane, Frederick Heldrich, Jr., and J. Edmund Bradley, "The Use of BAL in the Treatment of Acute Lead Encephalopathy," *Journal of Pediatrics* 42 (Apr. 1953): 409–13.
* The physicians in this study excluded patients who died within 12 hours of admission, arguing that such cases would not offer a fair test of the new treatment. Had these patients been included, overall mortality in the preintroduction (control) group would have been consistent with earlier studies that found up to a 60 percent fatality rate.

In the early 1950s Riker Laboratories of California released a second chelator, calcium ethylenediamine tetraacetic acid (CaEDTA), which it marketed under the names Versene and Versenate.[21] Within two years, the first studies of CaEDTA in pediatric lead poisoning appeared. In 1954, Randolph Byers published a report of six cases that confirmed the effectiveness of the new drug. One child, "who was suffering from coma and convulsions when treatment began, was sitting up feeding herself 36 hours [later] . . . and began to talk in 48 hours." A researcher in Chicago reported that of nine patients suffering from lead poisoning at the cerebral level, the six who began treatment with Versene before the onset of severe encephalopathy survived. By the end of the decade, CaEDTA had come into wide use in pediatric cases.[22]

Despite coordination among the pharmaceutical company, doctors, and the Lead Industries Association, there were some problems in the introduction of CaEDTA, especially in its oral form. An effective oral chelator approximated as closely as anyone dared imagine a "magic bullet" for plumbism. But early trials in pediatric cases were marred by overly eager administration of the compound. Randolph Byers reported the case of a child who was admitted to Children's Hospital with "mild" lead poisoning. Animal tests had suggested that EDTA delivered by mouth actually enhanced absorption of lead from the gut, so the hospital's protocol called for "cleansing" the child's digestive tract before administrating oral Versene, then under trial. In this child's case, the staff administered the CaEDTA first, and the boy's blood lead soared, rendering him unconscious and convulsive, and leaving him "with damage which was not reversible, to the brain."[23]

Even more dangerous was the temptation to use CaEDTA as a preventive medicine. In the late 1950s, Riker vigorously promoted Versene to occupational physicians. Riker recommended the intravenous form for "dramatic short-term treatment," while the oral form was for "convenient prevention" or therapy for mild cases; Riker's advertising suggested doses for both adults and children. Robert Kehoe and other industrial hygienists condemned the prophylactic use of CaEDTA and enlisted the LIA's assistance in discouraging "charlatans" who pitched it to lead manufacturers as a "wonder drug."[24]

Radiographs and urinalysis had given doctors basic diagnostic tools for detecting pediatric lead poisoning since the 1930s. With the development of BAL and CaEDTA, treatment caught up to diagnosis, leading one medical journal to declare "the physician of 1957 has sufficient knowledge to diagnose accurately and treat successfully most children or infants with lead poisoning." The ability to treat even the most deadly forms of lead poisoning encouraged doctors to seek out and treat new cases. A psychiatrist at George Washington University recalled that his staff "began to accumulate children with lead poisoning, as we were looking for them for this research."[25] No longer could a lack of medicine excuse a lack of concern.

Increasingly, physicians and public health workers saw their moral obligation extending beyond caring for the poisoned children who showed up stuporous in the city hospital. "All therapy is wasted," a pediatric radiologist from Detroit urged, "if a child is sent back into the environment where trouble originally started." While specialists removed the lead from the child's body, workers should remove the sources of lead in the child's home. And "the public health–oriented physician will also not miss this opportunity to look for other cases in the same milieu." This environmental awareness pointed directly to the need for primary prevention. The biochemist who had established George Washington's treatment regimen admonished the researchers, "You've got to stop these kids from eating that stuff that is bringing on the lead poisoning."[26]

Each widely publicized infant death due to lead added to the LIA's public relations problems. The lead industry's responses to this growing threat can be seen in the careers of Felix Wormser and Manfred Bowditch. After becoming secretary-treasurer of the LIA in 1928, Wormser collected clippings from newspapers around the country and wrote countless letters to newspaper reporters and the doctors they mentioned. He also wrote to lead researchers such as Joseph Aub and Robert Kehoe, hoping they would provide "scientific" confirmation of his assumption that the claims were fraudulent. A typical letter began "Another alleged case of lead poisoning in an infant attracted my attention," and concluded "I do not see how he can call this a genuine case of lead poisoning, do you?"[27]

Public relations occupied increasing amounts of Wormser's attention, and he continued to meet criticism of "his metal" directly and vigorously. In 1944, he wrote to the LIA's member companies of his skepticism about the rise in reports of pediatric lead poisoning and warned of the potential for bad publicity. He recommended educating the public about what he would later call "The Facts and Fallacies of Lead Poisoning." J. H. Schaefer, a vice-president at the Ethyl Corporation and a member of the LIA board, offered Wormser his company's help in formulating a public relations policy for the association. In the meantime, Schaefer warned that the lead industries would have to get their "house in order with regard to industrial hazards. . . . We are never in a sound position to take our story to the public until we have demonstrated our ability to handle our own affairs within our own industries." And, echoing advice Kehoe had been giving Wormser for years, the association should not try to disprove any particular case, "because certainly there are a sufficient number of legitimate cases of lead poisoning of children and it is hard to prove to the public what is truth and what is fiction." Repeated denials would only erode public trust.[28]

Schaefer's advice, with its implied criticism of LIA member companies' record on hygiene, must have rankled Wormser. Ethyl had almost complete control over the conditions under which its product was made, distributed, and sold.[29] Further, with Robert Kehoe running the Kettering Laboratory, they had a corner on research and a convincing record of high standards for hygiene. The LIA, on the other hand, was a large and loose affiliation of scores of companies and could not exert such tight control. But the growing problem of pediatric lead poisoning demanded a vigorous response.

Wormser was a businessman, not a doctor, which may have led him to ally the association with respected research institutions rather than to establish its own research center. This policy worked well in the years when almost no one else was interested in lead poisoning. But with a growing public relations crisis, the LIA would need stronger allies and more expertise. In February 1946, Wormser met with eight of the nation's leading experts on lead poisoning to plan a symposium to be cosponsored by the American Medical Association. Wormser complained that American doctors exhibited "an extraordinary lack of knowledge about lead poisoning" and urged the AMA to correct the situation. Harvard physiologist Philip Drinker agreed, complaining "that the metal had been kicked around unnecessarily and that it was a good time to correct this situation."[30] The symposium was held in early October 1946. One of its highlights was Wormser's fiery "Facts and Fallacies of Lead Exposure," which outlined his long experience with erroneous reports of lead poisoning. Aub concluded that the symposium had been a great success, and that with Wormser's lecture the meeting "ended up with everyone admiring your point of view and therefore that of your association."[31]

The practical impact of the symposium was negligible, but nine years later the LIA allied itself with another powerful group of medical professionals, the American Academy of Pediatrics, with more concrete results. This campaign would be coordinated not by Wormser, but by a career occupational specialist hired to run the LIA's health programs. The association must have seen that "damage control" was going to require the full-time attention of such an executive. It had also become apparent that Wormser's background in engineering and business made him a less effective intermediary between lead manufacturers and physicians than the situation required. In January 1948, soon after Wormser rose to the presidency of the Association, the LIA's new secretary-treasurer, Robert Ziegfeld, appointed Manfred Bowditch director of health and safety.[32]

In direct contrast with Wormser, who had almost single-handedly managed the LIA's public affairs since before the Depression, Bowditch was the consummate "organization man" called for in the business climate of the late 1940s. In the thirty years before he joined the LIA, he had served as manager of a metals and chemical brokerage, plant hygienist at two General Electric facilities, instructor in occupational hygiene at Harvard's School of Public Health, and founder and director of the Massachusetts Division of Occupational Hygiene.[33] Although he was not a doctor, Manfred had close ties to the Harvard Medical School through his father, the noted physiologist Henry P. Bowditch. And though his service with the Commonwealth of Massachusetts brought him into conflict with manufacturers, he retained and strengthened his alliances with industry. When, in 1945, proposed cutbacks in the Department of Labor threatened Bowditch's position, many of the companies he regulated rushed to testify to his abilities. Andrew Fletcher, vice-president of St. Joseph Lead Company, warned that other organizations were waiting "to employ Mr. Bowditch in responsible and well-paid capacities."[34] Three years later, Bowditch took one such well-paid position in the Lead Industries Association.

Robert Kehoe congratulated Bowditch on his appointment, and praised the LIA for its renewed interest in improving "hygienic conditions in the affiliated lead industries." He warned Bowditch that the need for improvement was "very much greater than you ever suspected" because the lead industries, "by and large, have not taken full advantage of the available information that now enables satisfactory control to be achieved." He concluded with the warning he had been giving Felix Wormser and others for years: "I hope that the lead industries will not wait until the state of society and the attitude of organized labor compel them to do what they could have done with great profit to themselves and to the community, at a much earlier time."[35]

Bowditch proved to be a creative and industrious executive, and under his guidance the lead industries expanded their activities in both public relations and basic research. One of his first projects was an industrial film, *The Safe Use of Lead in Modern Industry.* An early draft of the script, written by the Jam Handy Organization, opened with scenes of modern American life: "Batteries in streamlined automobiles, a steel frame of a sky scraper being covered with red lead, a neat Colonial residence receiving a coat of white paint, modern enameled kitchen equipment and utensils. . . ." Then the cameras would turn to "close shots of some of the hazardous jobs," including painters eating sandwiches or smoking with paint-smeared hands. As the voice-over provided statistics on how many cases of lead poisoning resulted from "eating lead," the screen would show "a child with face smeared as it sits on the floor near the base board which has some paint chips peeled off." The writers specified that the early scenes "should, if possible, be played deliberately under old fashioned, obviously dangerous conditions so as to serve as a marked contrast to modern factory scenes to be shown in a summary at the end of the film."[36]

Bowditch also prevailed upon the LIA to include health issues in its 1952 book *Lead in Modern Industry.* After twenty-four chapters extolling the thousands of applications for lead and its compounds, the book turns to "The Safe Handling of Lead and Its Products." In what amounts to hardly more than an appendix at under three pages long, the authors admit that "lead if improperly used and handled may be a health hazard." But quickly the book assures the reader that "its properties have been recognized for so many years" that it can now be used in complete safety. Workers and children with pica do sometimes get sick, but "lead intoxication [not "poisoning"] can be cured and recovery is usually complete, leaving no disability." Readers wanting more information were told to send seventy-five cents to the LIA for a brochure.[37]

According to Bowditch, the very existence of this chapter 25 was more than a small victory. In a cover letter he enclosed to accompany Joseph Aub's complimentary copy of the book, Bowditch cautioned, "Before yielding to the quite natural impulse to comment caustically on the meager section . . . please consider that the volume represents the thinking of our 70-odd member companies and that the book which it supersedes, published in 1931, made no mention of the toxicity of the metal which is our bread and butter."[38]

The LIA's efforts at educating the public about lead poisoning were understandably circumscribed. But its direct sponsorship of research burgeoned, as Kehoe had recommended. It participated in testing EDTA and funded pediatric research at Boston Children's Hospital through grants to Randolph Byers and at Johns Hopkins through grants to J. Julian Chisolm and others. For the first time, between 1951 and 1957, the LIA also funded several studies at Kehoe's Kettering Laboratory.[39]

Despite Manfred Bowditch's efforts, however, the LIA could not quell the rise of "plumbophobia." The rumblings from city health departments, the fears of the concerned parents who wrote to government agencies, the periodic outbreak of negative publicity accompanying "alleged" cases of lead poisoning drummed on. In a report of his department's activities in 1952, Bowditch laid out the seriousness of the situation, which continued "to be a major 'headache' and a source of much adverse publicity." His figures from nine cities showed 197 reported cases of childhood lead poisoning in 1952, with forty fatalities. Bowditch felt certain that "many could be shown to be of doubtful validity." But since "detailed investigation of any such large number of cases . . . would be prohibitively expensive" (LIA's recent expenses in Boston and Baltimore alone totaled more than $15,000), Bowditch recommended instead that the LIA continue to invest in research directed at "securing more accurate diagnosis." The stakes were high, for if cases "of doubtful validity" continued to garner adverse publicity, the LIA faced "the likelihood of widespread governmental prohibition of the use of lead paints on dwellings."[40]

Since the 1930s the epidemiology of childhood lead poisoning had defined it as a disease of poverty, assuring that the job of studying it would devolve upon city health departments—the government agencies whose primary duties involved practical action on the problems of the poor—and large urban hospitals, in whose emergency rooms the lead-poisoned children of the poor showed up every year. Decades before the federal government assumed a role in coordinating screening and treatment programs, these two institutions made remarkable progress in identifying cases and reducing the rates of fatal lead poisoning.

Since the early 1930s, Baltimore had led the nation in locating, studying, and combating childhood lead poisoning. Through the activism of Huntington Williams, the Baltimore Health Department developed strong ties with researchers at the Johns Hopkins University and the University of Maryland, where pediatricians had demonstrated continuing interest in lead poisoning since the time of Kenneth Blackfan's research in the 1910s.[41] This alliance made Baltimore the nation's pioneer in implementing mass screening and abatement programs. (Ironically, these organizations' successes did not bring praise for their detective work but instead earned Baltimore the erroneous label of the national capital for childhood lead poisoning.) As a result of the health department's activities, in 1942 Baltimore reported nearly one-quarter of the nation's lead-poisoning fatalities, with an incidence of plumbism about fifty times higher than that for the balance of the United States.[42]

Over the course of the 1940s, greater understanding of the epidemiology of lead poisoning encouraged Baltimore's policymakers to incorporate environmental and economic components in their plans to fight it. Health Department maps drawn from twenty years of case finding showed that the areas with the greatest rates of plumbism coincided with those for tuberculosis and venereal diseases. "Lead alley" and "lung block" shared the same address. Accordingly, Baltimore in 1941 enacted a housing ordinance giving the Health Department authority to make landlords abate hazardous conditions in their properties. World War II halted most Department casefinding and abatement activities, however, and landlord resistance made the ordinance essentially unenforceable. The city council diluted it in 1958 to require lead abatement only in homes where a child had been poisoned. In the first year of the new ordinance, the Health Department ordered one hundred landlords to remove flaking paint from apartments. All complied with this modest demand, although two did so only after being taken before a judge and fined.[43]

Each new case of lead poisoning sparked further case finding. Every year, the Health Department sent more blood samples to Maryland's Bureau of Occupational Diseases for analysis. Between 1940 and 1950, the number of pediatric blood samples tested increased fourfold. Not surprisingly, increased testing revealed more lead poisonings: the number of reported lead-absorption cases rose from fifteen to fifty-nine. In most years, the ratio of confirmed "cases" to patients tested remained close to 25 percent (see Table 3).[44] Baltimore's published reports left undefined its criteria

FIGURE 9-1
Huntington Williams Inspects a Baltimore Apartment
Baltimore City Health Department.

for a diagnosis of lead poisoning. In 1950, children with blood-lead levels greater than 70 μg/dL and who exhibited pica were called "probable" cases, so it is likely that Baltimore followed the occupational standards that defined lead poisoning as 80 μg/dL, over three times the level now associated with profound intellectual and behavioral deficits in young children.[45] A record number of cases in 1948 prompted the department to step up its campaign of education and hazard abatement. Testing increased as well, resulting in another record-breaking number of cases in 1951.[46]

The Lead Industries Association contracted with the Johns Hopkins Department of Physiological Hygiene to analyze all of the reported cases in Baltimore in 1948 and 1949. Anna Baetjer, a physiologist at the Johns Hopkins School of Hygiene and Public Health, examined the medical records of the cases of sixty children reported by Baltimore clinics and physicians and measured the lead content in paints found in the children's homes. Victims were classified under two categories: "positive" cases were those who "demonstrated pica" (i.e., they ate paint or dirt), had lead paint

TABLE 3 *Lead Tests and Results, Baltimore, 1936–51*

	Tests Conducted			Cases* (children only)	Deaths (children only)
Year	Adults	Children	Total		
1936	51	32	83		8
1937	88	43	131		2
1938	80	60	140	22	6
1939	112	68	180	13	4
1940	152	61	213	15	7
1941	201	78	279		3
†					
1947	169	78	247	22	3
1948	197	150	347	46	4
1949	231	242	473	34	2
1950	240	253	493	59	2
1951	130	323	453	77	9

SOURCE: Data for tests conducted and cases identified, *Baltimore City Health Department [Annual Reports for the Years Ending 1936–51]* (Baltimore: City Health Department, 1937–52); data for deaths, 1936–39, Baltimore City Health Department, "Lead Poisoning in Children" (Baltimore: 1949); 1940–51, Baltimore City Health Department, "Lead Paint Poisoning in Children" (Baltimore: 1968).

* I have included cases reported as "probable," which may cause discrepancies with other reports from Baltimore that count only "confirmed" cases.

† World War II apparently brought the screening program, or at least the reporting of it, to a halt.

in their homes, and had blood-lead levels of at least 100 μg/dL; "probable" cases were those whose blood-lead levels fell between 60 and 100 μg/dL. The sixty cases included nine probables and fifty-one positives; eight of the latter group died.[47]

If the LIA intended this study to fend off further unfavorable publicity from Baltimore, its plan backfired. Anticipating the coming deluge of questions and publicity and unsure of how to proceed, Manfred Bowditch sought help from Randolph Byers, who was still doing research under an LIA grant, and his long-time mentor, Joseph Aub. Bowditch sent Aub a set of questions the Baltimore study raised, saying "I hope you may find the time to discuss with me some of the matters referred to. . . . These young Baltimore paint eaters are a real headache and I need all the advice I can get."[48]

The Baltimore study stirred wide ripples in public health circles. Baetjer's report was the most detailed epidemiological study of pediatric lead poisoning to date, and it raised a number of important issues. The most perplexing was the marked summertime peak in lead-poisoning cases. Lawrence Fairhall of the PHS wondered to Bowditch, "Ordinarily I would expect that children confined to the house, as in winter time, would be more likely to be gnawing painted surfaces than those out in the open during the warm weather."[49] If "perverted appetite" caused lead poisoning, were children more perverse in the summer? Or perhaps the summertime peak suggested a "normal" metabolic reaction to "abnormal" environmental lead exposure.

With logic that would seem more appropriate coming from Felix Wormser, Aub suggested to Bowditch that perhaps the seasonal incidence of lead poisoning was actually an indication of "summertime complaints"—infectious diseases misinterpreted as lead poisoning.[50] But when epidemiologically minded doctors and public health officials turned this argument on its head, the seasonality issue presented another reason to screen for lead poisoning. How many summertime deaths in their cities, now attributed to vaguely defined "summertime complaints," were actually caused by lead poisoning?

On reflection, the study had even more ominous implications. Fairhall told Bowditch, "It occurs to me that if so many cases of lead poisoning in children occur in a city the size of Baltimore, the number for the entire United States in any one year must be enormous." May Mayers, medical chief of the New York State Department of Labor's Division of Industrial Hygiene and Safety Standards, wrote Bowditch that "for the first time, I am beginning to be really concerned that there may be many more children who develop lead poisoning and die of it throughout the country than we had hitherto suspected."[51]

A second report that came out of Baltimore's program would have an even more significant impact on doctors' perceptions of the scope of the problem. In a 1956 study of a new test to screen for lead through a simple urinalysis, researchers at the University of Maryland and the Baltimore Health Department took both blood and urine samples from 333 babies in well-baby clinics, children with no reported symptoms of lead poisoning. While the study cast doubt on the urine test's accuracy, the blood tests done on 148 children to monitor the new testing method revealed that over 40 percent had blood lead in excess of 50 μg/dL. Although this level did not put these children in the "positive" category in the mid-1950s, twenty-one had reached this "threshold" with blood-lead levels above the industry standard of 80 μg/dL; eight of the study children developed acute symptoms requiring hospitalization.[52] If two-fifths of the "well babies" in Baltimore had this level of lead absorption, what was the cost in chronic illness?

Although Baltimore had a head start on fighting childhood lead poisoning, a number of variations on that city's experience played out in urban centers in the eastern United States throughout the 1950s. Cincinnati shared with Baltimore two factors that encouraged lead-poisoning research: poor children living in old housing and research and testing facilities where lead poisoning was being studied actively. In Cincinnati, Robert Kehoe's Kettering Laboratory provided the facilities and expertise. Since the late 1920s, Kehoe had maintained ties to hospitals in the region. For his study of lead concentrations in "normal" populations, he had collected specimens of blood and urine from Cincinnati children. The Kettering Laboratory routinely analyzed tissue samples in cases of childhood lead poisoning from regional hospitals, frequently without compensation.[53]

Kehoe was aware of childhood lead-paint poisoning early in his career, and although he always held the lead-industry line in public, his internal correspondence vigorously criticized the LIA's record on public health. In the late 1940s he cooperated with Hugo Smith, a pediatrician at Cincinnati's Children's Hospital, to study childhood lead poisoning. Smith recalled that before 1949 the hospital's medical personnel "were not yet alerted to the disease." In that year, however, postmortem exams of three Cincinnati children diagnosed with polio revealed that lead had been the cause of death: "This experience jarred us violently," Smith wrote, and "we soon found that the disease was quite prevalent." From 1941 to 1948 doctors in the city had diagnosed only eight cases; over the next eight years Cincinnati doctors reported 113 lead-poisoned children, with twenty-six fatalities.[54]

In 1951, Smith and Kehoe began "a concerted cooperative program of study and follow up." They held conferences with hospital staffs, enlisted the aid of Cincinnati's health commissioner, and sought funds for an analytical survey along the lines of the Baltimore study. Kehoe suggested to Manfred Bowditch that the LIA might want to "have a share" in the proposed study. It would seem, however, that the LIA's funding of programs in Boston and Baltimore had satisfied its interest in pediatrics research: Bowditch replied, "Your group most certainly has our unqualified moral support; whether something more tangible may later be forthcoming can hardly be predicted now."[55] Apparently, nothing tangible materialized. But even without the LIA's sponsorship, Cincinnati's public health community remained alert to childhood lead poisoning.

New York City's doctors got their wake-up call in the summer of 1951, when the story of Celeste Felder, a 2-year-old from Bedford-Stuyvesant in Brooklyn, made the local headlines.[56] Celeste died from lead poisoning in Cumberland Hospital on August 23 at 4:15 A.M. This was not newsworthy in itself; New York–area hospitals would report fifteen lead-poisoning deaths in 1951 and 1952. What put Celeste in the papers was the tragedy of errors that marked her final days, as one doctor after another misdiagnosed her increasingly agonizing symptoms. Her story was only brought to light by the anger, frustration, and determination of her father, William Felder, a Brooklyn subway clerk.

Celeste's illness had started in July, when she began crying and refusing to eat. The family's private physician diagnosed intestinal grippe, and gave her penicillin. But her symptoms got worse. William Felder took his daughter to another doctor, who suspected a viral infection. He assured Celeste's parents that there was nothing to worry about. The next day, "she went out of her head" and fell unconscious. A physician at Kings County Hospital examined her and determined that she had an upper respiratory infection. He told Felder not to worry, gave Celeste more medicine, and sent her home. All the next day, Celeste shook with convulsions, which only abated when she lost consciousness that evening. An ambulance rushed her to Bushwick Hospital, where a doctor examined her throat and declared that she had "a bad case of tonsillitis." Despite the father's pleas, the doctor told him it "would be foolish and a waste of money" to admit Celeste to the hospital and insisted that he take her home.

By the time they got Celeste home, however, her screaming and convulsions had resumed. This time, Felder drove her to Brooklyn Eye and Ear Hospital, where, finally, a physician suspected that Celeste was lead poisoned. Unfortunately, Brooklyn Eye and Ear was not equipped to treat plumbism, so Celeste was moved to Cumberland, where the staff immediately began treatments. But it was too late.

Celeste Felder's case impressed the medical community enough that nearly a decade later, physicians were still talking about it when they advocated further medical training and research to improve diagnosis.[57] It is difficult to measure how influential the case was in prompting New York to initiate lead-poisoning case-finding programs. As this case demonstrated, there were doctors in the New York area already on the lookout for lead poisoning. A person just had to know where to find them.

Over the next two years, the New York City Board of Health improved its reporting of lead-poisoning cases.[58] Awareness of lead poisoning in New York hospitals nevertheless remained spotty for at least four more years. One of the first epidemiological studies of lead poisoning in the New York area, published in 1956, found that just three hospitals reported over 60 percent of the region's pediatric cases. The "lead belt" in Manhattan's Lower East Side corresponded closely with the service area of Bellevue Hospital, whose staff was "well aware of early symptomatology of lead poisoning." The study also praised for its vigilance Kings County Hospital, one of the hospitals that had turned Celeste away five years earlier.[59]

In 1955, the Department of Health further enhanced its case-finding program. It now questioned parents of suspected victims about pica and the condition of paints in the home and included intensive professional training for area physicians and a strategy of aggressive house inspections. As more doctors in more hospitals sought lead-poisoning cases, the familiar rise in cases and gradual reduction in fatalities followed. In 1954, eighty cases were reported, with twelve deaths (15%); ten years later, New York doctors reported 509 cases, with only seven fatalities (1.4%).[60]

Through the mid- to late 1950s lead-poisoning detection programs proliferated. St. Louis instituted free blood-lead testing in 1946, driving the number of reported cases up by an order of magnitude over the next decade. Chicago began actively searching for lead poisoning in 1953.[61] In Philadelphia, where the American lead-paint industry began in 1804, case-finding limped along with little in the way of a concerted effort until 1959, when Philadelphia Children's Hospital established screening protocols and initiated a two-year study.[62]

Without the regulatory power to remove lead paint from tenement walls, urban case-finding programs could not hope to eliminate lead poisoning. But those cities with aggressive case-finding programs significantly reduced the number of children who died from it. As each city began to screen, the same pattern developed. Initial screening detected only the most serious cases, giving the impression of an acute but limited problem. As the screening program reached deeper into the affected community, it identified greater numbers of children who benefited from treatment. As a result, the fatality rate among lead-poisoning cases fell—often dramatically. The availability of effective treatments no doubt speeded this process along, but it is noteworthy that Baltimore's screening program, which began many years before the advent of BAL and CaEDTA, showed the same general result. Early detection alone seemed to lower mortality (see Figure 9-2 and Table A-3). In addition to saving lives, the urban screening pro-

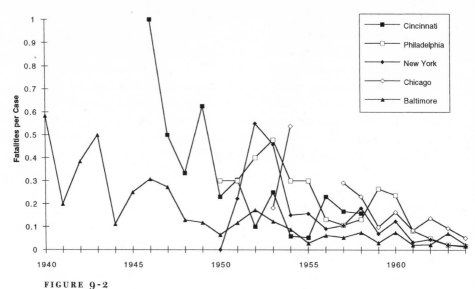

FIGURE 9-2
The Effect of Case-Finding Programs upon Lead-Poisoning Fatality Rates
See Appendix, Table A-3, for numbers of cases per year and source information.

grams forced the lead industries to acknowledge the extent of childhood lead-paint poisoning and to take some actions to reduce the dangers.

Late in the summer of 1953, a summer in which five children in Cincinnati alone died from lead poisoning, Robert Kehoe speculated about the national scope of the problem. From the Kettering Laboratory's data on lead-paint poisoning, a collection that was probably second only to that of the LIA, Kehoe estimated "that there are considerably more than one thousand severe cases of juvenile lead poisoning per year in the U.S.A.," with between 250 and 300 fatalities "at the minimum." Although he was writing to paint company officials, Kehoe did not hedge on the cause of these deaths: "There is no question about the principal source of exposure in our cities. It is usually lead-containing paint."[63]

The growing certainty about this forced the LIA to abandon blaming parents for repainting old cribs and furniture. Now, with a complete inversion of logic, it faulted parents for not realizing that most house paint contained lead and thus (contrary to decades of lead-industry promotion) posed a hazard to their children. Bowditch complained to Kehoe that "until we can find means to (a) get rid of our slums, and (b) educate the relatively ineducable parent, the problem will continue to plague us." Kehoe agreed that education was the solution, but he had a different pupil in mind. At an LIA meeting in 1953, he told the association's members that "a concerted and well organized program of education and control on the part of the paint manufacturers" was needed to get them to stop making lead-containing paints and putty for interior use. He warned: "If this is not done voluntarily by a wise industry concerned to handle its own business properly, it will be accomplished ineffectually and with irrelevant difficulties and disadvantages through legislation."[64]

The paint industry chose a voluntary solution. Key industry executives served on a subcommittee of the American Standards Association (ASA), established to set a voluntary limit on the amount of lead in interior paints. Sponsored by the American Academy of Pediatrics, the ASA's "Sectional Committee on Hazards to Children, Z66" included representatives from health associations, the federal Children's Bureau, insurance companies, lead-using industries, technical societies, and the Boy Scouts. The committee first met in March 1953 and established a subcommittee, Z66.1, to set standards for lead. In November 1954 the full committee approved the final standard.[65] The so-called Z66.1 standard required that any paint sold for interior use should contain less than 1 percent lead by weight in its dried film. The standard's stated purpose was to reduce the chances of poisoning "if, by chance, some [complying] coating should be chewed off and swallowed by a child."[66] John Foulger, director of medical research at DuPont, who served on the lead subcommittee, recalled that the 1 percent standard was based on "information we obtained from the paint industry." From this information, and estimates of lead metabolism from Robert Kehoe, the committee determined that a 3-year-old would need to eat 7 ounces of the new paint "to get lead poisoning. . . . It was considered," Foulger said, "that surely the parent would notice that amount of chewing."[67] This statement suggests two principles underlying the committee's decisions: protecting children from old paint was not on the agenda, and responsibility for lead poisoning was placed squarely on parents, who were expected to repaint their homes with new lead-free paint and be constantly vigilant where their children were concerned.

While no informed person would have argued that lead paint was nontoxic in 1955, the very definitions of "toxic," "lead paint," and "lead poisoning" were troublesome. To avoid the appearance of arbitrariness, the Z66.1 subcommittee claimed to have employed a complex calculus of technical and toxicological reasoning in establishing the 1 percent standard. But the simple arithmetic of economic interest had actually dictated the answer: 1 percent was the lowest the paint industry was willing to accept.

Although in the preceding decades the average lead content in paints had continued to decline, subcommittee chair A. G. Cranch of the Union Carbide and Carbon Corporation argued that "it would be almost impossible to produce a lead-free paint." Lead may no longer have been the chief ingredient in pigments, but many contained lead compounds, and lead naphthanate was still "the most satisfactory dryer."[68] Paint makers would have to scramble for some new low-lead paints, but the committee seemed to agree that 1 percent was a convenient compromise.

The seeming paradox of the LIA's support for Z66.1 disappears in the light of the changing market for lead products in the 1950s. From 1926 to 1955, the share of overall lead consumption held by white lead pigments fell by more than seven-eighths, largely because paint companies had introduced new pigments. Titanium dioxide, especially, had gained popularity as the cost of producing this superior pigment fell. From the perspective of the larger lead-paint manufacturers, titanium's progress was a friendly invasion. Many had begun investing in titanium mines and processing technology decades earlier, when the new pigment was a costly additive used for special purposes.[69] As long as Z66.1 allowed manufacturers enough leeway to add a dash of lead as needed, the LIA's members could harvest the public relations bonanza of having acted selflessly in the best interests of children.

The Z66.1 specification also contained a provision for labeling: "Coatings complying with this standard may be marked: 'Conforms to American Standard Z66.1–1955, for use on surfaces which might be chewed by children.'"[70] This voluntary provision, included as it was in an already voluntary standard, suggests that the paint industry hoped that in addition to protecting children, efforts by cities and states to require them to provide content labels would be forestalled.

Paint-labeling legislation had been a thorny issue for paint makers for fifty years, but the rules of the debate had changed as much as the players. Where lead-paint manufacturers once pushed for warning labels on any paint that was not 100 percent pure white lead, now they fought efforts to require labels that warned customers that a paint contained as little as 1 percent lead. But because of the decades of legislative stalemate between the "white-leaders" and the mixed-paint trade, and since most paint manufacturers had continuously experimented with formulas and varied their products according to changing prices for their raw materials, consumers in the 1950s knew less than their Progressive Era counterparts about what they were putting on their walls, their windows, and their baby furniture.

This troubled many parents, and some turned to the federal Children's Bureau for advice. In 1952, Albert Solomon, a counselor at the FBI academy at Quantico, wrote, "I am faced with the problem of repainting some children's furniture and toys for my infant daughter." Having read "at length" about fatal childhood lead poisoning, he sought advice on how to identify safe paints. In the early 1940s, the bureau had distributed a brochure entitled "Toys in Wartime," which defined "safe and unsafe pigments." But after a number of revisions, the bureau decided that existing data on many pigments' toxicity were inadequate and stopped distributing the pamphlet. Instead, Alice Chenoweth, a pediatrician at the Children's Bureau, sent Solomon a reprint on "Lead Poisoning in Young Children."[71]

But requests for specific information kept coming in—from parents, paint salesmen, manufacturers, and high school students doing research for their science classes. Eventually, Marian Crane, the bureau's assistant director for child development, became dismayed that "the questions that are coming to us are technical ones that we are not qualified to answer." She asked Lawrence Fairhall of the Public Health Service to take over answering such questions or to write a new statement about toxic pigments that the Children's Bureau could refer to or distribute. Fairhall agreed, assuring Crane that he understood "the difficulties which may arise because of inquiries from paint manufacturers, as that is not a group you are trying to assist with reference to safe children's furniture and toys."[72]

Despite this agreement, Children's Bureau staff continued to reply to letters from parents and manufacturers. Until the ASA established Z66.1, correspondents received much the same message—we no longer have that information, it is too uncertain—or were given the vague recommendation to avoid enamels in favor of flat paints, which are less soluble. After 1955, parents received the ASA's leaflet on Z66.1, a reprint of a report on "The Toxicity of Dyes," and, surprisingly, excerpts from the bureau's 1942 "Toys in Wartime."

Other than supplying this sort of information about toxic paints, the Children's Bureau remained detached from any effort to prevent lead poisoning. Similarly, no federal agency attempted to require paint labeling, preferring to let the cities set the rules. In 1951, Baltimore Commissioner of Health Huntington Williams sent the Public Health Service a copy of his report of his city's recent epidemiological study and an announcement of its new regulation banning for interior use paint that contained "any lead pigment." J. O. Dean, assistant surgeon general for the PHS's Bureau of State Services, recommended its publication in the service's *Public*

Health Reports, but he and the industrial hygiene staff at the agency warned against endorsing the paint ban "because of the public relations problem that would be sure to develop with the lead industry."[73]

In the absence of any federal interest, a number of cities attempted to establish labeling laws. In 1954, New York began requiring all paints containing more than 1 percent lead to bear the label "Contains Lead. Harmful if eaten. Do not apply on toys, furniture, or interior surfaces which might be chewed by children." Early drafts of the ordinance would have required POISON to appear in bold print on the labels of lead-containing paints, but the paint industry demanded that this provision be removed. In February, New York's state legislature considered paint-labeling bills modeled on New York City's ordinance. The LIA urged its members to protest, because such a bill's passage "would impose a heavy burden on interstate commerce" and would only duplicate the ASA standard (though voluntary, the LIA secretary called it a "regulation") that was soon to go into effect.[74] Three years after the promulgation of the ASA standard, Baltimore's health commissioner mounted a vigorous campaign to get manufacturers to stop producing and promoting lead-based paints for interior use. That city passed strict labeling laws in 1958 and 1959, and Health Department officials surveyed, investigated, publicized, and threatened to prosecute local manufacturers and retailers who violated the ordinances. Standard Z66.1 or no, lead paints were still being sold for interiors.[75]

The calls for paint-labeling laws and the development of city-based case-finding programs, tied as they were to local personalities and local politics, were sporadic, lacked any meaningful coordination, and could have ended in a dead end if not for the simultaneous rise of the poison-control movement. Between the early 1950s and 1961, when President Kennedy signed into law a bill making the third week of every March National Poison Prevention Week, pediatricians and public health officials developed an effective, nationally coordinated system for data collection, professional education, publicity, and gaining political clout.[76]

An early report on poison-control programs reflected both the new nuclear-age anxiety about chemical poisons and the postwar emphasis on the home. "There is a 'toxin,'" the authors exhorted, "that may be more deadly than that generated by the germs causing typhoid fever, tuberculosis, diphtheria, or leprosy and this toxin has already spread to almost every household in the United States." Children were the chief victims. A survey of American pediatricians in 1950 revealed that over half the emergencies they handled were the result of accidental poisoning. According to census statistics, between 1940 and 1950 over four thousand children under age 5 died from poisons. And prospects for the future did not look better, as one of the founders of the first poison-control centers prophesied: "Poisoning now looms as a major public health problem. Moreover, at the rate the new chemical compounds are being synthesized and distributed with the aid of modern merchandising methods, an even greater threat to health from poisoning is likely in the future."[77]

In 1952, Edward Press of the American Academy of Pediatrics' Accident Prevention Committee led a drive to establish a pilot poison-control center in Chicago, the first in the United States. With the cooperation of the five local medical schools and affiliated hospitals, the AMA, the Food and Drug Administration, and the National Safety Council, the center collected lists of all known household toxins, established treatment protocols, and standardized reporting methods. National support and centralized coordination were critical in promoting the poison-control center idea: by 1956, thirty-eight cities had established centers; by 1983 there were over six hundred. And in 1957 the federal government came on board officially, when the FDA established the National Clearinghouse for Poison Control Centers.[78]

Why did poison-control centers grow so rapidly, in contrast to the sluggish pace in the growth of city-based lead-poisoning case-finding programs? The answer lies in the relatively limited population served by the latter. According to accepted epidemiological wisdom, lead poisoning affected primarily poor children living in bad housing. By contrast, the specter of household poisonings hung over urban, suburban, and rural children of all economic classes. Popular magazines in the 1950s alerted housewives to the menace in their kitchens and medicine cabinets and reported the growth of the poison-control centers. In 1957, *Saturday Evening Post* readers learned of "apple-cheeked and tousled" little Lenny Behrens, who one Wednesday morning in his "quiet Brooklyn home" tippled furniture polish in his playpen while his mother did "her kitchen chores." When Lenny fell unconscious that night, his father took him to Kings County Hospital, where a doctor told him it was nothing worse than a common cold. The next morning, after Mr. Behrens had gone to work, Mrs. Behrens "started to put the play-pen toys in order," when she found the empty bottle of polish. "Mrs. Behrens remembered then, with shock, that she had forgotten to put the furniture polish back in the closet after Tuesday's cleaning. She had probably left it on the floor, under the bed, where she had last used it. Lenny could have reached it through the playpen bars." She rushed Lenny to the hospital, where a pediatrician called the local poison-control center for specific information on the contents of the polish. The answer came back that one-third of an ounce of the oil in the polish could be fatal for an infant. Tragically in Lenny's case, the information arrived too late: he died that Saturday.[79]

The similarities in the cases of Lenny Behrens in 1957 and Celeste Felder in 1951 demand a brief comparison of the two. Both children died from poisons picked up around the home. Both died at least partly because doctors failed to consider nonbiotic, environmental causes. But their stories, as narrated by the public press, reveal very different attitudes about blame, risk, and faith in science. Celeste Felder's story was told from the perspective of the shocked, attentive father; the responsibility for her death rested clearly on the incompetence of local physicians and hospitals. But in the later story blame shifts to the mother, who not only left the poison where her child could reach it, but forgot about it until it was too late for the doctors to save him.[80] What the two stories share is an abiding faith that medical science can solve the most obstinate problems. The narratives were in effect jeremiads, pointing out the medical community's failure to maintain its standards while impelling them to do better.

There can be little doubt that the advent of poison-control centers saved many lives. And the fact that they brought the direct involvement of a federal government agency paved the way for the Centers for Disease Control's leading role in coordinating lead-poisoning programs beginning in the 1970s. Initially, however, the poison-control movement's impact on finding and eliminating lead poisoning was counterproductive. Subsumed under the rubric of household poisons, lead appeared to be less troublesome than aspirin, which according to federal mortality records killed almost twice as many children each year.[81]

As lead came to be defined as just another household poison, lost was the distinction between a child's momentary impulse to swallow a bottle full of aspirin or furniture polish and the slow, ongoing process of mouthing painted surfaces around the home. This explained why most lead poisonings went undetected while almost every case of aspirin poisoning was correctly reported from the outset. Another outcome of the immediate and obvious proximity of cause and effect in aspirin, kerosene, or barbiturate poisoning was the recourse to direct preventive remedy. Medicines in bottles with childproof caps quickly replaced outdated containers in the bathroom cabinet. But all the paint labels and voluntary standards for new coatings would not remove a square inch of old lead from the walls.

Subsequent experience showed that the handful of city-based lead-screening programs and the advent of poison-control centers were inadequate. Still, some health practitioners in the late 1950s were prepared to declare victory. Recalling pediatrician John Ruddock's assertion in 1924 that "the child lives in a lead world," radiologist Paul Woolley summed up progress over the intervening years: "A final phase in the story might be the observation that Ruddock's lead world has been shrinking until now it is almost a microcosm composed of neglected surfaces from which paint is crumbling and peeling." Lead poisoning remained a persistent problem to be sure, but one for which adequate measures had been taken. It was simply one of many tragedies of the ghetto and was likely to get the same attention that other "ghetto problems" received. As one researcher had concluded in 1940, "like the poor, lead poisoning is always with us."[82]

The efforts to combat lead poisoning in the 1940s and 1950s suffered from cultural blinders about what could be done for the poor and tended to calcify the old definitions that limited "at risk" populations to lead workers and the urban poor. But a generation of pediatricians learned their trade in hospitals where every summer doctors and nurses fought—often in vain—to save what began to be recognized as lead-poisoned children. These doctors came to curse the economic and political stalemate that perpetuated the situation. In the 1960s they would forge strong alliances with politicians, lawyers, and—most important—the people whose knowledge of lead poisoning came not from a medical book or a legal brief but from personal experience, usually of loss. Lead poisoning might have remained a "disease of poverty," but in the years of civil rights, the Great Society, OSHA, and Rachel Carson, the fact that most lead victims were poor or nonwhite or laborers would be an increasingly inadequate excuse for neglect.

TEN

The Screaming Epidemic

On a Tuesday evening in late November 1969, dozens of families in New York's barrio received a very unusual visit. Knocking at doors of tenements along 112th Street were uniformed members of the Young Lords (the Puerto Rican counterpart to the Black Panthers), accompanied by medical students from New York Medical College. Working in two-person teams they went door to door, asking if any young children lived in the home and collecting urine samples in order to test for lead poisoning. Two hundred test kits had been obtained after thirty Young Lords, along with nurses and medical students, staged a three-hour sit-in at the Health Department. It is not known exactly how many samples the teams collected, but by the following Monday, the night's work had turned up two positive cases, including a 2-year-old with a history of convulsions.[1]

If you live in a lead trap, you can't win.

DR. AGNES LATTIMER,
MICHAEL REESE HOSPITAL,
CHICAGO

The Young Lords' impromptu screening program was certainly unique, but in most important ways it resembled many of the more conventional programs that sprang up in the late 1960s to combat childhood lead poisoning in the inner cities. In many large urban centers in the northern states, community activists mobilized local resources and forged alliances with medical professionals to make childhood lead poisoning a major thrust of community action programs. Often they had to overcome indifference and ossification at city hall and in hospitals. Always they faced critical shortages of money and manpower.

In 1966, a Chicago pediatrician stated what seemed a simple truth: "Lead poisoning is linked closely with poverty and poor housing." Less than a decade later, a new epidemiology based on a revised definition of lead poisoning would significantly alter this appraisal by greatly enlarging the "at risk" population. But in the 1960s, activists blamed this perceived link for the slow pace at which screening and treatment programs were established. Lead poisoning was a poor person's disease, and poor people's diseases did not get the attention afforded those of middle-class children, as Hyman Merenstein of the New York Downstate Medical Center pointed out: "When we used to have ten polio cases, the whole city rose up in arms, but when 30,000 kids are affected with lead poisoning, nobody notices."[2]

But people had in fact begun to notice lead poisoning, not despite but because of the fact that it was linked to poverty. The War on Poverty and Great Society programs directed unprecedented resources at the problems of the poor and their often substandard housing. Lead poisoning came to be seen as a fundamental but treatable symptom of poverty, a favorite target for eradication. Lead-poisoning control was by no means a Great Society program, however; to suggest so reverses the flow of influence. Although federal money eventually sustained the movement, the impetus for change ran from the community to the city and beyond.

This local initiative stemmed from the interactions of five distinct groups, each working on the problem from its own direction, but in remarkable concert. Local public health officials, often portrayed at the time as entrenched or insensitive bureaucrats, were not deaf to the call to reduce lead poisoning. In conjunction with the poison-control movement, most city public health departments had poison-control officers, who naturally played an important role in administering lead-poisoning programs. Critical in devising these plans was a new generation of doctors who often trained in city hospitals or took jobs in federally funded community health centers where they gained plenty of exposure to childhood lead poisoning.

Two groups of professionals outside of the realm of public health also supplied needed expertise and political energy. Scientific advocacy groups joined forces with medical groups such as Physicians for Social Reform and became deeply involved at both the local and national levels. The motto of the Scientists' Institute for Public Information (SIPI) caught the spirit of this movement: "Scientists inform, citizens act." Helping citizens act were hundreds of lawyers, either alone or through groups such as the Welfare Rights Organization, Community Legal Services, and the Massachusetts Advocacy Center.[3]

Childhood lead poisoning would not have remained a compelling issue for all these lawyers, scientists, physicians, and public health officials if not for the new resolve of members of the communities ravaged by it. As housing activist Paul DuBrul put it, "We have already been told by the Health

Department that no money can be found for a testing program until the black community begins yelling 'Murder.'" And it did. As a Philadelphia Community Legal Services lawyer recalled, "These inner city residents saw that health department maps and graphs demarking the so-called lead belts defined the areas where they lived, not the Beacon Hills, Society Hills, or Upper East Sides of their cities." Lead became a focus, a symbol for all that was wrong with housing and medical care in America's largest cities. This awareness led to activism, which eventually brought a federal response.[4]

By the early 1970s Congress had granted the federal government great power to wield against lead poisoning, which it did with both tremendous success and heartbreaking insufficiency. The Lead-Based Paint Poisoning Prevention Act of 1970 authorized $30 million to help local agencies establish screening programs and to develop technologies to remove lead paint from America's homes. Despite the many difficulties in funding and administering these programs, in their first decade they awarded more than $80 million to local public health agencies to run lead-poisoning prevention and hazard abatement programs (see Table 4).

This chapter is about the origins of federal lead-paint legislation, but the hearings, lobbying, and speeches leading up to the bill's passage receive scant treatment here. The impetus for federal action lay in a number of dramatic city campaigns. This local experience also anticipated many of the problems that would beset the federal programs. Some of the limitations that slowed cities' responses were eased by federal coordination and funding, but although federal money and better training provided some measure of stability, local lead-poisoning prevention programs often lacked a clear mission or consistent methodology—sometimes concentrating limited resources on screening and detection, sometimes hazard abatement, and sometimes using lead poisoning as a wedge to secure funding for general community health services. Like many well-intentioned federal programs to help the poor, campaigns to fight lead poisoning often foundered on their insensitivity to multiple issues, misjudged or alienated their client communities, or were hamstrung by national politics and bureaucratic and federal-state turf wars.

The limited case-finding programs of the late 1950s and early 1960s had been inadequate to their task, but they did establish several important precedents. They showed that no matter what treatments were employed, the very act of looking for cases of lead poisoning reduced the number of fatalities. And even though lead poisoning took a back seat in the poison-

TABLE 4 *Funding History, Lead-Based Paint Poisoning Prevention Act, Fiscal Year 1971 to Fiscal Year 1982*

Fiscal Year	Authorization (millions of dollars)	Appropriation (millions of dollars)
1971	10.0	0.0
1972	20.0	6.5
1973	*	6.5*†
1974	63.0	11.0†
1975	63.0	9.0
1976	10.0	3.5
1977	12.0	8.5
1978	14.0	8.5
1979	14.0	10.3
1980	14.0	11.3
1981	15.0	10.1
1982	‡	‡

SOURCE: Susan Bailey, "Legislative History of the Lead-Based Paint Poisoning Prevention Program," in House Committee on Energy and Commerce, *Lead Poisoning and Children,* 97th Cong., 2d sess., 2 Dec. 1982, 3–12.
* The program was not reauthorized until fiscal year (FY) 1974; funding for FY 1973 came from a continuing resolution.
† An additional $4.5 million was appropriated for FY 1973, but the administration impounded it. The impounded funds were added to the official appropriations for FY 1974.

control movement, that movement assured that environmental hazards would receive greater attention in medical training. Young interns working in large city hospitals, now sensitized to domestic poisons, were more inclined than their predecessors to identify the many cases of lead poisoning that turned up every summer.

When in 1961 Philadelphia pediatrician Walter Eberlein proposed a grant for federal funding to establish a program to screen for lead poisoning, he probably hoped to duplicate or surpass Baltimore's successes in joining research to municipal action. Eberlein, who had been a fellow at Baltimore's Harriet Lane Home and had published a paper on lead poisoning with J. Julian Chisolm of Johns Hopkins, came to Philadelphia Children's Hospital in the late 1950s. Now, in addition to his other duties at the University of Pennsylvania Medical School, he was its poison-control officer.[5]

A preliminary study showed that the out-patient department of the teaching hospital, which annually provided "general medical care to approximately 14,000 indigent children, mostly Negro," had identified twenty cases of lead poisoning in the previous year, over a third of the total reported in the whole of Philadelphia. Since most of these cases had been diagnosed only after the children began to exhibit signs of encephalopathy, Eberlein introduced a protocol to enable earlier identification. In the next six months the department detected fifty cases, a 500 percent increase, and Eberlein sought to make this protocol part of a hospital-wide screening and training program.[6] This protocol and Eberlein's grant proposal highlighted the need for local communities to establish effective screening and treatment programs; they also illustrate some of the inhibiting factors: imperfect clinical understanding, the limitations of available diagnostic technologies, and the realities of city politics.

According to the protocol, the first key to finding lead poisoning lay not in symptoms but in the child's behavior: "The parents of all out-patient children age 1–7 will be specifically questioned about pica in their child." In medical parlance, "pica" denotes an appetite for nonfood substances, often taking the form of geophagy (dirt eating) in rural settings or, in the urban environment, of eating paint and plaster. Of the 103 lead-poisoned children treated at the Philadelphia Children's Hospital out-patient clinic from 1959 to 1961, eighty-three provided "a positive history of pica."

Under Eberlein's protocol, clinicians performed a hemoglobin count on all patients. A history of pica or a low hemoglobin count placed the child in the next tier of screening, which specifically targeted lead. A blood sample was sent to a local automotive battery manufacturer, which operated one of the three analyzers in the region capable of doing blood-lead determinations; for some years the company had been analyzing samples for the hospital at no charge. Since it might take days to get the results, clinicians also took urine samples to test for urinary coproporphyrins, a rough index of blood damage from lead exposure—not as reliable as a "blood-lead," but a test whose results could be read during the office visit. Patients with suspiciously high coproporphyrins were x-rayed to look for lead-lines in the long bones or lead in the stomach. Either sign put the child in the hospital for immediate chelation therapy. Regardless of the urine test results, children were brought back for treatment if their blood-lead determinations came in at over 60 μg/dL.[7]

The lack of accurate, quick, and cost-effective methods to measure the amount of lead in a child's body was a major obstacle to establishing large-scale screening programs. Eberlein estimated that equipment and staff to do blood-lead determinations of the 12,000 to 14,000 new patients in his clinic each year would cost between $150,000 and $200,000. The equipment was so costly that it was usually found only in the hygiene departments of large lead-using companies or well-funded science laboratories, and rarely at state occupational hygiene agencies. The handful of moderately effective city programs for detecting childhood lead poisoning before the mid-1960s relied on such resources.[8]

Screening cases by asking about a child's behavior or administering a urine test avoided expensive blood work, but these methods were not sufficiently reliable or accurate. Eberlein knew the costs of this compromise and hoped that his trial program would generate the data needed to "bring about suitable municipal action," including "new legislation, law enforcement . . . public education," and perhaps of greatest interest to him, "a city-supported laboratory for free lead determinations." Eberlein knew the importance of early diagnosis. As New York City's commissioner of health warned in 1966, "Lead poisoning in children may be compared to an iceberg, with the small visible portion being cases of lead encephalopathy and the major portion being the invisible and as yet asymptomatic patients."[9]

In most cities, lead poisoning's status as a disease of poverty left the iceberg submerged. Cultural assumptions about the poor shifted blame from the toxin to the victim, inhibiting the discovery of the true scope of childhood plumbism and postponing indefinitely its eradication. In his Harben Lectures of 1960, Robert Kehoe voiced the prevailing understanding of childhood lead poisoning as a symptom of a cultural defect:

> *The circumstances of life conspire to make this a serious problem of public health, however, in that young children, in the poorer areas or slums of our cities, live under unsatisfactory conditions of housing, and in addition are likely to be left too much to their own devices; some of them are unwanted and unloved, while the mothers of others are compelled to leave them in the care of children but little older than themselves, while they (the mothers) engage in employment outside the home. The children develop aberrant appetites, interests and habits of eating (pica), and tend to deviate psychologically in other respects from those with more favourable social and physical environments.*[10]

Lead poisoning, according to this view, did not occur among intact families in the suburbs where Mom was home looking out for her children. Kehoe did not mention race, but most of the known victims of lead poisoning were African American children, and the common association of blacks with "the poorer areas or slums" made Kehoe's cultural etiology compatible with the racialized picture of plumbism found in most contemporary reports. Race and culture loomed especially large in discussions of pica as the cause of lead poisoning.

Lead intoxication from paint killed children, but what caused the intoxication? Pica—an abnormal, if not pathological, behavior. As informed health experts used the term, pica was distinct from the "normal" hand-to-mouth or oral behaviors almost every child exhibits. One researcher defined it as "not merely random mouthing to satisfy oral curiosity . . . it is a persistent and purposeful pursuit of the substance," which might be dirt, plaster, ashes, or paint.[11]

But what caused this behavior? The answers invariably came back to poverty and ignorance. Researchers attributed pica for lead paint to a number of causes. Hyman Merenstein, a pediatrician at Brooklyn's Kings County Hospital out-patient clinic, took the direct approach: after sampling a small patch of lead paint, he declared, "It's surprisingly good. It has a sweet taste—kind of like a cordial candy—with a kind of alcoholic aftertaste." Another physician in the District of Columbia observed children sticking wet lollipops on the wall and peeling off a patch of paint, which they ate as a coating to the candy. More frequently, however, doctors assumed the child with pica was undernourished or suffered from mineral deficiencies.[12]

Other observers portrayed pica as symptomatic of "exotic" cultural norms passed to children by their mothers—deviant practices imported from the poor South, from slavery days, and from behaviors deriving from biological adaptation to mineral-poor soils in Africa.[13] African American culture received the majority of the blame in many accounts of pica. A team of physicians related the story of a Mississippi-born black woman raising her children in Cleveland's inner city. "Each time Mrs. L. was pregnant she developed a craving to eat gray clay" which relatives back in Mississippi mailed to her. She reported that her children were "very oral." After her youngest had been hospitalized twice with lead poisoning, she forbade the children from eating paint but continued to share her clay as a treat when they got cranky.[14]

Regardless of whether accounts of childhood pica evoked images of ignorant migrants from the backward, impoverished South, almost all laid the blame on mothers for neglecting their children or disciplining them too loosely. "Working mothers," a *Time* article cautioned, "and those busy with community causes or a new baby may give a child too little attention, so that he seeks gratification by the readiest means available." Daniel Haley, a paint-industry representative, told a Pennsylvania health advisory group in 1972 that "parental neglect is the root of the problem." The solution to a child eating paint? "Spank the offending child with a shoe, something that we've all had at one time or another as a child." In an otherwise sympathetic article, *Look* magazine pointed out the inadequacy of this simple approach. Reporter Margaret English highlighted the story of a St. Louis mother in whose apartment the lead paint was falling off in sheets. English noted a hole in the plaster about a foot from the floor near the kitchen. "That's the hole my baby eats out of," the mother reported. "I smack him for it, but he keeps doing it anyway."[15]

Although health professionals and journalists tended inaccurately to portray pica as the sole cause of lead poisoning, the association made sense, since most of the children who were diagnosed with lead poisoning had histories of eating nonfood items. And since so many young children then (as now) ate nonfood objects and so many had alarming blood-lead levels, clinics interested in increasing their identification of lead-poisoning cases could do so simply by asking parents about their children's oral habits. Aggressive programs employing behavioral screening coupled to follow-up for possible treatment could improve case-finding and reduce the incidence of symptomatic lead poisoning.

More encouraging to public health bureaucrats was the notion that if pica "caused" lead poisoning, education to address parents' ignorance of the dangers and lack of parenting skills could prevent it. In 1962, Baltimore practically abandoned its thirty-year program of screening children for lead poisoning and encouraging lead-paint removal from housing. "Instead of removing the paint from the child's reach," the health department's new marching orders read, "the new program is intended to impress the person who cares for the child with the importance of watching it and keeping it from nibbling the paint." Historian Elizabeth Fee judges that "the 'hard-sell' educational program was a complete failure," but mostly because decades of Health Department activism had already made Baltimore residents well aware of lead paint's toxicity, even if they could not avoid it.[16]

Public education programs could be effective in communities where the level of awareness was low. An informal survey of pedestrians in Brooklyn's "lead belt" in 1966 found that two-thirds did not know that ingesting paint was hazardous. Those who had access to routine medical care seemed to be better educated: among fifty parents bringing their children to a Brooklyn health center for various treatments, thirty-nine knew that paint and plaster were hazardous when eaten. Armed with this evidence and the aid of the New York City Department of Health, the health center began a campaign to spread the word by posters, door-to-door visits by community volunteers, leaflets, radio announcements, and a sound truck cruising the area to broadcast simple warnings in English and Spanish.[17]

Splashy public campaigns were most beneficial when tied to case-finding and abatement programs. In the early 1970s Dr. Vincent Guinee headed New York's Bureau of Lead Poisoning Control. His "pica-balloon" (*pica* mispronounced "peek-a," perhaps to evoke "peek-a-boo") television campaign informed children that "pica is when little kids put stupid things in their mouths and eat them, like paint chips," and counseled that "mommies and daddies can call the Health Department" for information about free blood-lead tests.[18] But for cities without adequate resources, education programs might simply take the place of more direct action. It cost a health department little to hold a press conference, run a few radio spots, and distribute leaflets in the "affected community." And when local doctors continued to diagnose serious cases of childhood lead poisoning, the cultural arguments about poverty and pica allowed overburdened health bureaucrats to throw up their hands in defeat.

Education could only go so far, however. Even if every child's caregiver heard the message that pica was a "bad habit" and lead was poisonous, the benefits of education would flounder on the realities of the highly toxic environments in which many children lived. Agnes Lattimer, director of ambulatory pediatrics at Chicago's Michael Reese Hospital, saw poor nutrition as the root of pica but realized that it was the virulent toxin readily available to "paint eaters" that caused lead poisoning. Focusing on behavior placed impossible and inappropriate burdens on mothers: "You have to have a home safe enough to put the child down in while you do the washing and ironing. You can tie him down, but then he doesn't get to explore his environment. . . . If you live in a lead trap, you can't win."[19]

Lead poisoning, as activist Paul DuBrul recalled, lent an undeniable reality to the old adage that slums kill.[20] Lead poisoning was a particularly attractive target for community activists and their professional allies. The problem, the opposition, and the solutions seemed so clear. Poor children lived in deadly conditions because the greed of landlords went unabated due to the lax or moribund enforcement of regulations by city housing and health bureaus. The answer seemed equally simple. City health departments should screen every child; city housing codes should ban lead paint; city code-enforcement officials should take a hard line on getting landlords to comply.

These solutions called for expanded government intervention and greater involvement by professionals from outside—somewhat ironic given the rhetoric of community cohesion, volunteerism, local participation, and local action that characterized many lead-poisoning campaigns. In fact, many of these campaigns seemed designed to draw the city's attention as much as to solve the community's problem. They aimed at revealing the true scope of the problem, in the hope that a crisis would bring action. Each demonstration that education and screening could be effective would goad the city into institutionalizing programs. Unfortunately, setting sights on arousing action at the municipal or state level often produced short-lived campaigns: an encouraging word from city hall, a modest reallocation of resources, or the almost inevitable improvement in statistics frequently cooled the community's fervor. Further action had to await another crisis.

Variations on this theme of vigorous, direct community agitation for a political or bureaucratic response played out in many cities and towns during the late 1960s. A brief review of the lead-poisoning control activities in three cities—Chicago, Rochester, and New York—suggests both the range of experiences and the common factors that demanded a unified, well-funded federal regulatory response.

Chicago was the second major city, after Baltimore, to institute a broad screening program. In the late 1950s, a number of researchers there had published statistics on lead poisoning, and Chicago's hospitals were reasonably alert to the disease. Each year from 1959 to 1965, the Board of Health reported between 154 and 218 cases, with a fairly steady decline in the case-fatality rate. But before 1966, the city did little more than collect data. In 1961, two physicians from the Poison Control Program conducted a trial screening program, using urinary coproporphyrins to test fifteen hundred children, but the follow-up program they proposed never materialized. In the summer of 1963, after Chicago hospitals reported fourteen lead-poisoning deaths, a child advocate group called on the governor to declare a state of emergency. Chicago's commissioner of health took a more modest approach: thirty building inspectors were to check the homes where poisoned children lived; any homes found to contain lead paint could be closed until the owner repainted.[21]

The following summer, community activists on Chicago's West Side began agitating for a less paltry response. One evening in August, the normal activities of an American Friends Service Committee–sponsored block club meeting were suspended when a woman reported that both of her young children, who had been suffering with fevers and convulsions, were now diagnosed with lead poisoning. At an emergency information meeting the next week, representatives from a broad range of Chicago's church and social agencies formed the Citizens Committee to End Lead Poisoning (CCELP).[22]

The Board of Health was prepared to assist with a trial screening program the new organization proposed. In 1965, with twelve lead-poisoning deaths reported through August, statistics threatened to be as bad as they had been two years earlier, and city hall, smarting from the newspaper coverage, accepted CCELP's offer to provide community outreach and hundreds of volunteer hours. That fall, "a dedicated group of teenagers" canvassed Garfield Park, collecting almost six hundred urine samples. The city analyzed the samples and administered additional tests to children whose results were positive for lead. The program identified four children for chelation therapy, and reported their addresses to the Building Commission for inspection. When the trial ended in November, it had demonstrated that untrained workers could help run a screening program. Until it stopped operating the next May, CCELP continued its work "as a reminder to the city."[23]

In the spring of 1966, Chicago's Board of Health began what would become the largest screening program in any city to date, staffing it in part with War on Poverty workers. Over the next twelve months, more than forty thousand urine specimens were tested, and in October 1966 these tests began to be augmented with a finger-prick blood test. Over seven hundred children exhibiting no clinical signs of plumbism were identified for treatment.[24] Between 1966 and 1971, Chicago tested some quarter of a million children from the city's most distressed areas, and the health department's new lead clinic evaluated or treated almost ten thousand. Just as case-finding programs brought increased awareness among medical professionals and reduced lead-poisoning deaths, screening well children for lead reduced the average blood-leads in the screened population. Of course, screening alone did not lower any individual child's blood-lead level, but testing a larger population of asymptomatic children brought many more children with lower lead levels into the calculation, thereby lowering the average. Chicago's case rate (the number of new cases per hundred children screened) plummeted: in 1967, 8.5 percent of those screened had blood-leads over 50 μg/dL; by the third year of the program, the rate had fallen to 2 percent.[25]

The Citizens Committee to End Lead Poisoning had scored a major victory with lasting consequences. But there were serious limits. Although Chicago's screening program was much larger than that undertaken by any other city, it still served only one-quarter of the children health officials considered to be at risk. And as researchers extended trial screening programs into the surrounding suburbs and working-class communities away from Chicago's "lead belt," they found pre-1967 case rates and alarming average blood-lead levels.[26] In areas untouched by these programs, children continued to turn up every summer in emergency rooms with severe headaches, constipation, or vomiting, some slipping into convulsive unconsciousness. Perhaps worse, thousands of infants grew into childhood under the constant burden of high blood-lead levels: "subclinical" or "asymptomatic," but still lead poisoned.

A second limitation that, if better appreciated, might have kept Chicago's community activists alert, was the Board of Health's lethargic approach toward eliminating the source of these poisonings. As late as 1972, Chicago had no laws outlawing lead paint, and enforcement of housing codes was notoriously lax. Community activists and residents would have to rise to form new alliances in the 1970s to take on this very different issue, to treat lead poisoning "as a social disorder rather than a disease which will be remedied through . . . medical care." Conceiving of lead poisoning as a housing issue raised thorny economic questions and assailed powerful interests. As Bruce Flaschner, deputy director of the Illinois Department of Public Health, complained, "The fault rests with many extremely influential people who have money and power to squash efforts aimed at reducing substandard housing."[27]

In contrast with the CCELP's emphasis on health screening, Rochester's first community-based lead activists focused on lead paint instead of lead poisoning, in part because Rochester's program was initiated by a chemist, not a physician. University of Rochester professor David J. Wilson was vice president of the Rochester Committee for Scientific Information. After learning of Chicago's program at the annual meeting of the Scientists' Institute for Public Information, Wilson saw that lead poisoning was a suitable project for his group. With the cooperation of Rochester's Urban League, he began a demonstration project employing "Project Uplift" teenagers to collect samples of peeling paint. Of the fifty-nine households tested, twenty-two contained lead paint. "What followed," according to Wilson, "was one of the more constructive applications of 'Black Power' to our local governments." The Urban League began a program to educate "parents living in the slums," and screened those children living in homes found to contain lead paint. Under pressure from the league, the City

Building Bureau received instruction in Wilson's testing techniques and promised that housing inspections in the slums would include routine tests for lead paint. At the next annual meeting of the Scientists' Institute for Public Information, Wilson presented an optimistic report about the Rochester program.[28]

Alex Matthews, a Rochester father, told a different story. In May 1969, Matthews, a cattle handler at a meat-packing plant, attended a "Project Uplift" health fair at Rochester's War Memorial Auditorium. He was struck by the similarity between his family's home and the hazardous conditions described in a booth on lead poisoning. Worse, his 2-year-old daughter Wanda was "like the children in the demonstration." Matthews asked the Urban League volunteers to inspect his house.[29]

In addition to bad wiring and defective plumbing, the Project Uplift volunteers found "leaded paint on window sills, lead in wall plaster, and lead in flakes floating down from the ceiling." In June, the city agency responsible for code enforcement told the Urban League that its inspectors had found no lead paint in the home. They had turned over a list of other violations to the city's Department of Urban Renewal, which owned the home and, they claimed, was equipped to fix the problems.

About this time, doctors at Strong Memorial Hospital diagnosed Wanda Matthews's poor appetite, constipation, and irritability as "some kind of virus." Two weeks later, she was unable to walk, and the emergency room doctors had run out of ideas. As Matthews recalled, he told the doctors, "You checked for everything else; why don't you check her for lead poison?" Wanda tested positive, but the doctors hesitated to admit her for treatment; perhaps the diagnostic results did not suggest a level of toxicity high enough to alarm them. When her father asked for a specialist, the physicians warned him that treatments were "going to run into a serious bill." A specialist was nevertheless called in, and he admitted Wanda, who by then "could hardly breathe and . . . was just out of this world."

The Matthews family faced the problem of where to live after Wanda came home. The Department of Urban Renewal, responding to the publicity Wanda's case was bringing, promised to eliminate the hazards. Some teenagers from the Summer Youth Opportunity program were sent, and they "literally slapped paint over walls, doors, even furniture." Professional painters were also sent in, but Wanda's doctor declared that the house was unsafe. After the Matthews family rejected temporary housing in a welfare hotel and another house as bad as their first, Urban Renewal offered them a large, well-built home close to the school their older children attended. Urban Renewal assured the family that "our two Negro employees checked it themselves" and guaranteed the house was lead free. But this house, too, contained peeling lead paint.

Eventually the Matthews family was lodged in new, safe, low-rent housing operated by the city's Housing Authority. Wanda was saved only by luck and her parents' determination to do right by their poisoned daughter. Getting treatment and a safe place to live for Wanda Matthews taxed to the limit the community resources available to poor blacks in Rochester. The city was certainly not up to the task of abating lead-paint hazards. Wanda was fortunate—she lived, although her mother had to quit work to devote more time to her special needs—but her story points to the need for broader action, better training, and larger resources than the isolated city-based programs of the late 1960s could offer. As dozens of cities repeated aspects of the experiments in Chicago and Rochester, the calls for strong state and federal intervention became more insistent.

New York's experience would generate the national headlines and activist fervor needed to seed the coming regulatory downpour. The city's record on lead poisoning was much like that of other cities. Until the late 1960s, a few Health Department officials sustained, and made modest improvements in, programs to improve case finding and educate the city's physicians. In 1951, Mary McLaughlin, who became commissioner of health in 1969, instituted a small but aggressive case-finding program at a number of the department's Health Stations. In 1958 the department began intensive training of hospital staffs, while continuing to improve case finding. Largely as a result of earlier detection, the number of reported childhood lead-poisoning deaths continued to drop even as the number of reported cases rose.[30] The department invested considerable resources to case finding and treatment given its budget and other pressing public health issues such as drug addiction, venereal disease, and tuberculosis, which received greater media attention.[31]

Like other cities, New York was slow to enforce housing codes that might reduce children's exposure. The city's Health Code had since 1959 banned the sale or application of any paint containing over 1 percent lead. Even if it had been possible to enforce the ban, little effort was made in attempting to do so. Shortly before his death in 1966, Health Commissioner Harold Jacobziner boasted that every reported lead-poisoning incident resulted in a housing inspection by a Public Health sanitarian, and "in over 95 percent of cases of Code violations, the landlord complies within two weeks to three months from the time of notification." Tellingly, Jacobziner did not divulge the number of code violations issued.[32]

The drive to enlarge New York City's lead-poisoning control programs came, as it had in Chicago and Rochester, from a cadre of social and scientific activists, who joined with residents whose children were suffering, and national associations such as the Scientists' Institute for Public Information. Glenn Paulson, a 24-year-old graduate student from Rockefeller University and cochair of New York's Scientists' Committee for Public Information (SCPI), attended the 1967 SIPI meeting in St. Louis that had inspired Rochester's local chapter.[33] Early the next year, Paulson and other SCPI members planned a limited screening program modeled on Chicago's and called a meeting on the subject at Rockefeller University. Although most of the participants gave only a lukewarm response, Paulson enlisted the help of Paul DuBrul, the housing director of a Lower East Side settlement house. DuBrul, then 30 years old, had been active in New York liberal and radical politics for years. His contacts, commitment, and energy proved to be priceless.

The scientists' committee was unable to interest the city in participating in a trial screening. Assistant Commissioner of Health Donald Conwell explained that funding constraints precluded the department's participation, but confided that outside pressure might make it possible to get funding later. In August, DuBrul and other SCPI members joined with health workers and residents of Brooklyn slums to form a new coalition, Citizens to End Lead Poisoning (CELP). DuBrul framed an "Action Program" for the group based on the guiding principles that "existing agencies only respond in the face of crisis; the crisis exists; we have to draw attention to it." To do so, CELP's program called for case-finding programs, pressuring the city to improve code enforcement, and organizing programs at the local level to foster education on the risks of lead paint.[34]

Through the summer and late fall, CELP met with modest success. Young volunteers collected 409 urine samples from children living in the South Bronx, and students at Einstein Medical School, using equipment and space they borrowed from the school, identified eighty-nine children for further study. Pressure on city hall seemed to be having some effect. Joseph Cimino, director of New York's poison-control center, agreed with SCPI's characterization of childhood lead poisoning as "a silent epidemic"; indeed, he thought the committee's estimate of the problem was "conservative." Instead of the 9,000 to 18,000 New York children SCPI estimated were at risk, Cimino put the figure at over 25,000, at least 5,000 of whom were in serious danger. He claimed that the city was "moving relatively fast on this compared to its movement on other problems." In August the Health Department formed a task force to act as liaison to other city departments concerned with housing and environment. "What the department says looks good on paper," a SCPI spokesman retorted, "but action is needed. And they have nothing going now."[35]

The following year saw plenty of action. As SCPI and CELP stepped up their activities, lead poisoning began attracting more local media attention. In March, a year after SCPI's inauspicious first lead-poisoning conference there, Rockefeller University hosted another gathering, this time sponsored by five scientific and health associations including SIPI and the New York Department of Health. Glenn Paulson spent six months arranging the two-day meeting, chaired by René Dubos, one of his professors at Rockefeller and also president of SIPI. In his closing remarks at this conference that stressed both the economic and the environmental sources of childhood lead poisoning, the Pulitzer Prize–winning microbiologist issued a stern warning: "The problem is so well defined, so neatly packaged with both causes and cures known, that if we don't eliminate this social crime, our society deserves all the disasters that have been forecast for it."[36]

Volunteer screening programs and a number of lead-poisoning deaths in the New York area also received a good deal of publicity. In April 1969 a local politician held a press conference to denounce the Public Health Department's handling of the death of Janet Scurry, a 2-year-old from the Bronx. In May the *New York Times* reported the death of a Newark infant, who along with her three siblings was hospitalized for lead poisoning after eating lead paint and plaster. A door-to-door survey of Brownsville tenements, conducted over the summer by Neighborhood Youth Corps volunteers and supervised by Brooklyn College and a local hospital, found a frightening incidence of lead exposure: forty-five percent of the children tested showed signs of abnormal lead exposure and were referred to a clinic for blood tests. Of these children, almost half were admitted to the hospital for chelation therapy.[37]

New York's press covered most of these stories with a remarkably measured tone. Then, in the middle of September, the *Village Voice* broke ranks with a front-page exposé, "Silent Epidemic in the Slums," written by editor Jack Newfield. With a muckraker's outrage and flair for the dramatic, Newfield highlighted the Scurry story as a lead-in to his tale of government ineptitude and the heroism of DuBrul and Paulson—two "young, white, middle-class radicals" who had been "waging a lonely crusade, bereft of money, manpower, or organizational support, to pressure the city and the health establishment, and to alert parents."[38]

Newfield's criticisms cut a wide swath through much of New York's liberal establishment. City hall was taking no appreciable action against lead poisoning, and neither were a number of organizations that Newfield assumed should be concerned. The NAACP had no program; the United Federation of Teachers did not return his call; religious leaders in East Harlem seemed indifferent, with one even complaining that lead poisoning "was a phony issue, that *asthma* was a bigger community problem." The media were uninterested, the medical community deeply divided over solutions, the federal government slow and detached, and only a handful of local politicians seemed willing to make lead poisoning a campaign issue. But the standout villain in Newfield's shop of horrors was the Public Health Department.[39]

It is easy to overstate the impact of such searing negative publicity, and risky to draw causal inferences from temporal proximity, but it appears that the *Village Voice* articles hit home. Only weeks after Newfield's first exposé appeared, Mary McLaughlin, by then health commissioner, announced that she was transferring $150,000 from other programs to increase the number of staff members devoted full time to lead-poisoning control and to buy blood-testing equipment. McLaughlin also announced amendments to the city's health code that would permit landlords to cover lead-painted walls with wallboard or other sheathing and empower the department to call out emergency repair crews—at the landlord's expense—if offending apartments were not repaired in five days. It is not clear whether the publicity merely hurried changes that McLaughlin had begun putting in place in July, when she took over the department. She called it "a crash program," but suggested it had taken time, "pinching here and begging there," to come up with the necessary funds.[40]

McLaughlin's efforts did not keep the department out of trouble. The more Newfield pursued the issue, he said, the angrier he got "and the worse certain individuals in the city Administration look[ed]." Contrasting Chicago's program with New York's, Newfield noted that Mayor Lindsay had to shoulder much of the blame for the larger city's shortfall, "but it is more directly the fault of the little-known individuals he appointed"— notably Mary McLaughlin. Critics accused the administration of institutional racism, with Newfield charging that "if lead poisoning affected white, middle-class children, it would be covered on the front page of the *New York Times*. . . . But 30,000 undiagnosed cases of lead poisoning, living in Bed Stuy, El Barrio, and the South Bronx, is not news." In December, Paul Cornely, president of the American Public Health Association, repeated the charge that racism explained the city's indolence. But two months later he retracted this statement, explaining that "I had apparently been given a wrong set of facts concerning the New York City Health Department's activities—a very wrong set."[41]

Critics frequently threw off a quick contrast of the massive public health response to polio and the meager efforts to fight lead poisoning. In the mid-twentieth century, middle-class children composed a significant percentage of polio's victims—the specter of infantile paralysis haunted summer camps in the Poconos as well as playgrounds in the ghetto. The massive financial and intellectual investment in eradicating polio surely stemmed in large part from the demographics of its victims. But the way the disease was moralized is as significant. At the beginning of the century, polio was still as thoroughly a childhood "disease of poverty" as lead poisoning was to become.

The middle class's embrace of domestic hygiene had prevented their young children's exposure to the virus in its resistance-bestowing infantile form, thereby transforming polio into an epidemic plague for young adults and children of all ages.[42] Hence, polio sufferers were portrayed as innocent victims of a "natural" pathogen emanating from below. Lead's victims, on the other hand, were (to the less sympathetic) victims of parental neglect or backward culture or (to the more sympathetic) of a culture of poverty or institutional racism. No vaccine for either "cause" was in the offing. Chelation therapy was often depicted as a cure, but in fact it was only a treatment. The cure for lead poisoning was to cleanse the environment, but the enormity of that was, and continues to be, overwhelming. Racism no doubt fostered apathy in some politicians and health officials; institutional racism retarded the appropriation of funds and played a part in setting health departments' priorities. But the institutional racism argument itself provided an excuse for well-meaning individuals to throw up their hands in frustration with "the system."

A different "set of facts" might not excuse New York's apparent lethargy, but it could at least partially explain it. McLaughlin might have wanted to do much more, but lead-poisoning programs had to compete for limited funds with higher priority Health Department programs of long standing, such as drug addiction and venereal and other infectious diseases. And new programs concocted by department bureaucrats could quickly founder on the hostility of overworked or unconvinced staffs in branch offices. Since effective action on lead poisoning involved detecting lead in housing and forcing landlords to remove it, the best efforts of the Health Department could be defeated unless full cooperation was forthcoming from other city agencies.

Technological limitations also slowed progress. Much of the harshest criticism of McLaughlin's program in late 1969 arose over the department's failure to institute mass screening. A medical equipment supplier, Bio-Rad Laboratories, had developed a urinary screening test that was simple to use and inexpensive. Hoping to get New York City to test-market the product, Bio-Rad offered 40,000 kits, and Lindsay officials, in the heat of October campaigning, suggested the city would use them. Come November, the city had picked up only a few hundred of the kits and had yet to announce plans for using them.[43]

The lead activists devised a test program, however. As described earlier, the Young Lords, with assistance from students at New York Medical College, would distribute literature and collect specimens. Metropolitan Hospital promised two hundred kits, but at the last minute withdrew the offer. The activists staged a sit-in at the Health Department and finally, after meeting with two assistant commissioners, received two hundred kits and instructions on their use. On the following Tuesday, five days after Thanksgiving, the Young Lords swept through tenements along 112th Street requesting urine samples.[44]

According to Henry Intill, a salesman at Bio-Rad, the city had not developed a way to administer the tests because it simply did "not want a large scale screening of children . . . because then the city will be obligated to do something about the situation." The Health Department claimed that although it had considered trying the Bio-Rad urinary screening test, it was committed to the more reliable blood test. Although trials in Chicago showed Bio-Rad's product to be fairly reliable, the test seemed to produce too many false-negative results, raising the likelihood that some lead-exposed children would go untreated. The department planned to increase the number of blood tests it administered, and as part of the lead-poisoning program initiated in October had secured new equipment and staff sufficient to test five hundred children a week. Where the department had tested blood from 5,000 children for lead in 1968, it tested 87,000 in 1970 and 100,000 in almost every year after that. And as in almost every screening program, the case rate fell precipitously.[45]

These improvements marked the end of the first stage of New York's lead activists' campaign and must be counted as a major victory for community action. Citizens to End Lead Poisoning had created a political crisis by activating local interests and generating tremendous publicity. Its efforts had produced results: a dedicated but beleaguered city department had suddenly been given the funds and the mandate to act. McLaughlin requested $1.2 million from the 1970 city budget for a new lead-poisoning program; in January, the city turned over $2.4 million and established a Bureau of Lead Poisoning Control, headed by media-savvy epidemiologist Vincent Guinee.[46] And in late 1969 New York City's housing code was changed to reflect CELP's demand that the Health Department be required to order lead abatement in any home where a child had been poisoned. Before this the code had permitted, but not required, this action. The department sought the activists' advice in other matters as well.

Although new coalitions would rise to fight city hall on lead poisoning, CELP had completed its work. A few years later, one of the committee's supporters noted its passing. "Part of the problem that you get into is that you get diluted out. Paul [DuBrul] is now with the Borough President, I'm now with the [City's Health and Hospital] Corporation, Glenn [Paulson] is with the whole world's environmental problems, Jack's [Newfield] with the prisons and everywhere else."[47]

Within a year of New York City's dramatic response, Congress was hammering out the final version of a national law to promote case finding and abatement. Conventional wisdom credits adverse publicity, beginning with Newfield's *Village Voice* articles, for shaming New York City's health officials and their political overlords into taking decisive steps. But it does not appear that a similar wave of national publicity galvanized support for federal action. Nationally distributed magazines gave the subject very little coverage: the *Reader's Guide to Periodical Literature* lists only eleven feature stories on pediatric lead-paint poisoning in the two years prior to the enactment of federal legislation.[48]

The media response suggests that childhood lead poisoning was a local story repeated in many cities across the nation rather than a national issue playing out at the local level. Local media covered lead poisoning but emphasized the dramatic local political issues and the responses of scandalized city bureaucrats. The Philadelphia headlines for 1970 read like those in New York from a year earlier, as local activists roused the city into purposeful action. The story was similar in Washington, Boston, and other cities.[49] Local action on lead poisoning made good headlines for domestic consumption, but lead poisoning itself, as Jack Newfield correctly observed, was a hard sell: "How do you show a *process,* how do you show indifference, how do you show invisible, institutionalized injustice in two minutes on Huntley-Brinkley? How do you induce the news department of a television network to get outraged about nameless black babies eating tenement paint, when the public health profession, school teachers, housing experts, scientists, the NAACP, and the politicians haven't given a damn?"[50] A number of city and state politicians felt the sting of local controversy and let their federal representatives know about it. Although there was little adverse national publicity to stir wide support, the fear of it may have been an important factor in enlisting federal legislators' interest.

The move for the federal lead-poisoning legislation came in a flurry of bills proposed by federal lawmakers representing three cities with strong, vocal lead activists: New York, Philadelphia, and Boston. William Ryan, Democratic Representative from Manhattan, introduced lead-poisoning bills in 1969 and 1970; the House considered these in 1970, along with a similar bill submitted by Philadelphia's Representative William Barrett, who presided over the hearings. On the other side of the Capitol, Massachusetts Senator Edward Kennedy introduced a bill in 1969, and Pennsylvania Senator Richard Schweiker drafted another in 1970. In October 1970 the House passed a bill that cobbled most of these proposed laws together, and after hearings in November the Senate passed an amended version, which the House accepted on New Year's Eve. In mid-January, President Nixon signed into law the Lead-Based Paint Poisoning Prevention Act (LBPPPA).[51]

The LBPPPA created three programs, with funding of $30 million over two years. The Department of Health, Education and Welfare (HEW) was given $15 million to make grants to cities establishing lead-paint abatement programs and $10 million to establish screening and treatment programs; HUD was given $5 million to survey the scope of the lead-paint hazard and establish methods for abatement. Finally, the Act empowered HEW to prohibit the use of "lead-based paint" in federally constructed or rehabilitated housing. Sixty years earlier, any good "white-leader" would have puzzled over the law's definition of a "lead-based paint" as one containing more than 1 percent lead.

It is important to note that the law delegated primary responsibility, even that pertaining directly to housing, to HEW. The House version of the bill divided the tasks more evenly between HEW and HUD, but Kennedy had convinced Congress that his version would produce greater accountability. Jonathan Fine, deputy commissioner of Boston's Department of Health and Hospitals, supported putting responsibility for abatement and health control under one department. When federal funds came from more than one source, the result was "confusion within the city, lack of coordination in implementation, and much delay."[52]

As implemented, however, the federal programs guaranteed that "lack of coordination and delay" would continue. City health and housing departments in the late 1960s jostled for tight funding and blamed each other for permitting the *real* cause of lead poisoning to go unchecked. Under the provisions of the Act, especially as it was amended over the years, HEW and HUD retained an unhealthy mix of autonomy and overlapping jurisdiction, creating at the federal level the same situation that caused so much tension at the city level.

Still, as was true at the city level, increasing the resources available to lead programs made a tremendous difference. New York's new Bureau of Lead Poisoning Control did not eliminate the friction between the Health Department and the housing agencies, but the funds the new program received, beginning with its $2 million start-up, upped the ante; friction could no longer excuse inaction. Similarly, although the LBPPPA was far from perfect, it marked the end of decades of half-hearted efforts on the part of federal health agencies that seldom produced more than little pamphlets or funds for basic research.

The testimony of HEW officials before the Senate committee considering the new lead-poisoning prevention bill made it clear just how limited the federal effort had been. As was true in many city health departments, a small number of specialists within the Public Health Service had struggled to sustain lead programs. In the late 1960s, Jane S. Lin-Fu, a Children's Bureau pediatrician, directed much of the government's effort. A large measure of this work consisted of educating doctors, health departments, and parents. In 1967 the Children's Bureau published Lin-Fu's booklet, "Lead Poisoning in Children." Between 1967 and 1970, the bureau distributed 16,000 copies to doctors' offices, health departments, and pediatrics departments in medical schools. The Government Printing Office reported selling an additional 12.6 thousand to community groups and doctors. The government's distribution of Lin-Fu's tract paled next to the efforts of the Lead Industries Association, which distributed 61,000 copies of the pamphlet as part of its free booklet, "Facts about Lead and Pediatrics."[53]

The publicity generated by city-based programs in 1969 and 1970 had prompted Surgeon General Jesse Steinfeld to convene an ad hoc committee to establish guidelines for municipalities that were developing screening and abatement programs. One of the most obvious signs of the lack of guidelines was disagreement over the very definition of lead poisoning. New York and Baltimore defined a "case" of pediatric lead poisoning as a child whose blood-lead exceeded 60 µg/dL; Chicago reported children with 50 µg/dL as positive cases; others probably clung to the industrial standard of 80 µg/dL established by Kehoe. The Surgeon General's committee set a three-tier standard. Children whose blood-leads exceeded

79 μg/dL should be considered "as unequivocal cases of lead poisoning." Since at this level severe symptoms could materialize at any moment, these cases should be "handled as medical emergencies" whether or not they showed symptoms. Children whose blood-leads were between 50 and 79 μg were to receive further medical evaluation, with monthly blood tests and an evaluation of the child's environment recommended.[54]

Lin-Fu lobbied the committee hard to include in the guidelines a third action level, which defined a blood-lead between 40 and 60 μg/dL as proof of "undue absorption of lead." A small but growing body of research literature dealt with the effects of chronic blood-leads in this so-called high-normal range. Despite the evidence of potential danger, Lin-Fu encountered stiff opposition on the Surgeon General's committee, largely because in some cities almost half the screened children had blood-leads in this range; hence, "this new standard would frustrate public health officials with an overwhelming caseload." According to Lin-Fu, the committee chairman reluctantly agreed to adopt this definition in the Surgeon General's statement only after she promised "to respond to all the letters of complaint that might be written about the new limit." Hoping to fend off professional opposition, Lin-Fu wrote a review article in the *New England Journal of Medicine,* and the Surgeon General's office received not a single letter.[55]

On 7 November 1970, the Surgeon General released an official statement on childhood lead poisoning which began, "The U.S. Public Health Service recommends that screening programs for the prevention and treatment of lead poisoning (plumbism) in children include all those who are 1 to 6 years of age and living in old, poorly maintained houses." Nowhere in Dr. Steinfeld's statement did there appear a hint that federal funding might help local communities implement his recommendations. Indeed, although the draft of a HEW booklet to help cities develop programs listed potential sources of federal funds, it cautioned that "the term 'potential sources' cannot be stressed enough. . . . In none of these programs are there specific allocations of monies either for medical case finding or treatment."[56]

In direct contrast to the activist tone of the Surgeon General's report, HEW came out against the LBPPPA, which would enable—require, in fact—the department to move forcefully toward the Surgeon General's stated goal. Under Nixon's "New Federalism," the department was not about to support a bill that mandated top-loaded categorical grants to be administered from Washington. The Public Health Service claimed that it already had the authority to award the grants the new bill mandated. The bill was unneeded, "since a successful program is dependent upon the resolve of States and localities to solve a local problem."[57]

The Public Health Service did have the authority, Kennedy agreed in Senate hearings, "but there is no evidence to indicate that the Public Health Service is prepared to expend any resources to fight this problem." The information HEW Deputy Administrator John Hanlon later supplied to the committee confirmed Kennedy's assessment. Nearly as an after-thought, appearing at the end of a document that highlighted three demon-stration programs funded by the PHS (for a total expenditure of $45,850) and made vague references to "undetermined" amounts spent under Ma-ternal and Child Health Services block grants, came a noteworthy admis-sion. In recent years, twenty-seven state, local, and private institutions had requested funds totaling $33 million to establish or enhance lead-poisoning programs. "To date," the summary concluded, "the Department has been unable to fund any of these applications." A Boston public health official testified that when he discussed his department's plans with an HEW administrator, he was discouraged from applying for a grant because of "the stringency of funding."[58]

Funding the Lead-Based Paint Poisoning Prevention Act remained a serious problem. At first the Nixon administration ignored it. As Jack New-field reported in a *New York Times* op-ed piece, "Although the bill author-ized $10 million for fiscal 1970–71, and $20 million next year, HEW secre-tary Richardson refused to ask Congress to appropriate the money. The President did not mention lead poisoning in his annual health message to Congress; Secretary Richardson did not mention it in his budget testimony before various committees."[59] During the next year, when authorizations under the Act doubled to $20 million, Nixon appropriated only $6.5 mil-lion. By the end of the decade, authorizations and appropriations reached a near balance (see Table 4). In 1982 the LBPPPA's appropriations were subsumed under the Maternal and Child Health Services block grant, part of Reagan's Omnibus Budget Reconciliation Act.[60]

The programs administered with LBPPPA funds prevented thousands of deaths and reduced the consequences of lead poisoning in hundreds of thousands of children. Jane Lin-Fu reported that from 1972 to 1975, HEW funded more than 70 programs each year, programs that screened over 1 million children and identified 64,000 at risk, 10,000 of whom re-ceived chelation treatment. By the early 1980s, federal funds had established some 100 lead-testing laboratories, 4 million children had been screened, 250,000 at-risk children had been identified for treatment, and 112,000 homes had been cleared of toxic paints and plaster.[61] The Act's ban on "lead-based" paints had an impact far beyond manufacturers of paints used in government-funded housing. Amendments in 1973 and 1975 lowered permissible lead contents, first from 1 percent lead in dried film to 0.5 percent, and then to 0.06 percent. In 1977, after lengthy legal, political, and scientific conflicts, the Consumer Product Safety Commission (CPSC) extended the ban to interstate commerce.[62]

Despite these successes, federal lead poison-control programs served but a fraction of the children at risk. By 1981 only twenty-five states and the District of Columbia were drawing federal funds under the LBPPPA. Cities that did have screening programs concentrated their efforts in the oldest, most economically stressed communities. Although plentiful evidence showed that significant percentages of children living outside of large cities had high blood-lead levels, suburban and rural children fell beyond the scope of most screening programs, as did many not living in the industrial northeast. East Coast and Midwest states made up over half of the states using federal funds for lead screening in 1981. Southern and western states were particularly slow to institute screening programs.[63]

Given this often haphazard, woefully incomplete effort, it is puzzling that in the decade following the onset of federal lead-poisoning programs, fatal childhood plumbism all but disappeared, and the number of children with high blood-lead levels dropped dramatically. No doubt, greater awareness on the part of physicians and parents played its part in these favorable developments. But it is now clear that the elimination of leaded gasoline from American cars was even more significant—that the presence or absence of the local health department's mobile testing truck was less important in determining the amount of lead in children's blood than was the kind of fuel the truck was burning.

ELEVEN

Facing the
Consequences of
Leaded Gasoline

A famous paint-company trademark shows the earth suspended beneath an overturned can of paint, its contents spilling over the globe in thick, clinging drips and globs. For many years, the slogan "Sherwin-Williams Paint Covers the Earth" accompanied the dramatic logo.[1] The image of the world being swallowed in a flood of paint seems at least metaphorically apt. Considering how much lead was in that paint and how much remains on walls, windows, and ceilings, inside and out, and weighing the ongoing danger that exposure to this paint poses to children and the billions of dollars it will cost to remove the hazard, the image is ominous—a paint bucket of Damocles. Metaphors aside, the logo was more accurate than its designers probably imagined. By the 1960s, the lead industry had succeeded in covering the earth—or at least the northern hemisphere. Lead paint did its part, but by far the most widespread source of environmental lead—and it was widespread, from Mexico City to the glaciers of Greenland—was the exhaust fumes of cars burning leaded gasoline.

Now it's the whole world. And it isn't the Grand Inquisitor's universal anthill that we have to worry about after all, but something worse, more Titanic—universal stupefaction, a Saturnian, wild gloomy murderousness, the raging of irritated nerves, and intelligence reduced by metal poison, so that the main ideas of mankind die out, including of course the idea of freedom.

SAUL BELLOW,
The Dean's December

Airborne environmental lead pollution differed sharply from both occupational lead poisoning and pediatric lead-paint poisoning. The latter were identified as arising from gross exposure to lead from a single source, usually in a specific interior environment. This presumed specificity allowed lead's advocates to shift the blame from the product to the victim or to neglectful supervisors (plant managers and dilatory hygienists or careless parents and greedy landlords). Despite the public outcry, the "silent epidemic of the slums" did not challenge the prevailing complacency about environmental lead any more than Alice Hamilton's exposés of occupational lead poisoning had in the Progressive Era. It was tragic that children were being killed and workers sickened, but neither pediatric nor occupational lead poisoning made environmental lead pollution of the atmosphere particularly frightening to the general public.

Then, in 1965, came the revelation that the average American's blood-lead level was one hundred times higher than was natural.[2] Lead-industry supporters claimed that even if environmental lead was more common than in the past, it was still too limited to be of regulatory concern. But the recognition of the ubiquitous threat of airborne lead came at time when those concerns were rapidly changing. The public health establishment was placing ever-greater emphasis on chronic illness. A nascent environmental turn in the sciences was bringing about a fundamental change in thinking about low-level hazards and a sharp revision of the understanding of "natural" levels of environmental toxins.

This chapter discusses the long road leading to the nearly complete elimination of leaded gasoline in the United States by the mid-1980s. It begins in the late 1950s, with the Ethyl Corporation's request to increase the amount of lead in gas. It ends in the late 1970s with the Environmental Protection Agency's phase-out of leaded gasoline and a discussion of the health effects of those reductions. Most important, this story marks the end of the lead industry's domination of scientific research into lead toxicity. Weaning American automobiles of lead required retooling the scientific community's entire outlook on the mineral, which up to that time had largely been dictated by the lead-using industries. Geochemist Clair C. Patterson, whose research unraveled Kehoe's principle of "normal" lead levels, stands as a representative of a new generation of scientists. Charged with environmental sensibilities and funded with major government grants, this generation put company-sponsored hygienists on the defensive and overturned the lead industry's most sacred medical precepts.

Although this transformation took place during the same years as the "silent epidemic," the two problems seemed only tangentially related. Throughout the 1960s there was little argument that childhood lead poisoning arose from decaying painted surfaces in ghetto housing, a clear-cut etiology left standing from the late 1930s, and the lead industry's occupational standards were subjected to no fundamental overhaul. The long-standing norms in pediatric and occupational lead poisoning offered no fresh insights into solving environmental lead pollution. But the successful reductions in airborne lead would bring a new understanding that transformed the epidemiology and politics of both occupational and pediatric lead poisoning from the mid-1970s to the present.

The full-sized family car, not the flower-bedecked Volkswagen Khombi, should be the automotive icon of the 1960s. For every counterculture freak taking a trip to ecstasy in a fuel-efficient bug, a hundred thousand workers, commuters, and housewives steered their four-door sedans and station wagons on trips to town or to the supermarket.[3] Since the end of World War II, the American automobile industry had cranked out ever bigger, heavier, more powerful cars whose engines raised the demand for higher octane fuels. Sales of leaded gasoline in the United States increased almost continuously from the introduction of the new fuel, though the maximum amount of tetraethyl antiknock fluid added to each gallon remained the same. In 1926, Surgeon General Hugh Cumming's committee investigating the public health risks of leaded gasoline recommended the maximum concentration of tetraethyl lead be fixed at 3 cubic centimeters per gallon, 2 cubic centimeters less than the Ethyl Corporation had marketed initially.[4] Although the Public Health Service had no authority to force compliance, Ethyl honored this standard for three decades.

Then in 1958, with auto makers demanding more octane for ever-more powerful V-8 engines, the Ethyl Corporation sought Surgeon General Leroy Burney's approval to raise the ceiling by a third, to 4 cubic centimeters tetraethyl lead per gallon of gasoline.[5] Officially, Burney had no more regulatory authority than his predecessor had in 1926. But this time, Ethyl would have much more difficulty pitching its line. Burney convened an ad hoc committee to evaluate the scientific literature on leaded gasoline and public health. It found that the most detailed report on the subject was over thirty years old: the hastily concluded study commissioned in the wake of the industrial poisonings at Bayway and Deepwater.[6] Although Ethyl and DuPont provided reams of technical data concerning industrial exposures, the committee noted a lack of epidemiological and environmental data. "It is regrettable that the investigations recommended by the Surgeon General's Committee in 1926 were not carried out by the Public Health Service. . . . If data were now available on body lead burdens, with 1926 as a baseline," the committee argued, "a more objective decision would have been possible."[7]

With the industry's sanguine data from its hygiene programs and no damning evidence to argue against the increase, the committee concluded that "the lead content of gasoline could be raised to a new level of 4 cubic centimeters per gallon and such change would not significantly increase the hazard to public health." More cautious than their 1926 counterparts, however, the committee members recommended approving the manufacturers' request only under three conditions. First, any increase should be made gradually over the course of five years. Second, Ethyl and DuPont would supply the Public Health Service with "the domestic consumption figures for tetraethyl lead." The third provision was more portentous: it called for a joint research project, involving the manufacturers and the PHS. The project would "provide more definite data on levels and trends of atmospheric lead contamination in selected urban areas and of the body burden of lead of selected population groups." This call was to produce the largest environmental study of lead contamination to date, the Three-Cities Survey.[8]

A Working Group on Lead Contamination was formed, comprising representatives from the petroleum industry, the producers of tetraethyl lead, car manufacturers, and the PHS. Beginning in June 1961 the group initiated a one-year study to collect air and human blood samples in Los Angeles, Philadelphia, and Cincinnati. The Philadelphia and California surveys were entirely government funded, but the Kettering Laboratory conducted the Cincinnati survey with funds from the American Petroleum Institute, DuPont, and the Ethyl Corporation.[9] Unlike the circumscribed research project undertaken in 1926, the Three Cities Survey cost millions of dollars and took thousands of man-hours. Since the working group could count on the scientific community to scrutinize the study, it was required to meet higher standards. The group remained keenly observant of the methods employed to collect and interpret data and was active in directing the policy implications arising from those interpretations.[10]

In the intervening years, air pollution had become a controversial field for both scientific inquiry and governmental regulation. The remarkable increase in public concern over air and water pollution after World War II cannot readily be explained. Such pollution had been a public health issue since the nineteenth century, but the postwar years brought a fundamental transformation in its meaning. To the industrializing America of earlier years, banks of towering smokestacks issuing clouds of black smoke signaled progress and prosperity. In the atomic age, they symbolized a way of life that millions of recent transplants to the suburbs wanted to leave behind. With the nuclear genie had come the crucial realization that what you cannot see can kill you. Coal dust and foul smelling fumes were bad enough, but modern chemistry and modern epidemiology brought odorless, invisible killers. Environmental consciousness arose with the "perception of a chemical world out of control." Early poison-control advocates focused solely on the domestic aspects of exposure to "the new chemical compounds . . . being synthesized and distributed with the aid of modern merchandizing methods." Historian John Burnham finds that the concerns propelling the poison-control movement coincided with and motivated the larger modern environmental movement. And Samuel Hays has demonstrated that many factors went into the mix to create modern environmentalism: changing economics, the demographics of the middle classes, a transformed federal government, and new attitudes about science and religion.[11]

But lead could hardly be categorized as a "new chemical." Complaints about nuisance smoke from lead smelters date back to Greek times.[12] But tetraethyl lead had transformed the equation, and when the Three Cities Survey came out, lead in the general atmosphere was taking center stage in many discussions about urban air pollution. The lead industries were especially active in this dialogue. Early in 1963, at the conclusion of the Three Cities investigation, the Kettering Laboratory sponsored a symposium on lead and lead poisoning. Similar to meetings sponsored by Kettering or the LIA in years past, the audience was largely drawn from lead-industry management and health specialists. Most of the papers dealt with prevention, diagnosis, and treatment in the occupational setting. But in his prefatory speech, Robert Kehoe noted the sharp contrast with former industry-sponsored symposia, when issues pertaining to "the external environment" were presented "less frequently." This symposium, he claimed, would deal with "certain primary and ancillary facets of the hygienic problem raised many years ago by the introduction of tetraethyllead as an antiknock additive into automotive fuel."[13]

The lead industries could not ignore the changes in the air. The situation was very different from that presented by lead paint. The marked contrast between the LIA's relatively cooperative stance on pediatric lead-paint poisoning and its intransigence in the tetraethyl lead controversy can be readily explained. In the light of the evidence coming out of city health departments since the 1950s, the lead industry could not deny that its one-time flagship product, lead paint, was a serious hazard to young children. Denial was useless and counterproductive. Instead, through the 1960s the LIA promoted an image of its members as representatives of a concerned industry and responsible corporate citizens. The association funded some of the most important pediatric lead-poisoning research in the 1950s, helped establish a ban on lead-based interior paint in 1955, and did little to obstruct the first federal legislation to start reducing the threat and treating its victims.

The members of the LIA could well afford this compassion. The public relations advantages in appearing to fight childhood lead poisoning far outweighed the economic losses. The great age of lead paint had long faded: in 1969, titanium dominated the pigment market. Lead pigments accounted for only 6 percent of all lead used in the United States, and almost all of that went to exterior corrosion protection and chrome-yellow street paint, none of which faced any intense regulation at the time.[14] Some paint manufacturers objected when proposed federal regulations would prevent them from using lead dryers; but the paints they were drying were based on titanium, not lead. In the 1980s and 1990s paint companies and the lead industry have faced the growing costs of lawsuits brought on behalf of lead-poisoned children. In the 1960s this crisis was still a remote concern, and the lead industries had not yet adopted the strategy of denial that has come to typify their public pronouncements and legal tactics.[15]

Tetraethyl lead, on the other hand, was a critical component of the lead industry's product line, and the industry defended it tenaciously. In 1959, American cars burned 160 tons of tetraethyl lead, up almost 70 percent from ten years earlier and about 15 percent of total lead consumption in the United States that year.[16] The lead industry would pay dearly if leaded gasoline were banned. And manufacturers could argue that unlike the now-discredited lead paint, tetraethyl lead was a strategic product whose loss would have serious repercussions in the petroleum and automotive industries. The prospect of losing the profits from leaded gasoline required a new strategy for handling questions of public health risks, a strategy best characterized by denial, distortion, and vigorous denunciation.

Ultimately it was a fundamental difference in perceived risks that defined the industry's very different public responses to lead paint and airborne lead. Flaking lead paint killed young children. Fact. The LIA could only quibble over the degree of its culpability and responsibility for removing the lead-paint threat. But little or no evidence implicated lead from car exhausts in the deaths of children or adults. As long as the old definition of lead poisoning as an acute disease with marked symptoms held, occupational standards for toxicity defined the risks and Ethyl and DuPont could continue to assure any government agency that their product was completely harmless in the concentrations found in the urban environment.

In an influential series of lectures before the British Royal Institute of Public Health and Hygiene in 1960, Robert Kehoe made explicit the lead-industry gospel that had guided its policies (and lead toxicology in general) for almost forty years. "The natural environment of this planet is such," he proclaimed, "that the occurrence of lead in the tissues, body fluids and excreta of its human inhabitants is inevitable."[17] This exposure had had evolutionary consequences, he argued:

> *It appears . . . that an equilibrium is established at an early age between the human organism and its usual or normal environment in the United States, whereby the stream of lead absorbed into the body from the environment is balanced by a counterstream of lead issuing forth from the tissues and from the body via excretory routes. No doubt this equilibrium is disturbed from time to time by the variability of the environmental conditions, and for a time the absorption of lead exceeds the excretion, or vice versa. There is good reason to believe, however, that the end result of the operation of these variables is the maintenance of an essential balance between absorption and excretion over the span of life of an individual.*[18]

Together with the threshold for effect, the concepts of "balance" and "normal" exposure forged a seemingly unassailable defense of current and foreseeable lead exposures.

To some extent the Three Cities findings confirmed the status quo. Air samples taken near the cities' commercial centers contained up to three times as much lead as surrounding rural areas. But there were significant levels of lead at all sampling spots; hence it could be argued that lead was a "normal" part of the human environment. The study concluded that most of the airborne lead in the cities was from leaded gasoline. But the study found no more airborne lead than in 1946, when lax smoke abatement regulation permitted high-lead coal to be burned without adequate emission controls.[19]

The study's findings on the biological impact of these levels of airborne lead were similarly ambiguous. Blood samples from 2,300 individuals in the three cities confirmed that city dwellers had higher concentrations of lead in their blood. But those who lived in the country were not lead free, and not a single subject's blood-lead exceeded the industrial threshold of 80 μg/dL, which seemed to prove that leaded gasoline posed no threat to public health.[20]

Twenty years earlier, the status quo might have gone unchallenged. But in 1965, the fact that eleven of the 2,300 had blood-leads in excess of 60 μg/dL alarmed several members of the Working Group on Lead Contamination as well as a growing number of skeptics in the medical and scientific community. But no matter how vocal or unified their opposition, they would not be able to overturn the prevailing nonchalance as long as Kehoe's three tenets held: lead absorption was natural and the body was adapted by evolution to deal with its leaded environment; the lead that industry added to the environment did not tax the body's natural ability to balance intake and output; and no illness resulted from lead absorption below established thresholds. City dwellers may have higher blood-lead concentrations than fishermen in the South Pacific. But if a Borneo aborigine far removed from automobile exhausts and lead paint had lead in his blood, then surely environmental lead has always been a natural part of human existence.[21]

Geochemist Clair C. Patterson did not set out to disprove such comforting theories.[22] Initially, his research interests were the origins and development of the earth's mineral composition. Just out of graduate school in 1953, he used stable lead isotopes from meteors to establish the earth's age as 5.6 billion years—2 billion years older than previous estimates. His work required unprecedented "clean techniques" to keep lead contaminants from tainting his samples. From this laboratory annoyance grew an abiding interest in industrial lead pollution, which resulted in his opposing applied scientists such as Kehoe. In Patterson's view, their efforts at reducing lead poisoning, despite their seeming success, actually increased the risks "because they fostered the increased production of lead."[23]

In the early 1960s, Patterson and research partner T. J. Chow were exploring the history of the earth's crust by studying trace elements in the oceans. They found that far more lead was entering the oceans from rivers than had been leaving by sedimentation.[24] Presumably, the source of these rising levels was industrial lead production. But to prove this thesis required an accurate record of the extent of lead pollution dating back to preindustrial times.[25]

Patterson and his colleagues turned to undisturbed polar ice, the layers of which can be dated like tree rings. They took core samples from remote areas in Greenland and Antarctica and analyzed them for date and trace metal content. The lead content in the Greenland samples was consistent with the historical record of industrial lead production, showing a "marked parallel with the increase in lead smelting and in consumption of leaded gasoline": from 1750 to 1940 the lead concentration rose 400 percent; from 1940 to 1965 it rose a further 300 percent (see Figure 11-1). Any suspicion that this was natural could easily be put to rest by comparing the Greenland data with the Antarctic cores, which showed very low concentrations of lead and were stable over time. The highest concentration found in the

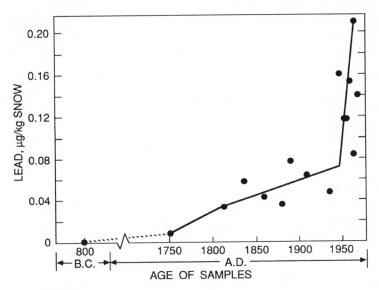

FIGURE 11-1
Lead Concentrations in Polar Ice of Northern Greenland
EPA, *Air Quality Criteria for Lead* (Research Triangle Park, N.C.: EPA, Environmental Criteria and
Assessment Office, 1986).

Antarctic ice was about the same as the lowest Arctic sample, from 800
B.C. The pollutants generated in the industrializing northern hemisphere
were confined by atmospheric conditions around the equator that act like
a wall separating the two hemispheres.[26]

The polar ice studies showed that atmospheric lead levels rose along
with industrial lead production, but they did not demonstrate a concurrent
rise in human blood-lead concentrations. Patterson took a unique
approach to proving that modern humans carry unnaturally high body-
lead concentrations. In an influential article published in 1965, he outlined
a remarkable counterfactual study. He carefully estimated how much lead
an adult would absorb if exposure involved only "natural" lead pollutants—those leaching from the ground or sea salts, or resulting from volcanic activity, natural wildfire, and even meteorites. He calculated that an
average body burden in a 70 kilogram (kg) adult living in such a pristine
world would be about 2 milligrams (mg) (appx. 0.00007 oz. per 154 lbs.).
Studies of different modern population groups had found body burdens
between 90 and 400 mg, and Patterson estimated the average as 200 mg
per 70 kg person, 100 times higher than his estimated "natural" levels. In
1965 the average American's blood-lead level was around 20 μg/dL, a full

25 percent of the 80 μg "threshold" for clinical plumbism. Patterson extrapolated that "the average resident of the United States is being subjected to severe chronic lead insult."[27]

Before 1964, Patterson's research exemplified the objectivity and dedication to rigorous scientific method expected of pure scientists and he only published his findings in journals and books dedicated to pure science, such as *Geochimica et Cosmochimica Acta* and *Earth Science and Meteoritics*.[28] But the trends his data suggested alarmed Patterson, and his annoyance with the scientific and medical communities' unquestioning acceptance of the standards and outlook set by the lead industry led him to seek a more public forum. He stepped out from the sanctuary of publications in his field and submitted his manuscript "Contaminated and Natural Lead Environments of Man" to *Archives of Environmental Health,* a leading journal for doctors and engineers in the fields of occupational hygiene and pollution control—the very heart of the lead technocracy and applied environmental science Patterson deplored.

The magazine's editor, Katharine Boucot, went out on a limb for the article, which in scope, tone, and methodology diverged radically from the *AEH*'s norms. (More typical fare appeared in the February 1964 issue, which featured papers from a lead-industry–sponsored conference hosted by Robert Kehoe written by specialists from the Kettering Laboratory, Ethyl Corporation, and the Lead Industries Association.[29]) The published version of Patterson's article was filled with overheated rhetoric and moralism, despite considerable effort by Boucot and six anonymous readers to tone down Patterson's "colorful" eighty-page first draft.[30] Most prominent among the readers Boucot enlisted were Robert Kehoe and Joseph Aub. Aub wrote Patterson that he had read the first draft several times and still found it "difficult to criticize and discuss." Kehoe suggested that the first draft was so filled with "extraneous material" as to allow it to be easily dismissed as alarmist claptrap. As he assured another lead researcher, "I'm not sure of the intellectual qualities of a youngish man (or any other man) who goes outside his field of training and competence with such passion and self-assurance."[31] Still, Kehoe saw the potential damage Patterson's paper could do to Ethyl's reputation. He warned an executive of the corporation that if not handled properly, Patterson's information could lead to the outlawing of tetraethyl lead and other lead products. Patterson did not reply to Kehoe's invitation to meet him to discuss his findings. The pure scientist "would have none of it," Kehoe bitterly recalled, "expressing the opinion that he had no intention of meeting or knowing me, or of discussing anything with me."[32]

When Patterson's article appeared, the National Science Foundation, under whose auspices Patterson was conducting his Antarctic research, issued a press release outlining the Antarctic project and discussing Patterson's Greenland findings and his theories as to the probable cause of rising lead pollution. A *New York Times* reporter interviewed Patterson, and the ensuing story, which announced an "alarming" increase in lead pollution, appeared on page 1.[33] The response from the lead and petroleum industries was immediate—and predictable. The *Times* reported the American Petroleum Institute's assurance that "all 'accepted medical evidence proves conclusively' that lead in the environment presents no threat to public health." The LIA reminded the press of the Three Cities Survey, which showed that even the highest lead exposures it found were "well within the accepted range of lead levels for humans."[34]

The tone of calm authority in these pronouncements belied the deep unrest Patterson's study caused inside the lead-using industries, both before and after the paper's release. Copies of early drafts had circulated through the corporate offices of Ethyl, DuPont, and the LIA. Six months before publication, an Ethyl Corporation chemist named Cooper critiqued Patterson's findings. For the most part, he confined his arguments to technical quibbles with Patterson's geochemical findings regarding the composition of the earth's crust. Patterson's most critical shortcoming, according to Cooper's memorandum, was in the area of biological evidence. Patterson posited theoretical geologic evidence for lower lead levels in the past but offered no evidence from preindustrial man or animals to prove a link. "One should think," Cooper argued, "Patterson would have felt the need for data from at least one specimen of early animal remains."[35] An Ethyl executive, A. S. Hawkes, suggested that Patterson's failure to use fossil evidence meant he was a "crackpot."

A tone of disbelief and indignation pervades much of the internal lead-industry correspondence about Patterson. If Patterson had not been such an outsider he would not have betrayed such an important industry. Cooper despaired, "Lead is, of course, one of man's most useful raw materials. We are very much surprised that a scientist of such considerable experience would mount such an attack on it from so very little evidence." This outsider status offered Ethyl a possible defensive strategy. Blacked out but still legible at the bottom of his memo on Cooper's findings was Hawkes's conclusion: "Patterson is playing fast and loose with chemistry and physiology. Since he is an authority in neither field, he should be vulnerable in both."[36]

Kehoe, too, was offended at Patterson's apparent lack of appreciation for the authority of lead-industry researchers such as himself. But he recommended Ethyl executives take a different approach toward damage control. Kehoe admired certain aspects of Patterson's work, which he argued "makes a very considerable contribution" to knowledge about "the distribution of lead over the surfaces of this planet." The lead industries should seek "multidisciplinary information and advice" instead of "flying off the handle everytime some untutored or learned jackass has a vision of things outside of his knowledge and experience." Instead, Kehoe argued, the lead industries should rely on the Kettering Laboratory's reputation and decades' worth of findings to "uphold the truth with evidence and also with the poise and dignity that are essential." Casting himself as lead's savior, he claimed that if Patterson's findings had been published twenty years earlier, "ignorance and apprehension" would have prevailed at great cost to the lead-using industries. In March, months before the article was published, Kehoe counseled patience, as he had many times before: "I urge, as vigorously as I can, that there be little or no obvious effort on the part of those who have an important stake in the lead-using industries, to discredit Dr. Patterson, for such effort will be a boomerang."[37]

Kehoe lost the argument—even, it would appear, with himself. Prominent (and first) among the uncharacteristically large number of letters to the editor in the year following publication of Patterson's article was a four-page critique by Kehoe, focusing on Patterson's lack of experimental evidence and dismissal of data from epidemiological studies of lead poisoning conducted for occupational hygienists. Despite what he saw as serious flaws, Kehoe supported the *Archives'* decision to publish, though perhaps only for strategic reasons. He reminded readers that since "Dr. Patterson's paper has been circulated in private communications, and presented to at least one official agency . . . it was essential . . . that it be brought into the open for professional appraisal."[38]

Those appraisals were for the most part very negative. Essentially they followed three lines of argument. First, some critics questioned Patterson's methods, his interpretation of data, or the logical gymnastics he underwent in establishing a baseline "natural" body burden of lead. Kehoe repeated the criticisms made by Ethyl chemist Dr. Cooper. He was more emphatic in criticizing Patterson for dismissing his own 1930 study of lead levels among people living in supposedly "preindustrial" conditions in the Mexican plateau. Robert Ziegfeld, LIA executive vice-president, blasted Patterson for vastly overestimating the percentage of lead-containing ceramics, glass, and solder that went into consumer products. Even those who supported Patterson's message questioned his methods. Dartmouth physiologist Henry A. Schroeder, whose proclamations against lead pollution would grow stronger over the following years, argued that modern body burdens of lead were probably half those estimated by Patterson. Even so, Schroeder warned, the present levels were "toxic in the sense of increased mortality and decreased life span."[39]

Second, many critics also commented on Patterson's transdisciplinary obtrusion. "Dr. Patterson has ventured," Kehoe complained, "without caution, humility, or appropriate critique, into what is clearly, for him, an alien area of biology." One of the reviewers of the early manuscript echoed Kehoe's comment and concluded that Patterson had "taken some interesting geological data . . . and derived some rather curious biological inferences."[40] In defense of his admitted transgression, Patterson replied that the growing awareness of "the scope and seriousness" of pollution problems had changed the discipline of toxicology. The urgency of the problem required "probing by research scientists in many fields." Industrial hygienists and "men in basic research disciplines in universities" would increasingly find themselves competing for the Public Health Service's attention. Therefore, he concluded, "It behooves industrial toxicologists and physicians in occupational health not to respond to this challenge as though their heritage was being stolen, but to reflect whether they have renounced it."[41]

Patterson was neither the first nor the last disciplinary interloper to populate the modern history of lead poisoning. The insights and outrage of energetic outsiders impelled almost every major turning point. Although trained as a bacteriologist, Alice Hamilton, for example, brought a sociological and epidemiological sensibility to her life's work in toxicology, a discipline that in her time concerned engineering and economics more than medicine or sociology. Boston pediatrician Randolph Byers had little training in toxicology, and his influential study of the late effects of noncerebral lead poisoning was coauthored with Elizabeth Lord, a psychologist. In recent years Herbert Needleman, a psychiatrist, and Richard Wedeen, a kidney specialist working for the Veterans Administration, changed the very definition of lead poisoning by demonstrating the mental and physiological effects of chronic low levels of lead absorption.[42]

Finally, and far more troubling to Patterson's critics than his credentials or his methods, was his assertion that current levels of lead pollution were bringing "severe chronic lead insult." To those who had long labored in the world of lead toxicology, either this conclusion suggested that Patterson had misread Kehoe's research, or—worse—it signaled his refusal to accept the prevailing understanding of lead metabolism. Joseph Aub thought Patterson would be less concerned with the levels of lead exposure he found if only he would remember that "when there is absorption the beautiful protective mechanisms of the body perform their assigned roles, retention and excretion." Rutherford T. Johnstone, a California physician, saw Patterson as a barbarian storming the gates of knowledge (located in Cincinnati). In a letter to *Archives of Environmental Health,* Johnstone proclaimed, "Surely the foundation upon which the Kettering Laboratory rests must have felt a quake following the accusation that 'a thorough understanding of the mechanism of lead metabolism does not exist today.' If Patterson's statement is true, then a lot of time, guinea pigs, mice, men, and money have been for naught."[43] And UCLA pharmacologist Thomas J. Haley published an article-length rebuttal entitled "Chronic Lead Intoxication from Environmental Contamination: Myth or Fact?" His answer, unequivocal and italicized: "*The supposed chronic lead intoxication from environmental contamination is a myth, not a fact.*"[44]

Kehoe argued that Patterson had been taken in by the "simple apprehension" that since lead was undoubtedly toxic at higher levels, "it may be, and therefore probably is, harmful in any quantity." Kehoe reminded readers that "the facts of physiological research and clinical medicine" predict "that no case, not even an occasional mild case, of lead poisoning will occur, unless the physiological threshold value is exceeded." But Patterson contended that the common acceptance of a threshold value for damage bred complacency and blinded clinicians to lower levels of poisoning. The "threshold for damage concept" relied on "the axiom that a worker must be either perfectly healthy or classically intoxicated with lead but cannot be neither." He feared that neurological and intellectual functioning might be widely and immeasurably impaired at far lower levels than those that bothered occupational hygienists, whose primary job was to keep workers healthy enough to produce more lead products.[45]

Harriet Hardy, a toxicologist from the Massachusetts Institute of Technology, where Patterson had worked on "Contaminated and Natural Lead Environments of Man," recalled that Patterson hoped his article would "have the effect that Rachel Carson's work did in stimulating [her] opponents to defend their position." Arie J. Haagen-Smit, one of Patterson's colleagues at the California Institute of Technology, also compared Carson and Patterson. While not all of Carson's book was good science, he said, "we scientists never get anywhere," and just as Carson "put something in" the discussion of pesticides, "by golly, Clair Patterson put something in there."[46]

Direct references to Patterson's article quickly disappeared from the popular press, but in professional circles the publication fanned a slow-growing debate into a firestorm of controversy.[47] Long before Patterson's article appeared, the Public Health Service had scheduled a three-day symposium on environmental lead contamination. The lineup of speakers suggests that the meeting had been set up to highlight the recently released Three Cities study. But Patterson, half a world away in Antarctica, still stole the show. Although a number of papers, such as Kehoe's "Under What Circumstances Is Ingestion of Lead Dangerous?," sought to hold the old line, lively discussions and exchanges with the press focused on the possible threat of low-level lead insult.[48]

Patterson criticized the Public Health Service for scheduling the symposium when he could not be there. The organizers in turn invited Patterson to send a delegate or additional data. And although another chemist from Patterson's department at Cal Tech attended, Patterson's most vigorous defender was Harriet Hardy. The MIT toxicologist had studied beryllium poisoning among bronze and other metal workers and had coauthored a textbook on occupational medicine with Alice Hamilton. Now she was studying the effects of low-level lead exposure. Her paper, with the deceptively neutral title "Lead," indicted the threshold-limit concept for ignoring the possibility that certain populations, including children, the elderly, and pregnant women, were subject to toxic effects from lead at much lower levels than the industrially based standard currently employed.[49]

The symposium revealed the deep cracks beginning to erode the Kehoe-Kettering paradigm. During a heated question-and-answer period following Kehoe's presentation, Dr. Hermann of the Harvard School of Public Health argued against the concept of a strict threshold. When Kehoe predictably asserted that his balance studies had proven the reliability of the 80 microgram threshold, Hermann questioned Kehoe's entire research program. He criticized the fact that so much research had been done in one place, and called for Kehoe's balance experiments to be repeated elsewhere. The 72-year-old scientist, unaccustomed to such questions, stepped back from his customary confident bandying of facts and admitted, "I seem to be a bit under the gun." His only defense was that his opponents must have misunderstood him. In boiling down his thirty years' experience into twenty minutes, he argued, the complexity of his findings had got lost.[50]

Much more had been lost. Although Kehoe's work continued to be cited

by both friend and foe, friends more frequently accompanied their praise-
with confirmatory citations, while foes more frequently cited his research
as the best work of a bygone era.[51]

What Patterson "put in there" was the universality of the threat of low-level
lead contamination. His pessimistic theory in effect asked the question, if
everyone is getting a little stupider, who is going to know? As Saul Bellow's
character Albert Corde put it: "Now it's the whole world . . . universal
stupefaction, a Saturnian, wild, gloomy murderousness, the raging of irritated nerves, and intelligence reduced by metal poison, so that the main
ideas of mankind die out, including of course the idea of freedom." Patterson was explicit about the potential historic significance of lead poisoning: "It is interesting and not at all unworthy to consider," he mused, "how
the course of history may have been and is now being altered by the effect
of lead contamination upon the human mind."[52]

Another paper appearing in an occupational hygiene journal in 1965,
also written by a nontoxicologist, highlighted the possible effects of lead
exposure on the course of history. Colum Gilfillan, a scholar of the sociology of invention, published an essay entitled "Lead Poisoning and the Fall
of Rome." In it he boldly suggested that plumbism "is to be reckoned the
major influence in the ruin of the Roman culture, progressiveness, and
genius."[53] Gilfillan's thesis may have overreached his evidence, but his conclusions are not without merit, and further research by several scholars
has provided substantial support. More significantly, his work gave lead's
opponents a powerful tool in the symbolism of an invisible threat capable
of toppling empires. As Bellow's weary dean put it, "It wasn't the barbar-
ians, it wasn't the Christians, it wasn't moral corruption . . . the real cause
was the use of lead to prevent the souring of wine. Lead was the true source
of the madness of the Caesars. Leaded wine brought the empire to ruin."[54]

With the empirical foundation Patterson provided and the powerful
rhetorical wallop of Gilfillan's thesis, environmentalists could portray lead
pollution in an apocalyptic light. The lead industries, however, had a very
different vision of the end of the world as they knew it. Upon reading
Patterson's first draft, Kehoe cautioned the Ethyl Corporation to prepare
for "the need which may develop for heading off action which would eliminate the use of lead in a variety of ways."[55] Conflicts over automotive pollution had tended previously to focus on reducing carbon monoxide and
sulphur dioxide; after 1965, lead played an increasingly important part in
the debates. With the enemy marshaling evidence against the metal, it
became crucial for lead's supporters to uphold the Kehoe-Kettering tenets:

that lead was a natural component of the environment and hence the human body; that the body had mechanisms to cope with its burden of lead; that lead absorption was harmless below a measurable "threshold" blood-lead level; and that the public's exposure to lead was no threat to health.[56] Although Patterson's work cast serious doubt upon all four principles, it would take years, and many empirical studies, to get lead toxicology to abandon them.

For the next ten years, the lead industries struggled to maintain the ability to set agendas, define risks, and influence their potential regulators. Their decades-long relationships with the automotive and petroleum industries were critical in maintaining lead's place in the industrial economy. For the most part, this momentum derived from the lead-using industries' convincing portfolio of scientific data and expertise developed to reduce the costs of lead poisoning in their workplaces. Patterson and his generation of scientists saw lead's momentum diminish over the decade, and the use of leaded gasoline was eventually prohibited in the United States. But rather than seeing them as Davids who by means of pluck and creativity single-handedly killed the leaden giant, it is more accurate to see them as gadflies who roused the sleeping federal regulatory giant and prodded it into action.

The story of the legislative, judicial, and administrative battles over the elimination of lead from gasoline is longer and more complex than can be fully covered here, and has in any case received adequate attention elsewhere.[57] Yet even a brief summary of those battles reveals a profound shift in the perception of lead's hazards, which undermined the industry's ability to define those hazards or keep them confined to categories of pediatrics and occupational medicine. The outcome of the leaded gasoline controversy—that is, the effective ban on leaded gasoline in the United States—went far toward clearing the nation's air, and its people's bodies, of a ubiquitous toxin.

Before the mid-1970s much of the progress in reducing airborne lead had little to do with reducing public exposure. Had he wanted to, the Surgeon General could have regulated the amount of lead in the nation's gasoline as early as 1963 under the provisions of the Clean Air Act. Of course, the prevailing understanding of risk would have made such a restriction impossible to justify. Even with proof that levels of lead in urban air increased with the number of cars burning leaded gas, the lead and petroleum industries could argue convincingly that average blood-lead levels were nowhere near the 80 microgram threshold. Kehoe testified before

Senator Edmund Muskie's 1966 Air and Water Pollution subcommittee that increases in car traffic would indeed add to the amount of lead in the atmosphere. But, Kehoe claimed, traffic congestion in the largest cities had reached its limit years ago. Traffic could only be increased by building superhighways, which would have the desirable effect of leveling exposures: if vehicles were "more widely and more uniformly distributed the high results will come down, the lowest results will go up."[58]

No matter what the distribution, testified John Kimberly, executive director of the LIA, in 1970 before a New York City committee considering a city-wide ban on leaded gas, "there is no evidence that lead in the atmosphere, from autos or any other source, poses a health hazard."[59] Lead's boosters actually argued that, if seen in the right light, continued use of leaded gasoline was good for the public's health. In an rare discussion of a health issue in the LIA journal, *Lead,* Kimberly warned that "pollution hazards could actually be increased if lead were to be eliminated from gasoline" because lead substitutes might be even more toxic.[60]

It was lead's threat to pollution-control devices, not to humans, that first encouraged serious efforts to ban leaded gasoline. By the late 1960s auto makers had developed the catalytic converter, an active chemical device in the exhaust system that converted carbon monoxide to carbon dioxide and water. The portions of the 1970 amendments to the Clean Air Act that addressed auto emissions relied on this technology in setting timetables and standards. According to John Quarles of the Environmental Protection Agency, Detroit saw in the catalytic converter a way to buy time and satisfy regulators while they developed cleaner running cars. But to oil refiners and the lead industry this very effective device had one serious drawback: even small amounts of lead in gasoline "poisoned" the catalyst, eventually rendering it completely ineffective.[61]

The Clean Air Act amendments also directed the EPA to set ambient air standards for such pollutants as lead. To those in the EPA who believed, as Director Ruckelshaus proclaimed in 1971, that airborne lead from tetraethyl gasoline posed "a threat to public health," the catalytic converter offered a strategy for reducing lead despite the industry's ability to counter credible medical evidence with credible medical evidence. In 1972, in the wake of the controversial EPA-sponsored study *Airborne Lead in Perspective* by the National Research Council, the agency proposed two separate regulations covering leaded gasoline.[62] The first required that all gasoline producers offer at least one grade of unleaded gasoline by 1975 in order to protect catalytic converters, which would by then be required equipment on new cars. The second proposed regulation, designed to protect humans, would give refiners four years to reduce by half the amount of lead in the entire gasoline pool. The first regulation was implemented in 1973, but besieged by industry opponents, the EPA hesitated to enact the second. Only a successful lawsuit by the Natural Resources Defense Council (NRDC) forced the agency to set a restriction schedule. In December, the EPA issued regulations calling for a five-year phased reduction to begin in 1975. The amount of lead in the gasoline pool would be reduced from the 1973 average of 2.0 grams per gallon to a maximum of 0.5 grams per gallon by 1979.[63]

Phasing out leaded gasoline was a hard sell in the mid-1970s. Tetraethyl lead was very effective at squeezing extra power out of a gallon of gas, and no one had successfully marketed a safe, economical alternative. Fuel shortages, rising petroleum prices, and new economic realities kept environmentalists on the defensive, and the EPA's phase-out limped along from compromise to delay. Administrators no doubt dragged their feet, but the delays must be seen in the context of the agency's mandate to set scientifically valid standards to protect health without weighing the costs. Under the Kehoe paradigm, setting such standards would be simple arithmetic. But in the uncertain and politically charged world of lead toxicology in the 1970s the criteria were fuzzy, and standard setting became a source of conflict between industry, environmentalists, and the EPA, as well as between environmental activists and more politically involved "realists" within the agency.[64]

As often occurs when politics throws the *sabot* into the bureaucratic works, lawsuits forced the machinery back into motion.[65] The process of deleading American automobiles was slow and rocky: in the mid-1980s over 40 percent of gasoline sold in the United States still contained lead, although in a far lower concentration than had been the average twenty-five years earlier. Still, between 1971 and 1980, the amount of lead consumed annually in gasoline dropped by 46 percent, to the lowest level since 1951. Over the next five years, tetraethyl sales plummeted by 60 percent.[66]

The health consequences of this sudden reversal in the trend toward greater environmental lead pollution will probably never be assayed, but the serious ongoing poisoning problems in countries where leaded gasoline continues to be burned suggest the human impact.[67] The Second National Health and Nutrition Examination Survey (NHANES II) conducted between 1976 and 1980 provided graphic evidence of the potential health benefits of deleading gasoline. The survey found that average blood-

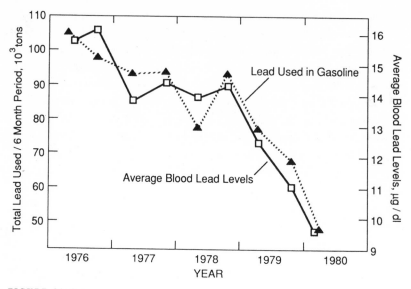

FIGURE 11-2
Parallel Decreases in Blood-Lead Levels and Amounts of Lead Used in Gasoline, 1976–80
Data from NHANES II, after EPA, *Air Quality Criteria for Lead* (Research Triangle Park, N.C.: EPA, Environmental Criteria and Assessment Office, 1986).

lead levels had declined by nearly 40 percent over the period, from 14.6 to 9.2 μg/dL. When researchers presented the NHANES data plotted alongside consumption of leaded gasoline over the same period (Figure 11-2), the parallel was striking. No other factor, including reduced lead in tin cans and pesticides, improved screening and treatment, or lead-paint abatement, bore any significant correlation to this nationwide detoxification.[68]

This reduction in average blood-leads had little to do with efforts to remove lead paint from substandard housing. The important work done by inner-city lead-screening clinics, and the increased awareness on the part of pediatricians and hospital staffs, no doubt saved thousands of lives and enhanced the effectiveness of treatments. But while all the programs mandated or funded through the Lead-Based Paint Poisoning Prevention Act and subsequent legislation represented hard-fought victories and important steps in the right direction, their impact on bioavailable lead in children's environments pales in comparison with the government-mandated reductions in permissible lead in gasoline. By the standards of the early 1960s, the elimination of leaded gasoline produced a healthy margin of safety in preventing "lead-paint poisoning."

Of course, by the late 1970s those standards were no longer in place. As the amount of lead in the environment and in human blood decreased, clinicians found more evidence of the effects of low-level exposure. Faith in the old thresholds continued to erode, and with each reduction in "acceptable" blood-leads, further studies revealed ever more subtle and widespread damage below that level, eventually calling into question the very existence of a threshold for damage from lead. And if there were no threshold, then all children were at risk; therefore, it came to be argued, all children should be tested.

TWELVE

The Rise and Fall of Universal Childhood Lead Screening

By the mid-1970s, lead-poisoning programs in city health departments had made remarkable progress toward many of the goals established in earlier years. And then, as if to snatch defeat from the jaws of victory, the very definitions of lead poisoning changed. A new epidemiology redrew the boundaries of the "lead belt." No longer confined to inner cities, it now took on global dimensions, and lead's victims were found to be all of us. The Centers for Disease Control confirmed this new definition. No mere "silent epidemic of the slums," childhood lead poisoning was "the most prevalent environmental threat to children in the United States." The new definition promised to return to John Ruddock's universal conception of lead's threat—that "the child lives in a lead world"—a conception that had been eroding ever since he asserted it in 1924.[1]

> *It is wasteful of all our capital—the time and goodwill of parents and health care providers as well as health care dollars for our children—to screen every child irrespective of lead exposure or plausible risk.*
>
> NANCY M. TIPS, HENRY FALK, AND RICHARD JACKSON, "CDC'S LEAD SCREENING GUIDANCE: A SYSTEMATIC APPROACH TO MORE EFFECTIVE SCREENING"

The most startling example of this transformation was the change in the popular image of the lead-poisoned child. When *Newsweek*'s editors wanted to highlight "Disturbing New Evidence" about the threat to "Your Kids" for a 1991 cover story, they did not show a black child surrounded by peeling paint and cracked plaster, as was typical in articles about lead poisoning in the 1970s. Instead the magazine's cover depicted Miranda Perrone, a red-haired 2-year-old from Milwaukee, standing next to a newly stripped wooden sideboard in her family's dining room. Miranda's was a case of what has been called "yuppie plumbism," a child exposed to dangerous amounts of lead dust during home renovations.[2]

In 1991, the CDC joined with antilead activists in urging that all American children be screened. But less than a few years later, when public awareness of this more general threat was at its peak, when it looked most promising that state and federal agencies would mandate screening to identify every child at risk, new studies found that average blood-lead levels had fallen. Even by the criteria set by the CDC in the early 1990s, lead poisoning seemed far from the pandemic lead's opponents had foretold. Some argued that the new battle—so recently entered upon—had already been won. The information seemed to provide a salient argument for abandoning universal screening, and the CDC did precisely that.

This chapter traces the transformation from the activism of the early 1970s to the optimism of the early 1990s to the present, when the universal approach has crumbled under cost-benefit analysis and the resurgence of the "comforting" impression of childhood lead poisoning as a nearly exclusive disease of poverty—sound epidemiology that may lull the public back to the complacency of earlier years.

As understood in the late 1960s and early 1970s, lead poisoning was an attractive cause for social reformers. Those affected, poor and often nonwhite, could be portrayed as victims of their particular circumstances: bad landlords, lack of access to health services, indifferent or incompetent public officials, or systemic racism. The vast majority of cases treated by medical practitioners or quantified by epidemiologists strengthened the association between poverty and plumbism. And although sympathetic reports found blighted environments to be the ultimate cause of lead poisoning, their descriptions often revealed underlying assumptions about the contribution of cultural or personal behaviors associated with poverty. Both contemporary epidemiology and the rhetoric of activists secured the image of lead-paint poisoning as a disease of poverty and ignorance.

The first federal law to address childhood lead poisoning appeared to dismiss such narrow definitions. Although it was lead poisoning in the ghetto that pushed Congress to pass the Lead-Based Paint Poisoning Prevention Act of 1970, the law itself proposed what seemed a more universal approach: "to eliminate lead-based paint poisoning by . . . eliminating the causes of such poisoning." The law's call for "intensive community testing programs" also suggested universal screening, but its reliance on "units of general local government" all but assured that the focus would remain on "children living in the 'lead-belts' of our city slums." In a similarly expansive pronouncement, the 1973 amendment to the LBPPPA required HUD to "eliminate as far as practicable the hazards of lead-based paint poisoning with respect to any existing housing which may present such hazard." But again, since this mandate extended only to "HUD-associated" housing, the ideal of universal screening was to take a back seat to the visible crisis of lead poisoning in the inner city.[3]

During the first ten years of federal programs to eliminate lead poisoning, activists continually attributed bureaucratic indolence to the fact that it was poor children, and often those from racial minorities, who were being poisoned—children who, in the words of Philadelphia attorney Jonathan Stein, "lacked the power, political and economic, to obtain the government infusion of funds to establish a prevention program."[4] These critics highlighted the association between plumbism and poverty in order to shame government officials into taking more aggressive action or to spur like-minded activists to direct their action to the inner-city child at risk. Ironically, this criticism reinforced the image of particular susceptibility and may have slowed growth in concern about universal lead exposure.

For the most part, epidemiology recapitulated expectation in the early 1970s. The National Committee for the Prevention of Childhood Lead Poisoning complained in 1975 that HUD had not been able to fulfill its mandate to determine "the nature and extent of lead-paint poisoning problems" because HEW had not developed a comprehensive database. Because all projections and computer simulations relied on inner-city data, the advocacy group warned that "the true extent of the impact of lead poisoning remains a mystery."[5] Although it was true that very few of the early screening programs looked outside the "lead belts," the few that did suggested a far broader problem than projected in government forecasts. A 1972 study of southern Illinois cities outside Chicago found that almost one-quarter of 6,000 children tested had "potentially dangerous" blood-lead levels.[6]

As discussed in Chapter 10, the first years after the passage of the LBPPPA saw a tremendous upsurge in government-funded screening, prevention, hazard-detection, and abatement programs. Funded research into the mechanisms of plumbism, its epidemiology, and treatment flourished as well, producing a steady accretion of biomedical expertise. J. Julian Chisolm, Jr., a pediatrician at Johns Hopkins, was the foremost researcher investigating childhood lead poisoning in the early 1970s, publishing many of the most influential studies and frequently testifying before legislative hearings on behalf of the American Academy of Pediatrics (AAP) and other health organizations.[7] In 1977, Chisolm published a short article with the provocative title "Is Lead Poisoning Still a Problem?" in which he addressed only acute lead encephalopathy. He found that although "current methods of chelation therapy have reduced the mortality from this severe form of the disease," children who survived frequently suffered from "severe behavior abnormalities, . . . blindness, aphasia and hemiparesis." Clearly, Chisolm argued, lead poisoning was still a problem. Chisolm noted that the cause for the decline in cases of fatal and cerebral plumbism was unknown. Perhaps "some reduction in substandard housing" was responsible, or improved nutrition among the at-risk population. He also considered increased awareness on the part of physicians, better testing equipment, and improved screening programs. Significantly, Chisolm did not mention the possible contribution of declining environmental pollution from leaded gasoline. The controversies then raging over lead and air pollution seemingly did not register. Although Chisolm firmly believed that lead poisoning was still a problem, it involved the old trinity of "silent epidemic" causation: "environment (mainly housing), the child, and the parent."[8]

But even within this narrow framework, the picture of childhood lead poisoning was changing. In addition to the more obvious ongoing problems resulting from acute plumbism, Chisolm and others were very interested in the effects of subacute or asymptomatic absorption. By exploring the mechanisms by which lead impaired blood metabolism, researchers expected to improve screening methods. This work found consistent relationships between blood damage and blood-lead levels well below those associated with clinical symptoms, casting doubt on the notion of a clear threshold for damage and raising questions about the effects of asymptomatic lead absorption in other body systems. Even without the confirmation later studies would bring, these studies on blood metabolism prompted the first two downward revisions of the Public Health Service's definitions of "undue lead absorption," from 60 μg/dL to 40 in 1971 and 30 in 1975.[9]

Lowering the threshold of concern rapidly accelerated the dissolution of the particularist ghetto model for childhood lead poisoning. Those whose blood-leads exceeded the old standards fit the conventional image of poor children ingesting old flaking paint in substandard housing. But under the new categories, such victims were merely the worst off of a vast population of lead-poisoned children whose members could be found almost anywhere.

The great irony of this geometric rise in the population defined as at risk is that it occurred at a time when the average American's exposure to lead from all sources was beginning its fall to today's historic low. Changing technologies in paints, food preservation, and automotive fuels reduced the amount of bioavailable lead in the environment. But as the "background" level of lead exposure dropped, researchers were able to identify the mineral's role in health problems at or below previous years' average blood-lead levels. Consequently, as Americans' exposure to lead dropped dramatically, their "exposure" to the troubling issues of low-level lead absorption rose just as dramatically. The new definition of lead poisoning sent the potential price of "deleading" America skyrocketing, but the social costs of doing nothing appeared to more than justify those expenditures.

As discussed in the previous chapter, data on average blood-lead levels in the United States from NHANES II, published in the early 1980s, provided much-needed encouragement for opponents of leaded gasoline. And reducing environmental lead did produce a marked reduction in average blood levels. The data for childhood lead poisoning were less encouraging, however. The study's authors projected that 4 percent of Americans between the ages of 6 months and 5 years—approximately 675,000 children—had blood-lead levels over 30 μg/dL, the CDC's definition of "elevated" at the time. This estimate far exceeded projections from local screening programs.[10]

Taken one way, the NHANES II findings seemed to support the particularist viewpoint. The percentage of poor black children with elevated blood-leads was twenty-six times higher than that for white children of more middle income. Inner-city blacks were more than fifteen times as likely as rural whites to have elevated blood levels.[11] The findings also convey how the conventional view had determined screening efforts: almost one-quarter of inner-city black children had been tested for lead prior to their NHANES II interviews; 12.5 percent of all black children had been tested, compared with fewer than 3 percent of white children.[12]

Further, from an earlier year's perspective, the NHANES II findings would have suggested that lead poisoning would soon be vanquished. Only a handful of subjects had blood-leads higher than 70 μg/dL, and average blood-lead levels for children were "only" 16 μg/dL, barely half the CDC's "elevated" level and only one-fifth the old threshold for lead "poisoning." But from the perspective of those who had spent a decade studying the effects of low-level exposure, the findings conveyed very bad news indeed and suggested an urgent need to screen all children for lead in order to avoid permanent damage from "invisible" plumbism.

The work of Pittsburgh psychiatrist Herbert Needleman was foremost among the studies that called for this transformation. For over twenty-five years, Needleman's influential research has turned up harmful effects from ever-lower lead exposures. Joseph Stokes, who supervised Needleman at Children's Hospital in Philadelphia during the 1950s, observed that he "showed noteworthy interest in new studies and also demonstrated unusual originality in his approach both to clinical problems and to the obtaining of new scientific data." In 1971, Needleman applied his original approach to the study of lead poisoning. Hoping to develop a means of gauging long-term exposure, he and his colleagues measured the lead content in baby teeth collected from 761 Philadelphia schoolchildren and showed a clear association between lead absorption and environmental lead pollution and poor housing.[13] Dentine lead provided the analytical tool for what would become Needleman's most influential, and controversial, lead-poisoning study.

In 1979 Needleman's group published a paper implicating subclinical lead exposure in diminished IQs and school performance among Boston-area children. By and large, the 273 children studied were healthy; all but ten showed no symptoms of lead poisoning and had blood-lead concentrations below 40 μg/dL, the level the CDC then used to define toxicity. The study concluded that IQ and teacher evaluations were inversely related to the amount of lead in the children's teeth. A follow-up study eleven years later retested 132 of the original subjects, finding among those whose childhood lead burden had been the highest "a seven-fold increase in failure to graduate from high school, lower class standing, greater absenteeism, impairment of reading skills . . . and deficits in vocabulary, fine motor skills, reaction time and hand-eye coordination."[14]

Herbert Needleman and J. Julian Chisolm have both been influential in setting regulatory agendas, though in very different ways. Chisolm sought to measure directly the effects of lead on biological processes that were invisible to the clinician—distinct physiological mechanisms taking place inside the body. When he found these responses at lower exposure levels, he warned of possible consequences for health and behavior but did not concentrate on quantifying those consequences. His work focuses instead on improving the prevention, diagnosis, and treatment of childhood plumbism. Although he continues to influence both clinical and regulatory practices, his measured, pragmatic approach has always lent itself to gradual improvements, a steady erosion of old standards—evolutionary, not revolutionary progress.

Needleman, on the other hand, sought to overturn conventional definitions of childhood lead poisoning. In direct contrast with Chisolm's focus on mechanisms within the body, Needleman found (by painstaking and

sophisticated, albeit indirect, methods) statistical associations between behavior, academic accomplishment, and the amount of lead in children's
bodies. His work seeks to assess the impact of mechanisms of lead toxicity
in terms of human experience. Despite his reliance upon complex, indirect
statistical measures—many of which even his advocates would be hard-
pressed to explain, understand, or defend—Needleman's focus on the
human costs of lead absorption has proven extremely effective in promoting regulatory action. Chisolm's work revising the link between blood lead
and erythrocyte protoporphyrin might alarm knowledgeable public health
workers, but Needleman's correlation of elevated but asymptomatic lead
absorption with reduced IQ scores hits a far wider target (see Figure 12-1).[15]

The revised epidemiology of childhood lead poisoning percolated into
the public sphere in the 1980s. Concern among middle-class parents was
not entirely new. In almost every decade this century, a small number of
well-informed parents sought information about child-safe paints. In the
early 1970s, researchers began investigating childhood lead poisoning
among middle-class families whose homes were being remodeled.[16] Where
previously middle-class professionals might have lent moral or electoral
support to fight a killer of ghetto children, the reassessed risks of low-
level exposure cast lead's threatening shadow over their bright, healthy
offspring, making the push for screening and abatement a personal fight.

As lead-removal programs proliferated throughout the 1970s, so too did
concern over the health consequences of the dusty work of abatement. One
body of new research focused on the hazards to the "urban lead miners"—
the workers who scraped, burned, and swept away lead paints, often in
dusty and ill-ventilated conditions reminiscent of the Pullman car factory
Alice Hamilton had observed generations earlier. Abatement workers frequently exhibited sharp increases in blood-lead levels, and since no one
suspected workers of nibbling on paint chips, lead dust must have been
the cause of their plumbism. Other researchers investigated the lead levels
in children living in or near homes undergoing lead abatement.[17]

FIGURE 12-1

Examples of Chisolm's and Needleman's Data Analyses
A, Chisolm scattergram showing correlation between blood lead and erythrocyte
protoporphyrin (ALA, excreted in urine); B, Needleman's distribution curve of
verbal IQ scores among children with high and low lead levels.
A, Julian J. Chisolm, Jr., M. B. Barrett, and E. D. Mellits, *Journal of Pediatrics* 87 (1975): 1152; B,
Herbert L. Needleman, Alan Leviton, and David Bellinger, "Lead Associated Intellectual Deficit,"
New England Journal of Medicine 306 (11 Feb. 1982): 367.

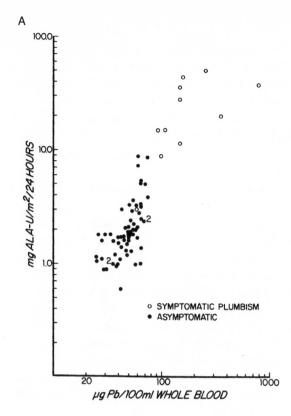

A

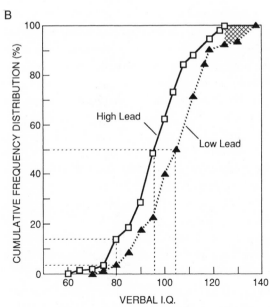

B

These occupational and pediatric studies focusing on lead dust hastened the move away from a pica-centered explanation for pediatric lead poisoning and broadened the search for at-risk children. Here, too, new concerns with low-level lead exposure transformed the agenda. The amount of lead in the dust on children's hands had been worrisome—but not alarming—when authorities labeled a pediatric blood lead of 39 µg/dL "below clinical concern." Once the consequences of Needleman's findings took root, however, the fine particles of dust assumed gargantuan significance, both for the way homes were to be deleaded and for setting goals for screening children for lead exposure.[18]

It also created an entirely different moralization of childhood lead poisoning. For most Americans, assigning blame for ghetto children eating the paint they picked off baseboards and windows had been relatively easy. Villainous landlords, "defective" children, parents mired in a "culture of poverty," or urban blight—all were factors that seemed safely remote. Dust was different. For one thing, exposure to lead dust required no dietary or behavioral pathology. More significantly, for a generation raised on the hygienic visions of June Cleaver and Mr. Clean, dust was more than merely a vexing nuisance, and to many a two-income couple, a dusty house signified failure to balance career and family.[19] Adding to this hygienic failure, the new allegations that dust threatened their children's mental health only multiplied parents' worries and guilt.

Ironically, many families first learned about childhood lead poisoning—an archetypal disease of poverty—as a direct result of their pursuit of that equally archetypal American status symbol: home ownership. Renovating or remodeling an old home introduced countless numbers to the childhood menace of lead dust. With the disclosure in 1990 that Millie, the nation's "First Dog" during the Bush presidency, had nearly succumbed to canine plumbism during White House renovations, no house seemed safe. By the early 1990s, home remodeling was a highly visible locus of concern over low-level lead exposure.[20] In a gentrifying neighborhood in a sun coast Florida city, the parents of a 2-year-old learned that despite their common-sense efforts to minimize the dust and fumes from remodeling their 90-year-old frame bungalow, their son's blood-lead teetered on the brink of the CDC's standard for elevated exposure. Their pediatrician's assurances could not erase the alarmed voices they heard when they called the lead-poisoning clinic at Boston's Children's Hospital. In Brooklyn, CBS news producer Richard Stapleton, concerned after finishing a story on childhood lead poisoning, asked his family pediatrician to test his son's blood. Matthew's results came back at 20 µg/dL, a shock for Stapleton and his wife, Andrea. "We had been renovating our new home while Andrea was pregnant with Matthew, and thus she was nagged by a mother's question, 'Did I do this to my baby?'" Some parents took up the cause of publicizing this newfound danger; thousands took their children in for blood tests.[21] More just went along sanding, scraping, and burning the old paint from their walls and windows.

Growing concerns over low-level lead exposure, the enlarged role attributed to lead dust, and the expansion of the at-risk population to include the children of politically empowered middle-class parents should have nearly assured adoption of universal screening. The optimistic rhetoric of many within the health community in the early 1980s suggested that what seemed a distant goal in the 1970 federal laws might soon become reality. In 1982, Jane Lin-Fu, now moved from the Children's Bureau to the Public Health Service's Office for Maternal and Child Health, recommended that all community child-health programs administer "routine periodic erythrocyte protoporphyrin screening of all children from one to five years of age," in conjunction with other health-screening activities.[22] The CDC's policy in 1985 recognized universal screening as an ideal goal, to be met as part of a proposed screening program for iron deficiency. The policy still reflected the particularist tradition, however: screening was "to identify children with significant exposure" and to concentrate on enrolling "the maximum number of children in high-risk groups." By 1991, the CDC's goals relegated particularism to a parenthetical qualification: "Because almost all U.S. children are at risk for lead poisoning (although some children are at higher risk than others), our goal is that all children should be screened, unless"—and here was the condition that more than any other has prevented universal screening—"it can be shown that the community in which these children live does not have a childhood lead poisoning problem."[23]

Unfortunately, in the 1980s, national policy on lead-poisoning prevention and national health care politics were moving in opposite directions. Universal lead screening would require millions of federal dollars designated exclusively for the purpose. But dismantling this very type of "categorical" program was a linchpin in Ronald Reagan's campaign to return spending authority to the states. Instead, federal money was to be allocated through block grants, which each state could spend as local exigencies demanded. In 1981 Congress placed lead-prevention programs under the new Maternal and Child Health Services block grants.[24]

Block grants, and the funding cuts that accompanied them, threatened the survival of existing state and local lead-screening programs, even in

states with high at-risk populations. In a typical example, St. Louis—
ranked third among the ten cities with the worst childhood lead-poisoning
problems—fired half of its lead-control staff in 1982, and the number of
children screened was four thousand fewer than in 1981, a drop of some
29 percent. Not only did block grants squeeze the budgets of existing programs, but they discouraged other cities from starting new lead programs,
eliminated the incentive for rural states to embark on lead-poisoning prevention, and made universal screening a nearly impossible undertaking.[25]
Any campaign for universal screening appeared quixotic, given the obstacles to large-scale projects in a political climate obsessed with dismantling the "New Deal Order."

The turning away from federally run categorical programs in favor of
locally run block grants benefited the lead industry by reducing the chance
that an activist federal bureau could mandate costly, universal screening
programs. The industry opposed revising the conventional picture of lead
poisoning; in the years since the LBPPPA it had fully accepted the particularist, "silent epidemic" view of childhood lead poisoning. In 1990, Jerome F. Cole, president of the International Lead and Zinc Research Organization, the lead-industry's research arm, declared to a Senate committee
that "there is no argument that old lead-based paint in deteriorated housing is a serious problem which may result in elevated blood levels." The
worst lead-poisoning problem remained in the inner cities, just as lead-
industry–funded studies since the 1950s had established. Chisolm's research had shown visible physiological evidence to support current thresholds for damage. Those researchers pushing for lower thresholds relied on
less tangible, statistical associations. Their findings, Cole argued, "provided no evidence that blood lead levels below today's guidelines (below
25 μg/dL) result in any harm to health." And since most of the children
whose blood-leads exceeded those levels were to be found in deteriorating
housing, the lead industry recommended staying the particularist course:
"Certainly, the most cost effective means of controlling the problem that
does exist is to target resources at the problems that do exist and that is
the prevention of blood lead levels above today's guidelines."[26]

Cole's summary dismissal of studies that argued for lowering permissible blood-lead levels is typical of the lead industry's efforts in recent years
to downplay, denounce, or discredit a research-government coalition it saw
as overrun with anti-industry activists and perhaps thereby regain some of
its earlier prominence in lead-poisoning research. Herbert Needleman has
frequently been the target of lead's efforts. In the late 1980s and into the

early 1990s, Needleman repeatedly squared off in acrimonious and potentially defamatory contests against lead-poisoning researchers associated with the lead industries. In a highly publicized proceeding concluded in 1992, he successfully fended off charges of professional misconduct brought by his employer, the University of Pittsburgh.[27]

Needleman was an attractive target for the lead industry, perhaps in part because the "unusual originality" and creative approach to "obtaining of new scientific data" noted by his old supervisor made his work more difficult to evaluate than conventional clinical studies. But many studies by other researchers, conducted with a wide array of methods, had already replicated Needleman's findings by the time of these investigations. If anything, his original findings seem moderate by comparison.[28] The charges against him, Needleman insists, stemmed not from flaws in his methodology but from the implications of his findings and his outspoken advocacy. Needleman had wielded his science unapologetically in public. In dozens of articles in medical journals and popular magazines, in legislative hearings and courtrooms, his activism had worked to undermine the political influence and economic interests of the lead industry.[29]

Highly publicized debates over particular lead researchers' methods or findings will no doubt persist. In the 1990s, the lead industry's enemies were multifarious, comprising factions from within every federal department, affiliates of major research institutions armed with federal research funds, powerful law firms pursuing massive class-action suits, and environmental and consumer action groups—a far more diverse, organized, and determined opposition than at any time in history. Toxicologist Ellen Silbergeld reflects this new activism that bridges the academy, government, and private activist movements. Throughout a career as a toxicologist that has taken her from academic fellowships at Johns Hopkins to positions at the NIH and the EPA, she has combined environmental activism with her professional life. She and her husband, Consumers Union attorney Mark Silbergeld, frequently appeared before the same congressional committees on childhood lead poisoning. In 1990, she helped found the Alliance to End Childhood Lead Poisoning, the country's largest antilead advocacy association.[30]

The late 1980s and early 1990s were heady times for this opposition. Their spokespersons had the ear of the federal government; they had basic science studies that revealed the harm lead caused below previous levels of concern and clinical and epidemiological studies that indicated the human costs of these low-level effects. And with NHANES II data showing nearly universal exposure to these low levels, they could argue that the nation had a clear mandate for universal screening and aggressive abatement programs.[31]

The reconstitution of childhood lead exposure as a pandemic thief of intelligence demanded their persistent efforts, no matter the political climate. And the ubiquity of the environmental hazard demanded more than one line of attack. Lead's opponents had noted the largely unintended health benefits resulting from the leaded gasoline phase-down and therefore were able to employ successfully the new definitions of lead hazards to further reduce the bioavailable lead in the environment. In 1986, the EPA lowered the permissible lead contamination of drinking water by a factor of ten, prompting passage of federal laws that sharply curtailed lead in solder, plumbing fixtures, and drinking fountains. The amount of lead in foods canned in the United States plummeted in the early 1990s, after the FDA secured a voluntary industry-wide ban on lead solder for cans.[32]

These successes involved reducing exposures to new sources of lead in the environment. The old lead—on walls, in old plumbing, and in the soil around houses and highways—remained intractable. The 1970 lead-paint legislation had proposed a two-front approach: screening and treatment of children, and testing and abatement of lead in housing. And while reductions in environmental lead had achieved many of the earlier health goals, they did nothing to delead the nation's housing stock. If anything, the lower lead levels in the general public took some pressure off abatement programs; indeed, the new fears of low-level lead exposure complicated efforts to mandate safe standards for abatement, and drove the per-unit costs above what landlords and government cost-benefit analysts deemed reasonable.

The Residential Lead-Based Paint Hazard Reduction Act, passed in 1992, finally took a universalist approach to the housing issue, but addressed the problem in a suitably post-Reagan, hands-off fashion. In addition to authorizing funds for enlarged HUD-sponsored abatement efforts in low-cost housing, the act required the seller of any home financed with federal assistance to notify the buyer of known lead hazards, and while it provided incentives to abate those hazards, it did not require abatement or provide for any government oversight.[33]

The principle of universal environmental lead exposure propelled both the Residential Lead-based Paint Hazard Reduction Act of 1992 and the regulatory actions against lead-contaminated food and water. But the proponents of universal blood-lead screening met with no comparable success. The percentage of young children screened each year hovered between 10 and 20 percent; a survey of pediatricians in Virginia found that fewer than one in eight screened all of their patients for lead, and one in four did not screen patients at all. Only 27 percent of San Francisco Bay-area pediatricians surveyed in 1993 screened all of their patients under six. Those in private practice or in health maintenance organizations (HMOs) were least likely to screen (19% and 0%, respectively).[34]

In the early 1990s, both the CDC and the American Academy of Pediatrics endorsed universal screening.[35] Ironically, reductions in new sources of environmental lead made the push for universal screening more difficult: with average blood-leads falling, the rhetorical power of lead as a universal menace began to wane. Opponents began publishing their doubts about low-level toxicity or questioned the cost-effectiveness of testing all children.

Much of the criticism of universal screening came from health professionals from western states. Oakland, California, pediatrician Edgar Schoen of the Kaiser Permanente Medical Care Program, then the nation's largest HMO, assailed the CDC's strategic plan for universal screening, treatment, and abatement. The plan was too costly, even given what Schoen asserted was the CDC's "gross underestimate" of the projected costs. And it was probably unnecessary, since the evidence of neurological effects of low-level lead exposure cited by the CDC for its 1991 redefinition of lead poisoning suffered from "problems of small effect, confounding variables, and questionable outcome measures." Alluding to the investigation into Needleman's work then underway in Pittsburgh, Schoen asserted that research into low-level lead effects was plagued by "faulty methodology, flawed statistics, and overstated conclusions." Although Needleman had had "great influence on other investigators in the field," Schoen did not mention that those investigators, by and large, had corroborated Needleman's initial findings.[36]

Others criticized universal screening as a waste of money in particular geographic regions, such as much of the west, where housing stock was relatively new. George Gellert, of the Health Care Agency of Orange County, California, reported that of 5,115 low-income children tested there, only 371 (7.25%) had blood-lead levels above 10 $\mu g/dL$, and only six (0.12%) had blood-leads higher than 25 $\mu g/dL$ (and for five of these children the chief exposure was from cooking utensils or lead-containing folk remedies, not paint). Unlike Schoen, Gellert did not question the CDC's findings regarding low-level effects, only the value of universal screening in areas such as his, where less than 8 percent of homes were built prior to 1950. Gellert estimated the costs of locating the six children with the highest blood-lead levels at $19,139 per child, money that Gellert argued would be better spent on "more compelling child health problems, such as immunization, nutrition, or injury." Other methods to locate at-risk children, such as interviewing parents about occupational exposures, folk remedies, and food preparation might be just as effective and more economical. "Knowledgeable community professionals," Gellert argued, "should be able to set local priorities based on specific local data and regional conditions in a manner that will best protect the public health and serve their communities."[37]

Schoen's and Gellert's views on screening probably resembled those of most American pediatricians, who, even if they screened for lead, found few children with elevated blood levels and worried that the low risk did not warrant the anxiety, pain, and expense of the venipuncture required for the most reliable lead test. In early 1994, the AAP published a forum on recent risk-assessment studies of lead poisoning in its journal, *Pediatrics*. In his concluding commentary, Stanford University pediatrics chief Birt Harvey reviewed the case against universal screening. He declared that it was "a time for action," but by this he meant that since most of the states were preparing—but had not yet begun—to implement the 1991 CDC screening guidelines, it was time to revise those guidelines before precious funds were wasted. "With government and industry demanding limitations on the cost of health care," he warned, "expenditures for new or expanded services must be justified, and necessary services must be provided in the most cost-effective manner."[38]

Harvey noted the uncertainty regarding the effects of blood-leads below 20 µg/dL and the high costs per case of screening in regions where average blood-lead burdens were low. He agreed with the goal of identifying all children with elevated levels, but measuring every child's blood, he argued, was not only a costly screening method, but was unnecessarily invasive, since the most accurate test required venipuncture. Simply asking parents the right questions was as effective as the frequently inaccurate finger-prick test for identifying those children whose blood should be tested further. And why spend millions of dollars to label as "poisoned" thousands of children with blood-leads between 10 and 25 µg/dL, when the medical and environmental interventions for such low levels were of such questionable value? Chelation therapy is generally not indicated with this degree of absorption, and the effectiveness of aggressive environmental controls—frequent wet-mopping of floors, dusting of surfaces, and a fastidious regime of hand washing—is doubtful.

Critics of universal screening cited studies showing dramatic differences in average blood levels—region to region, or between cities and suburbs within the same region. Those who favored universal screening argued that the NHANES II study of the mid-1980s gave a truer picture of the prevalence of elevated blood-leads. But in 1994, the National Center for Health Statistics and the CDC released the blood-lead findings from the third National Health and Nutritional Examination Survey (NHANES III). The new data revealed that the low prevalence of elevated blood-leads found in western cities now more closely resembled national norms, that there had been a change since the epidemiological studies undertaken in the eastern lead belt. A decade of aggressive restrictions on new sources of bioavailable lead products had brought a 78 percent drop in average blood-lead levels—from 12.8 to 2.8 μg/dL—since the midpoint of the NHANES II study. The change in childhood exposure was even more remarkable: where NHANES II had estimated that almost 90 percent of American children had blood-leads above 10 μg/dL, by the late 1980s and early 1990s the rate had fallen to under 9 percent (see Figure 12-2).[39]

Voices quickly rose to declare victory and cast off the heavy veil of "plumbophobia." In a 1995 issue of *Atlantic Monthly,* science journalist Ellen Ruppel Shell praised America's "triumph over lead," calling it "a stunning example of the strength of activism over vested interests." She questioned the obstinacy of universal screening proponents, characterizing them as zealots who had "picked up on the chant" that low-level lead poisoning was hazardous from Needleman and lobbying groups such as the Alliance to End Childhood Lead Poisoning.[40]

The heartening news in the NHANES III reports all but guaranteed that the CDC's 1991 call for universal screening would fail. In 1995, the CDC began reviewing its recommendations for screening and abatement, bowing to pressure from local health agencies and insurance companies, economic realities, and a shifting consensus within the agency regarding whether universal screening would—or should—be implemented. Late in 1997, the CDC published new guidelines that called for "targeted" screening and case finding focused on children in high-risk areas or groups, or on those identified as at risk "by means of a personal risk questionnaire." Local and state authorities, not CDC officials, were to set priorities and criteria for blood screening. "In some places the plan will call for screening all children in a jurisdiction, while in others the plan will call for screening children in selected areas and from selected populations."[41]

The CDC reassured the public that the guidelines were in no way a retreat from screening children; in fact, screening would increase if localities followed the new recommendations. And because the targeted approach

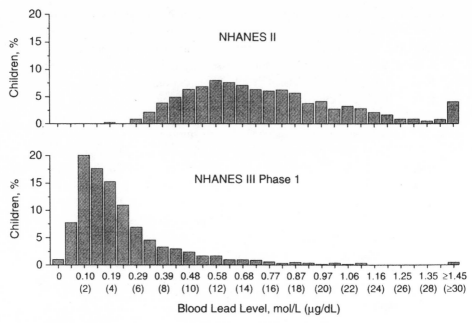

FIGURE 12-2
Blood-Lead Levels for U.S. Children Aged 1 to 5 Years as Shown in the National Health and Nutrition Examination Surveys
Top, NHANES II, 1976–80; bottom, Phase 1 of NHANES III (1988–91).
James Pirkle et al., "The Decline in Blood Lead Levels in the United States: The National Health and Nutrition Examination Surveys (NHANES)," *Journal of the American Medical Association* 272 (27 July 1994): 285.

would turn up far more of those children with truly worrisome blood-lead levels, the obviously improved cost-effectiveness would overcome "ongoing friction" among care providers that had "the potential to detract from or to jeopardize necessary screening."[42]

The new CDC guidelines are sensible, sound, and well reasoned, based as they are on credible epidemiological data, humane principles, and sophisticated cost-benefit analyses. Unfortunately, their particularist view of the problem, abetted by the CDC's determination to turn authority over to local agencies (also a sensible choice, given the current political climate), threatens to turn the clock back to the early 1980s, when only well-informed parents and poor folk living in enlightened towns had their children tested as a preventive measure. Even if perfectly implemented, the guidelines will leave many children to fall through the cracks. There will always be individual anomalous cases: the child who moves to a "lead-free" town from a region with older housing or other sources of high exposure; the child whose family uses lead-containing folk remedies or cooks with improperly fired lead-glazed ceramics. Presumably, these are acceptable losses, their risks balanced against the far greater cost-effectiveness of targeted screening.

Certain targeted screening programs will also miss whole groups of children. Some local health agencies will use average prevalence rates within areas determined by zip code or census tracts to determine what—if any—screening activities to implement. This may be effective where the condition and age of housing is fairly homogeneous, but in the many towns where tight pockets or widely scattered sprinklings of older housing exist in "safe" areas, many young inhabitants will be erroneously declared risk free by the power of the arithmetic mean. In towns where race or ethnicity are used as risk factors, there is the chance of reverse discrimination in screening.

Far greater is the risk that the new guidelines' abandonment of universalism and de facto acceptance of race specificity will perpetuate the status quo—that is, that the number of screening programs nationwide will stay roughly the same. In all probability, childhood lead poisoning will remain a defining environmental justice issue for the foreseeable future. In 1994, EPA staffers Lynn Goldman and Joseph Carra wrote that the picture of lead poisoning found in NHANES III data required that "the US strategy must begin to focus more than ever on poor, nonwhite, and inner-city children." Ellen Shell agreed: "Only by treating lead poisoning for what it is—largely a disease of the poor—do we stand a chance of beating it."[43]

Treating lead poisoning as a disease of the poor makes good epidemiological sense, but historical precedent suggests it is a dangerous choice. Substantial progress in reducing lead poisoning has only been made when lead has been considered a common risk. Seventy years ago, when conventional wisdom held that lead poisoning was confined to the workplace, "universal" blood testing within lead factories prompted simple but effective environmental improvements, improvements few had been willing to make when only the "wops and hunkies" in a factory were thought to be susceptible to poisoning.[44] The concept of lead as a "ubiquitous threat," as posited by Clair Patterson in the 1960s and borne out by Herbert Needleman and others since the 1970s, prompted the dramatic reductions in lead-contaminated air, food, and water that brought average blood-lead levels in the United States to their historic low. While it is true that poor children have shared in the widespread reductions in environmental lead over the last twenty years, they remain at greatest risk of exposure to the lead that remains. And with no comparable improvement in universal access to health care, a return to particularism amplifies the dangers of that exposure.

A recent article in a Florida newspaper reported the stories of two children poisoned by lead. Their experiences highlight the most serious dangers inherent in the CDC's targeted approach to lead screening.[45] One, a 2-year-old boy, lives in an old house in a historic section of St. Petersburg. His lawyer father and stay-at-home mother noticed that he sometimes played with chips of paint on the windowsills: "He didn't eat them. He just played with them, sometimes throwing them out the window." Just to be safe, the boy was tested, and his blood-lead was revealed to be over 30 μg/dL. His parents then eliminated all of the lead paint in the home. The child's blood-lead slowly fell, but his parents worry about the long-term consequences. Under current guidelines, it is unlikely that this child or any other children in his neighborhood would have been routinely tested.[46]

No more than a few miles away, T. J. Ellis lived with his mother, Clifford Anderson, in a small apartment in a sprawling public housing development in one of the city's oldest and poorest neighborhoods—just the sort of place one would expect screening programs to target under any guidelines formulated since the 1970s. According to his mother, T. J. frequently ate the paint that fell off the walls. He "would put the chips of paint in his mouth and it would melt just like candy and there would be gray stuff all over his mouth." At first a happy and healthy boy, when he was 15 months old his behavior changed and his language development seemed to slow. "He was a completely different person," Anderson reported. The first time she took T. J. to be tested, his blood-lead was 41 μg/dL. When a second test two weeks later showed his level had jumped to 65 micrograms, he was admitted to the hospital for chelation therapy. Clifford Anderson found a lawyer and sued St. Petersburg's Housing Authority, which settled for $200,000. At the age of 6, T. J. had "significant language, learning and behavioral problems" that experts hired by Anderson's attorneys attribute to the lead he ate.[47]

Like most cities its size, St. Petersburg had no screening programs in 1985, when the CDC called for local boards of health "to identify children with significant exposure," or in 1991, when the CDC declared "that all children should be screened." St. Petersburg introduced a screening program of sorts in 1998. With a grant from the CDC, the county health department provided a free capillary blood-lead test for any child brought to a screening site. In its first year about twenty parents each week were taking up the offer. Nothing in the new guidelines promises to increase participation, let alone make screening a mandatory part of child care, as commonplace as other childhood screening tests and inoculations.

Universal lead screening is controversial now only because of what some call the nation's "triumph over lead." Pediatrician John C. Ruddock would have celebrated the end of the child's "lead world" as he knew it. Still, reducing the massive lead exposures typical in the past has brought a new understanding of the hazards of what remains, accompanied by fear, ever more costly solutions, and unprecedented animus expressed by the competing factions. Universal lead screening is only one forum for the debate: other battles continue to rage over deleading homes and making lead factories safe for workers and their children. Although there is far less lead around us and in us, it weighs more heavily upon us.

THIRTEEN

Regulating
"Low-Level"
Lead Poisoning

Over the course of the twentieth century, the fight to end lead poisoning took place on several fields. In the age of Alice Hamilton and Robert Kehoe, plumbism was considered almost exclusively a problem of occupational hygiene. In the age of Jane Lin-Fu and J. Julian Chisolm, the locus of concern shifted from the factory to the slum, and lead poisoning was primarily considered a problem for children in lead-saturated environments. In the age of Ellen Silbergeld and Herbert Needleman, the shadow of pediatric lead poisoning has spread from the ghetto tenement to the suburban nursery and the classroom. At the same time, the reach of occupational lead poisoning has spread from beyond the assembly line to the birthing room.

> *The continuing overexposure of American workers to lead and the persistent occurrence of occupational lead poisoning is a national scandal. It is not necessary. It is entirely preventable. The question is not one of technology or of feasibility, but rather of national will to act upon the abundantly available medical data.*
>
> PHILIP J. LANDRIGAN, "LEAD IN THE MODERN WORKPLACE"

This social reconstruction perfectly reflects plumbism's participation in the "epidemiologic transition" that shaped public health in the late twentieth century. The earlier emphasis on acute and epidemic diseases, usually biotic in origin, is giving way to concerns with chronic, endemic health problems. Biotic agents still play a role, but the new epidemiology presumes multiple causes, many of which are associated with "lifestyle" and man-made environmental toxins. Diagnosing these diseases no longer follows the algebra of Koch's postulates wherein a particular, identifiable, and isolable causative agent produces a unique illness with predictable and reproducible symptoms. Treatment—even that term is giving way to "management"—requires not a single prescription but a lifetime commitment to a complex of therapies.[1]

Lead's own epidemiologic transition since the late 1960s is a bold example of this larger transformation. Gone are the days when hundreds of children died every summer from lead poisoning.[2] With frustrating regularity, children still show up at clinics and emergency rooms with the abdominal symptoms of acute lead poisoning, but many more "asymptomatic" children receive chelation therapy to prevent the unseen and unpredictable effects of chronic exposure. Millions of children have learned to live with the remaining dangers: their parents warn them of lead dust instead of germs when nagging them to wash their hands or to stay out of the dirt; they hear lead paint get the blame for their families' difficulty in finding a new apartment; a lead-poisoning test is added to the pricks and prods of their preschool checkups.

Through this transition, one link between lead poisoning and the broader world of diseases has held fast: it continues to be an underrecognized cause of a range of other illnesses, although the nature of these disorders has changed with the times. When acute and fatal forms abounded, plumbism was more likely to be recognized as a biotic assault than an environmental insult. Thousands of lead-poisoned children went to their graves diagnosed with tuberculosis or influenza. In even more cases, chronic lead poisoning hastened their deaths when TB or the flu struck their weakened bodies.

In those days, the public health response to lead poisoning was analogous to that for biotic diseases: a search for *the* pathogen, the *one* cause; quarantine in the absence of a cure; and one-shot treatments when the cure was found. For pediatric lead poisoning, eating paint was identified as the one cause. Literal quarantine would have sought to remove healthy children from the danger, an impossible task if the ubiquity of the threat was taken seriously. Instead, the quarantine was psychological only: since this disease's victims were identified as poor inner-city children, the sick were already isolated from the healthy, the disease already contained in the ghetto. Chelation was the presumed cure for lead poisoning, but as effective as it was, it was no vaccine. Environmental cleansing, not medical treatments, ended the "silent epidemic" of acute childhood plumbism.

Now, after passing through the epidemiologic transition, lead is most often associated with classic "lifestyle" diseases such as certain kidney and heart ailments. And as public health has grown to encompass mental health, education, and behavioral issues, lead has been implicated in each.

The solid evidence of lead's effects on neurological development has embroiled lead activists in the adjudication of problems in education, mental health, and even the criminal-justice system.[3]

As the disease known as lead poisoning changed in the final years considered in this study, so too did the prescriptions for its elimination. The regulatory regime that developed to cope with the new lead epidemiology bears little resemblance to its predecessor. This concluding chapter presents three situations where this contrast may be seen in high relief: pediatric plumbism and the abatement of lead paint; setting and upholding safety standards for lead-industry workers; and—bridging the pediatric and occupational—the issues surrounding female lead workers and industry-enforced "fetal protection policies." In each case, the distinction between low-level and high-level risks has become blurred, producing regulatory impasses that allow lead poisoning—in both its old and its new forms—to persist. An enormous infrastructure has developed to eliminate lead from the workplace and the environment, but now that workers' and children's lives no longer appear to hang in the balance, this regulatory and research apparatus gets bogged down in questions of multiple perspectives and competing rights. Hence, today's lead-poisoning issues tend to end up in the courtroom.

In an empty tenement apartment, a crew of workers, covered head to toe with protective clothing and wearing respirators, removes and replaces old wooden window frames and molding, scrapes and strips old paint, and covers the walls and remaining woodwork with special "encapsulating" paints. The windows and doors are sealed with plastic to protect children in adjoining apartments from the dust being stirred up. All the debris, including that accumulated by a high-efficiency particle air-filtered vacuum, is sealed into drums and loaded for disposal in a hazardous waste site. On the way the truck passes a small crowd of neighborhood children watching as a wrecking ball reduces a similar tenement to dust and rubble, which is loaded by bulldozers into dump trucks and hauled without ceremony to the city dump.[4] It would seem that standards for abating lead paint are applied somewhat inconsistently.

Over the 1970s, blame for pediatric lead poisoning shifted, from poorly supervised children with "perverted appetites" who ate deteriorating lead paint and plaster to the lead in every urban child's environment. Lead could be found in car exhausts, playground sand, tap water, and—more troublesome—normal house dust. Alertness to multiple sources of exposure and greater awareness of low-level hazards made development of safer, more thorough abatement techniques imperative. Leaving intact lead paint on the wall perpetuated the low-level threat, but doing so seemed safer— and certainly less costly—than removing it.

Advocates of lead abatement argued that "so long as there is lead in the environment and children continue to put their hands in their mouths, the hazard of lead remains."[5] In 1990, Herbert Needleman offered a "radical proposal" to reduce that hazard: "If one were to construct three maps, one indicating where lead is present in overabundance, one where there is a scarcity of decent housing, and one where jobs are scarce, the three maps would be identical. In this unfair and dangerous maldistribution of hazards, one can also locate the resources to dedicate to an effective program of primary prevention of lead intoxication." Needleman proposed training inner-city residents in lead abatement and housing renewal. And unlike programs some cities undertook in the 1970s, where summer students were employed to do slipshod cover-ups, these workers would be trained, supervised, and paid "a decent wage."[6]

It is ironic that more than any other lead researcher, Needleman established the new epidemiology that now complicates any abatement program such as the one he proposed. Reducing lead exposures to the lowest measurable levels is a laudable public health goal. But programs initiated to move even an inch toward this objective can stall because the abatement efforts themselves might violate the underlying principle of protecting against the lowest levels of exposure. No issue demonstrates this paradox better than attempts at removal of lead paint.

Most of the concern over abatement centers on reducing collateral damage—to neighbors, to the children who return to the renovated home, and to the abatement workers themselves. Safe abatement techniques involve costly equipment, special training, careful supervision, and considerably more time than conventional remodeling.[7] Consequently, estimates of the cost of safely removing the lead from America's walls run into the tens of billions of dollars. A new industry is blossoming to provide training, equipment, and abatement services, most bearing little resemblance to the community-based programs Needleman's progressive solution envisioned.[8]

The direct financial burden for lead abatement has fallen hardest on landlords, who have often fought city, state, and federal agencies to avoid, postpone, or reduce the cost of compliance. Increasingly, however, regulatory pressures to test for lead and abate the hazards have fallen on homeowners. In 1992, Congress passed the Residential Lead-Based Paint Hazard Reduction Act, which requires testing and full disclosure of all lead hazards when any government-financed residence is sold.[9] The pressure is mounting for a final accounting and a workable assignment of responsibility for the tremendous costs involved in deleading. This task has spawned another growth industry—lead-poisoning litigation. As lawyers and judges continue to debate how the financial burdens are to be assigned, the workers who scrape, vacuum, and cart the old paint away will bear the physical costs of abatement, and if recent history of federal efforts to regulate risks to lead workers is any guide, little protection will be forthcoming.[10]

What the Lead-Based Paint Poisoning Prevention Act of 1970 did for children, the Occupational Safety and Health Act of the same year did for lead workers: it strengthened government's hand in forcing industry to protect them, only to see implementation slow to a crawl in rule-making procedures and interdepartmental squabbling, often in controversies over the definition and reduction of low-level risks.[11]

The act created two administrative bodies: OSHA, the Occupational Safety and Health Administration, within the Department of Labor; and NIOSH, the National Institute of Occupational Safety and Health, within the Department of Health, Education, and Welfare (now the Department of Health and Human Services). The two bodies were meant to be complementary: NIOSH would conduct research and develop safety standards, which OSHA would mandate and enforce. But in practice, according to critics, NIOSH is too remote from the realities of the workplace to develop practicable solutions, while OSHA is overly beholden to industry for data and woefully underfunded to perform its mandate.[12]

In order to get the lead industry to improve safety, OSHA needed rules and standards to enforce. Consequently, NIOSH conducted an intensive review of the medical literature and in 1973 published its findings and recommendations in the "NIOSH Criteria Document." The agency agreed with the traditional view held by lead-industry officials—that is, there is a usable "Threshold Limit Value" (TLV) for blood-lead; it even agreed with the lead industry that 80 μg/dL was a safe TLV.[13] Despite agreement on a threshold, NIOSH's recommendations for enforcement by "environmental" monitoring—measuring the lead content in the factory air—aroused the immediate ire of the lead industries.

For decades, lead companies had relied on routine blood, fecal, or urine tests to detect potential cases of overexposure. So-called biological monitoring was cheap and, as a Lead Industries Association spokesman defended, "moves you closer to what you are really trying to protect and assess." Lead industries cited several problems with environmental monitoring, problems they argued added up to greater risk to the average worker. Air monitors were more sensitive to large airborne lead particles than to smaller ones, while the reverse is true for the human body. Moreover, air monitoring was not to be constant; as a result, short bursts of high exposure would likely be undetected. But most important according to the industry, air monitoring did not account for variability in individual workers' susceptibility, their exposure to lead outside the workplace, or (it never ends) their lack of adherence to industry-established standards of work hygiene that would put them at risk in an otherwise safe environment. The human body, as industry saw it, was a more effective monitor than machines. "Biological monitoring," wrote a lead-industry spokesman, "employs the worker as the environmental sampler, rather than depending on mechanical sampling devices."[14]

Labor's viewpoint was different. Speaking at an LIA-sponsored conference on occupational lead exposure, Sheldon Samuels of the Lead Council (an interunion health advocacy group) blasted "the myth of the TLV; the attempt to hide the economic base of 'acceptable' risk." Industry knew that lead workers were constantly at greater risk than was considered normal for the general population. He produced a letter from Jerome Cole, director of environmental health for the LIA, stating that "it is for all practical purposes impossible to keep blood lead concentrations in the lead industry within the normal range" (i.e., under 40 μg/dL). Eighty micrograms was the threshold, Samuels said, only because that was the lowest level industry was *willing* to meet, though it might be *able* to do better. Laborers did not expect the workplace to be "a womb"—they were willing to accept "reasonable" risks—but what constituted "reasonable" should undergo constant reevaluation.[15]

In 1979, OSHA established its Standard for Occupational Exposure to Lead, which, as might be expected, struck a compromise. The standard required lead workplaces to monitor air quality; plants with lead levels exceeding the established limit of 30 μg per cubic meter of air were required to establish a hygiene program and start biological monitoring by means of blood tests. In a significant break from tradition reflecting a reevaluation of what constituted reasonable risk, the OSHA standard lowered the permissible blood-lead level from its long-standing 80 μg mark to 50 and required "leaded" workers to be removed—with full pay—from lead exposure until their blood-lead fell below 40 μg/dL.[16]

The OSHA standard drew heavy fire from lead-using industries, from OSHA administrators who felt the standards were unenforceable, and—after 1980—from the White House. The lead standard flew in the face of President Reagan's drive to deregulate industry. Under Reagan, OSHA delayed enforcement and tried to get the Supreme Court's permission to abandon their own standard's requirements. Steel, automobile, paint, and other manufacturers sued OSHA, claiming that to enforce such high standards would bring undue economic hardship. The case went to the Supreme Court, which rendered its decision on 29 June 1981. Despite the pleas of the Reagan administration and OSHA, the court upheld the rules: lead industries and OSHA had a duty to protect workers from proven dangers, regardless of cost-effectiveness. Six months later, however, OSHA was still dodging responsibility, publishing a four-month study that showed it had neither the technology nor the funds to enforce its own rules. Later accounts from OSHA inspectors suggest that the problems continued.[17]

Many lead workers still face the merry-go-round of work, lead absorption, treatment, and reinstatement in the same contaminated workplace that got them sick in the first place. The situation is not that different from the revolving-door policy prevalent in lead workplaces at the turn of the century. Ninety years ago, a steady stream of fresh workers took the place of sickened employees—to be replaced themselves on the next go-round. Today, the same old faces make their way back through the unending cycle. Now, more than in Hamilton's day, many of those faces are women's.

In April 1984, the United Automobile Workers sued Johnson Controls, a large manufacturer of automotive batteries. Many of Johnson Controls' employees worked with toxic lead compounds in an environment where undue absorption of lead was not uncommon. The defendant was not under fire for endangering its workers, however. On the contrary—the lawsuit sought reinstatement for several employees into high-risk jobs denied them by their employer's "fetal protection policy." In effect since 1982, the policy barred fertile women from higher risk (and higher paying) jobs in the plant. The case went to the Supreme Court, which on 20 March 1991 ruled unanimously that Title VII of the Civil Rights Act of 1964 prohibited such policies.[18]

The Court's decision in the Johnson Controls case appeared to end a protracted battle that had been building since women first began testing the limits of the 1964 Civil Rights Act by demanding jobs denied them since the end of World War II. The ruling brought immediate cheers from women's advocates and civil libertarians. "This is the end of fetal protection policies," said Joan Bertin of the American Civil Liberties Union.[19] Despite such congratulatory rhetoric, however, the ruling is unlikely to settle the issues of reproductive hazards in the lead-using industries, let alone gender-based job discrimination. Seen in the context of shifts in lead epidemiology, the Supreme Court's decision looks anything but final.

The Civil Rights Act of 1964 disposed of the various Progressive Era "protective" laws banning women from dangerous jobs, and by the 1970s a growing number of women were taking their place on assembly lines in lead-using businesses. Initially, most of these industries did not raise insurmountable obstacles to female employees, and no doubt some welcomed the potential for a second, lower-paid workforce. Then came the rising concerns over pediatric lead poisoning, from the "silent epidemic" to the lurking dangers of low-level lead exposure. Whether or not they believed the data, industries did believe in the increased liability posed by lead workers who had the potential of becoming pregnant.

Some manufacturers sought to protect themselves from this liability by adopting a policy of education, advising new female employees of the risks to reproduction attributed to lead. Johnson Controls' battery division took this approach in 1977, when the company hired its first female lead workers. New female employees at Johnson Controls were required to sign a statement attesting that they had been informed about lead hazards.[20] After eight employees at the Bennington, Vermont, plant became pregnant during the first four years of the company's "advise and consent" approach, a more aggressive "fetal protection" policy was adopted, similar to those in place at American Cyanamid, the Olin Corporation, General Motors, DuPont, and other large users of lead.[21]

Under Johnson Controls' rules, several women were forced out of the highest paying jobs, and at least two underwent surgical sterilization. In April 1984, the UAW filed suit on behalf of several Johnson Controls employees in the Wisconsin District Court (the company's head office is in Wisconsin). The suit charged that the Johnson Controls policy violated Title VII of the Civil Rights Act of 1964. The District Court ruled in the company's favor, citing medical reports that "the fetus is more vulnerable to levels of lead that would not affect adults." The fetus must be afforded greater protection than the worker, and since no one was able to suggest affordable methods for reducing workers' exposure to a level safe for fetuses, the company's policy was justified by "the business necessity defense."[22]

Fetal risks played little part in the Supreme Court's considerations when it heard arguments in *Automobile Workers v Johnson Controls*. The court's unanimous decision to forbid fetal-protection policies focused on whether the ability to bear children could be used by a company as a legitimate reason for disqualifying a job candidate. In rendering the opinion, Justice Harry Blackmun wrote that restrictions can legally apply only to the ability of prospective employees to do the job they are hired to do. Fertile women are just as capable as infertile women—or fertile men, which brings up Blackmun's second objection.

Contrary to the lower court findings, the Supreme Court found that the policy "explicitly discriminates against women on the basis of their sex." The lower court had ruled that because the intent of the policy—protecting fetuses—was benign, its discriminatory effect was excusable. In countering this argument, Blackmun stressed the voluminous research indicating the risk to fetuses posed by the father's exposure to lead. If the company intended to protect fetuses, then it must also bar fertile men from working in lead-contaminated areas. Since the policy addressed only childbearing capabilities, it was "explicit facial discrimination," and illegal.[23]

In this decision we see again the paradoxical consequences of regulating low-level lead exposures. A large portion of the lead industry made itself over in the first half of the twentieth century, urged on by Progressive Era activists and sustained by professional hygienists, who assured them that dramatic improvements could be bought at reasonable costs. Conditions commonly found in the lead factories Alice Hamilton inspected brought her to agree reluctantly with "prohibiting the employment of women entirely in those occupations in which lead poisoning constitutes a considerable hazard." By 1964, however, conditions in most large lead plants surpassed the standards that Hamilton would have thought justified barring women.[24]

Reduced acute occupational lead poisoning permitted the lead industry to enlarge its female workforce. In the 1970s, new definitions of lead poisoning forced the industry to rethink its new impartiality and to enact fetal-protection policies. The first favorable judicial decisions upholding them acknowledged the importance of low-level lead exposure, but distinguished degrees of susceptibility. The Supreme Court's decision in the Johnson Controls case fully accepted the new construction of lead risks and took it to its logical extreme, all but equating one extremely low-level risk with another far greater one. Blackmun's argument held that potential genetic damage from the father's work was as great a danger to the yet-to-be-conceived fetus as was the lead passing from the placenta to the developing fetus. But the fetus carried by a woman working in a lead-contaminated environment faced a two-fold risk: the one-time risk of genetic damage at the point of conception (presumably the same as the risk that a male lead worker imposes), and nine months of steady exposure to lead in utero.

After the Johnson Controls decision was announced, Joan Leard, a union official working at the Bennington plant, told reporters, "We're no longer fertile, so it's not going to do us any good. But for the younger girls, hopefully the door will be open now."[25] The doors of hazardous job opportunity for women are open, but so is a large window of vulnerability for unborn children. Manufacturers are only likely to improve processes and ventilation sufficiently to make lead work safe for fertile workers if the threat of lawsuits forces them to. But Justice Blackmun acknowledged that if a manufacturer notifies fertile workers of potential dangers and meets OSHA lead standards, "the basis for holding an employer liable seems remote at best."[26]

It is doubtful that many women will try their luck against the odds as Blackmun states them. The ruling provides no incentive for recalcitrant lead manufacturers to make any substantial improvements in hygiene. Thousands of Americans will continue to work in "a daily atmosphere of lead," absorbing just enough of the toxin to bring on subtle damage.[27] Their children will not suffer from lead palsy or die from lead encephalopathy. No one will notice a ten-point loss in their IQs; no one will attempt to prove that their failure to thrive in school is related to a parent's job.

Two campaigns during the twentieth century made enormous strides against lead poisoning in the United States. Manufacturers, activists, and hygienists in the decades after the Progressive Era all but banished fatal plumbism from occupational settings. A similar contingent in the 1950s and 1960s brought fatal childhood lead poisoning into the light and began eradicating it. Both projects acknowledged and engaged the political-economic power of industry as well as the community's civic responsibility. The lead-poisoning problems that persist do so largely because of failure to employ or exert either.

Alice Hamilton's generation may have transformed American industry or merely shaped the inevitable processes of rationalization already under-way in a modernizing system of industrial capitalism. Either way, factory-inspection acts, workmen's compensation laws, the professionalization of occupational physicians, and other Progressive Era developments did not come from dissociating industry from society but were the product of conversation between members of a dynamic industrial order and a cadre of individuals deeply committed to social and cultural change, two groups whose memberships were seldom mutually exclusive. Hamilton was proud of her friendly relations with industrialists, boasted about enlightened capitalists she encountered in her surveys, and rallied her peers by describing "What One Stockholder Did" to change working conditions in Chicago's giant Pullman Palace Sleeper Company factory.[28]

The "discovery" and reduction of childhood lead poisoning, from about the end of World War II until the early 1970s, marks a second period in which the medical community and political activists appealed to industry's interests, and industry accepted a measure of responsibility (albeit limited by the scientific "truths" it had created about lead poisoning). In this period, the Lead Industries Association sponsored research in childhood lead-paint poisoning, invested in and helped direct tests of effective treatments, and cooperated with the American Academy of Pediatrics in establishing and implementing a voluntary ban on lead paints for interior use. In 1972, during his health subcommittee's hearings on proposed lead-paint legislation, Senator Edward Kennedy told the president of the LIA that his association deserved "great credit for the help it has provided to Congress in this area." Given the association's record at the time, there is little reason to question Kennedy's sincerity.[29]

The lead industry continued improving its worker-safety record from the Progressive Era to the 1950s. Its activities related to lead paint and the public health implications of that work have far less to commend them. Felix Wormser, secretary of the LIA during most of the interwar years, expended great efforts in denying and squelching discussion of lead's impact upon public health. His job grew considerably more difficult after World War II—a difference that cannot be explained by new legislation or regulatory power. Before the 1970s, no government regulations required lead manufacturers to take steps to prevent childhood lead poisoning. Instead, the pressure came from a society that was devoting unprecedented resources and attention to raising its children. More specifically, it came from growing concern among medical researchers and health departments. Initially these groups did less to press for legislative solutions than to secure the industry's cooperation. The evidence suggests that this cooperation was due in part to many industrialists' concern for the welfare of children. But the growing activism of doctors and public health advocates carried the threat of greater government interference, which no doubt encouraged preemptive action on the part of manufacturers.

Indeed, since 1970 the lead industries have faced considerable government intervention in the form of regulation, outright bans, negative publicity, and expensive litigation. Their response—understandably, but regrettably—has been to assume a defensive stance, to withdraw behind a barricade of computer printouts from industry-sponsored studies, and to deny any responsibility for problems stemming from lead in the environment—to deny, in fact, that there are any "real" problems at all. The public health community, increasingly beholden to the federal government, has been unable to establish, let alone maintain, a consistent stance or robust program to counter industry's intransigence. This failure is due in part to territorial disputes between government agencies over mandates and funding. It is also due to deep divisions between rival camps of lead-poisoning researchers, divisions over findings and funding. Policymakers' faith in scientists' ability to produce "objective" data has been shaken.

Instead of building alliances with lead industries, landlords, and governments in order to formulate solutions as their Progressive Era counterparts might have, today's reformers in the medical and environmentalist communities encourage the deepening rift. Twenty-five years ago, J. Julian Chisolm was one of the most prominent childhood lead-poisoning researchers and an effective advocate for lead abatement and treatment programs. He also openly accepted lead-industry grants for his primary research.[30] Many lead researchers today eschew such "corrupting" influences in favor of "untainted" foundation and government funding, and vilify those who accept industry support as toadies. The result has been a war of statistical one-upmanship, with each side's experts reinterpreting and exploding their opponents' findings and lambasting their opponents' obvious "interest" and bias. If complex multivariate statistics is the most effective tactical weapon in this war, charges of self-interest and bias have been the mustard gas, blowing back into the attacker's face as often as it hits the intended target.

The activists of the second and seventh decades of the century had no time to argue the merits of their data. They believed what their science told them and acted upon those beliefs. We still believe in—indeed, we place our ultimate faith in—science: science, armed with "objective" truth and miraculous technology, will ultimately triumph. But science no longer occupies that lofty realm where truth rings with algebraic precision. Differences over interpretation, accusations of bias, questions about the lack of precision or sophistication . . . myriad questions cloud science's window on truth and make it impossible to rely on science alone to determine policy.[31]

Senator Edmund Muskie ran head-on into this quandary in his 1966 air pollution hearings. After listening to Robert Kehoe and other lead-industry evangelists tell of the good news in the Three Cities Survey, he brought Clair Patterson into committee chambers to hear his jeremiad. One passage from the exchange is worth quoting in full:

> **SENATOR MUSKIE:** *Now in identifying the typical lead levels, as you have termed it, you use actual observations?*
> **DR. PATTERSON:** *Yes.*
> **SENATOR MUSKIE:** *Are those observations different from those that we have been hearing about during the days that these hearings have been going on?*
> **DR. PATTERSON:** *They are the same observations.*
> **SENATOR MUSKIE:** *So they are the same observations leading to different conclusions?*
> **DR. PATTERSON:** *Yes.*
> **SENATOR MUSKIE:** *You know, this is something we expect from lawyers and Senators but not from scientists.*[32]

Lawyers and senators have found the seeming fuzziness of scientific truth extremely problematic when dealing with admittedly multicausal health problems such as low-level lead exposure. At the end of the twentieth century, our atmosphere and bodies contained less lead than at the century's beginning. Today, we face a question that thirty years ago would have been laughable to anyone outside of the lead industry: Is lead overregulated?

Arguments of rights—workers' rights, fetal rights, employers' rights, property rights—have largely replaced the clear moral imperative to save lives that charged earlier campaigns in the workplace and the inner cities. The benefits of lead-paint abatement and further improvements in lead manufacturing processes no longer seem obvious to all interested parties. Owners of residential rental property and private homes chafe at the costs and red tape involved in guaranteeing lead-safe housing (or avoiding liability for not doing so). Likewise, fetal protection policies—which favor the unborn child's physical safety at the expense of the family's fiscal soundness—are far more problematic now than at the beginning of the century, when British physician Thomas Oliver asserted, "It is almost impossible for a pregnant woman working in lead to go to term and give birth to a healthy child."[33]

Like their opponents, advocates for more aggressive measures against lead increasingly rely on talk of rights. Lead pollution is now an important area in the growing field of environmental justice, which puts the specific issues of low-level lead exposure in context with the rights of the poor. It

often seems that lead exposure has become emblematic of the nation's ills—
in the words of Saul Bellow's dean, it is simply what we have "fixed upon
but stands for something else that we all sense."[34] Lead becomes a proxy
for the nation's education problems, for its failure to provide decent housing, and for its inequitable health care system.

Some would argue that if we just guaranteed all children adequate diets,
lead poisoning would never be a problem, and that remedial education for
lead-exposed children would compensate for any related IQ loss. By taking
care of children's basic needs, the argument runs, we can save tens of billions of dollars on screening, treatments for children with low-level exposure, and lead-paint removal. Lead activists will counter that as long as
there is old lead in the urban environment, there will always be children at
serious risk. They would argue that every child has a right to grow in an
environment free from powerful neurotoxins. Both sides have a point, and
the nation can afford to address both. Moreover, the nation cannot live up
to its highest values if it does not do both.

How is it to be done, and who is to pay? Obviously, in cases where lines
of accountability are clear, those responsible for present dangers must pay
to reduce them. Unfortunately, this simple equation only fits the occupational setting. The costs of maintaining healthy workplaces must be the
employers' responsibility.

With paint abatement the link between culpability and responsibility is
not so simple. The costs have tended to fall on individual property owners,
with some relief in the form of tax credits. This places an unfair burden
on owners who never used lead paint. Recent lawsuits have attempted to
use market-share liability to tag paint companies and the lead industry for
a portion of abatement costs. But there is an important distinction to be
made between cleaning up toxic dumps and abating lead paint. Lead was
not an unwanted chemical byproduct released accidentally into the environment. It was not dumped surreptitiously on the nation's walls. American consumers paid top dollar for 100 percent white lead paint. Professional painters, government contractors, homeowners, and landlords alike
called for white lead paint—inside and out. National Lead, Sherwin-
Williams, and hundreds of paint makers manufactured and sold lead paint
and must bear their share of the clean-up's cost. But how do we bill the
"white-leaders"?

Looking more carefully at the example of manufacturers' "clear" responsibility for the costs of industrial hygiene suggests not an answer, but
a more constructive perspective on assigning costs. A lead-battery manufacturer who takes on the burden of reducing lead exposure by investing

in safer processes will pass on portions of the higher costs to consumers. Further, the whole system will work only if government also invests its regulatory power. Holding all manufacturers to strict standards will prevent some from operating "cheap and dirty" and undercutting their more hygienic competitors.

In short, in both the question of factory hygiene and lead-paint abatement, at least three accountable parties must share the costs of the twentieth century's saturnine binge. The lead manufacturers, who have profited from selling poisons, must pay, as must the public, who traded health for whiter walls, more powerful cars, and spotless apples. Finally, the government must pay—with tax dollars and political fortitude—for its part in "covering the earth": for not joining its European counterparts in restricting white lead in the 1920s, and for failing to adequately test or regulate a gasoline additive whose effects competent critics accurately predicted from the start.

This book presents the history of what has and has not worked in regulating the risks of lead exposure and reducing lead poisoning. What has worked is cooperation, supervision, and moral suasion. When industries are left unwatched and unregulated, old abuses will persist and new ones will arise. But if we so distrust industry's capacity to act in any way other than blind self-interest, we risk forfeiting the opportunity to shape corporate behavior. Over her long life, Alice Hamilton entertained many unpopular sentiments and was branded a radical and a rabble-rouser. But as horrific as the industrial conditions she found were, as uncaring as many manufacturers seemed to be, she always sought ways to make capitalism and compassion compatible.

Lead, the element, is extremely malleable—it can be melted, recast, or worked to fit many needs. Lead, the commodity, is a phoenix—as one use for it disappears through obsolescence or health-dictated proscriptions, another use leaps from the smelter's ashes. As long as this is so, we may assume that some lead will make its way to the environment. Lead is not biodegradable—once disgorged from beneath the earth it remains a potential threat. The lead in the environment now will stay there until we remove it. The United States is lead poisoned. Somehow we must make the cure more profitable than the disease.

Reports on Lead Poisoning

TABLE A-1 *Reported Lead-Poisoning Deaths, United States, 1900–45*

Year	Deaths	Deaths per Million	Lead Consumed (tons)	Deaths per Ton of Lead
1900	61	1.98	269	0.23
1901	71	2.27	272	0.26
1902	94	2.95	332	0.28
1903	100	3.08	300	0.33
1904	81	2.45	320	0.25
1905	87	2.35	347	0.25
1906	97	2.37	377	0.26
1907	92	2.08	387	0.24
1908	95	2.00	336	0.28
1909	86	1.69	369	0.23
1910	136	2.21	379	0.36
1911	145	2.33	385	0.38
1912	148	2.36	388	0.38
1913	162	2.56	419	0.39
1914	149	2.28	453	0.33
1915	158	2.35	420	0.38
1916	190	2.65	461	0.41
1917	147	1.93	516	0.28
1918	123	1.52	543	0.23
1919	148	1.74	434	0.34
1920	120	1.36	605	0.20
1921	142	1.56	521	0.27
1922	137	1.46	683	0.20
1923	141	1.46	768	0.18

(continued)

TABLE A-1 *continued*

Year	Deaths	Deaths per Million	Lead Consumed (tons)	Deaths per Ton of Lead
1924	142	1.43	812	0.17
1925	142	1.38	856	0.17
1926	144	1.35	901	0.16
1927	135	1.22	841	0.16
1928	129	1.13	931	0.14
1929	134	1.13	949	0.14
1930	101	0.82	754	0.13
1931	111	0.90	568	0.20
1932	78	0.63	416	0.19
1933	117	0.93	453	0.26
1934	118	0.93	491	0.24
1935	130	1.02	543	0.24
1936	132	1.03	634	0.21
1937	77	0.60	679	0.11
1938	94	0.72	546	0.17
1939	97	0.74	667	0.15
1940	100	0.76	782	0.13
1941	70	0.52	1,050	0.07
1942	71	0.52	1,043	0.07
1943	58	0.42	1,113	0.05
1944	61	0.44	1,119	0.05
1945	45	0.32	1,052	0.04

SOURCE: Bureau of the Census, Mortality Statistics 1904–46 (Washington: GPO, 1902–47); U.S. Department of the Interior, Bureau of Mines, "Lead," in *Minerals Yearbook* 1924–45 (Washington, D.C.: GPO, 1925–46).

TABLE A-2 *Reported Lead-Poisoning Deaths, by Age and Race, United States, 1923-48*

Year	All Ages	Under Age 5				
		Number	% of Total	Black	White	% Black
1923	141	1	0.7	0	1	0.0
1924	142	2	1.4	0	2	0.0
1925	142	2	1.4	0	2	0.0
1926	144	3	2.1	0	3	0.0
1927	135	7	5.2	0	7	0.0
1928	129	5	3.9	0	5	0.0
1929	134	2	1.5	0	2	0.0
1930	101	8	7.9	0	8	0.0
1931	111	8	7.2	1	7	12.5
1932	78	13	16.7	2	11	15.4
1933	117	11	9.4	1	10	9.1
1934	118	17	14.4	4	13	23.5
1935	130	25	19.2	9	16	36.0
1936	132	28	21.2	4	24	14.3
1937	77	15	19.5	3	12	20.0
1938	94	20	21.3	4	16	20.0
1939	97	19	19.6	3	16	15.8
1940	100	28	28.0	11	17	39.3
1941	70	15	21.4	3	12	20.0
1942	71	13	18.3	6	7	46.2
1943	58	20	34.5	3	17	15.0
1944	61	15	24.6	2	13	13.3
1945	45	13	28.9	5	8	38.5
1946	46	14	30.4	4	10	28.6
1947	32	7	21.9	3	4	42.9
1948	46	18	39.1	7	11	38.9

SOURCE: Bureau of the Census, *Mortality Statistics,* 1931-50 (Washington: GPO, 1935-50).

TABLE A-3 *Reported Childhood Lead-Poisoning Cases in Five Cities, 1940–66*

Year	Baltimore Cases	Baltimore Fatalities	Philadelphia Cases	Philadelphia Fatalities	New York Cases	New York Fatalities	Cincinnati Cases	Cincinnati Fatalities	Chicago Cases	Chicago Fatalities
1940	12	7								
1941	15	3					1	0		
1942	13	5								
1943	10	5					1	0		
1944	9	1								
1945	8	2								
1946	13	4					1	1		
1947	11	3					2	1		
1948	31	4					3	1		
1949	34	4					8	5		
1950	31	2			1	0	13	3		
1951	77	9			18	4	23	7		
1952	29	5	5	2	20	11	10	1		
1953	49	6	23	11	24	11	20	5	33	6
1954	34	3			80	12	17	1	13	7
1955	35	1	22		115	18	19	1		
1956	48	3	38	5	99	9	13	3		
1957	56	3	28	3	85	9	18	3	24	7
1958	133	10	53	7	116	21	19	3	13	3
1959	66	2	49	13	171	12			193	19
1960	53	4	55	13	146	18			172	28
1961	48	1	109	9	181	6			178	15
1962	44	1	244	14	198	9			154	21
1963	42	3	136	3	338	7			203	19
1964	45	1	122	2	509	7			156	8
1965	32	0	134	5					218	18
1966	32	1	163	3					304	5

SOURCES: Baltimore: Baltimore City Public Health Department, "Lead Paint Poisoning in Children" (Baltimore: Author, 1968). Philadelphia: *Philadelphia Public Health Bulletins* 1954–70; *Archives of Environmental Health* 3 (Nov. 1961): 576; House Committee on Banking and Currency, *Hearings to Provide Federal Asssistance for Eliminating the Causes of Lead-Based Paint Poisoning: A Hearing before the Subcommittee on Housing*, 91st Cong., 2d sess., 22–23 July 1970, 211. New York: Harold Jacobziner, "Lead Poisoning in Children: Epidemiology, Manifestations, and Prevention," *Clinical Pediatrics* 5 (May 1966): 277–86; Mary Culhane McLaughlin, "Lead Poisoning in Children in New York City, 1950–1954," *New York State Journal of Medicine* 56 (Dec. 1956): 3711–14. Cincinnati: Hugo Dunlap Smith, "Lead Poisoning in Children, and Its Therapy with EDTA," *Industrial Medicine and Surgery* 28 (Mar. 1959): 148–51. Chicago: Arnold Tanis, "Lead Poisoning in Children, Including Nine Cases Treated with Edathamil Calcium-Disodium," *American Journal of Diseases of Children* 89 (Mar. 1955): 325–31; David Jenkins and Robert Mellins, "Lead Poisoning in Children: A Study of Forty-Six Cases," *Archives of Neurology and Psychiatry* 77 (1957): 70–78; *Science and Citizen* (Aug. 1968): 56 (1966 data estimated from November).

Notes

AEH	*Archives of Environmental Health*
AJDC	*American Journal of the Diseases of Children*
AJPH	*American Journal of Public Health*
BHM	*Bulletin of the History of Medicine*
CDC	Centers for Disease Control
EPA	Environmental Protection Agency
GPO	Government Printing Office
JAMA	*Journal of the American Medical Association*
JIH	*Journal of Industrial Hygiene*
MLR	*Monthly Labor Review*
MMWR	*Morbidity and Mortality Weekly Report*
NEJM	*New England Journal of Medicine*
PHS	U.S. Public Health Service

INTRODUCTION: WHAT'S LEAD IN THE BONE . . .

1. Most research into the effects of the chronic lead exposure common before the 1970s has dwelt on subtle neurological damage, but several studies now implicate low-level lead exposure in hypertension and kidney diseases; see Howard Hu et al., "The Relationship of Bone and Blood Lead to Hypertension," *JAMA* 275 (17 Apr. 1996): 1171–76; Rokho Kim et al., "A Longitudinal Study of Low-Level Lead Exposure and Impairment of Renal Function: A Normative Aging Study," *JAMA* 275 (17 Apr. 1996): 1177–81.

2. Since 1991, the blood-lead level for intervention in cases of pediatric plumbism has been defined as 15 micrograms per deciliter (15 μg/dL); see CDC, *Strate-*

gic Plan for the Elimination of Childhood Lead Poisoning (Atlanta: U.S. Department of Health and Human Services, PHS, CDC, 1991).

3. An early epidemiological study found that the blood-lead level in almost half of the children brought into neighborhood clinics in Baltimore exceeded 50 μg/dL; see J. Edmund Bradley et al., "The Incidence of Abnormal Blood Levels of Lead in a Metropolitan Pediatric Clinic," *Journal of Pediatrics* 49 (1956): 1–6.

4. *Plumbism*, from the Latin *plumbum*, or "lead." Lead poisoning has been known for millennia by many names, including plumbism or saturnism; painter's colic, potter's colic, colic of Poitou, or Devonshire colic; the dangles or the dry bellyache. For an almost encyclopedic history of ancient lead poisoning, see Jerome O. Nriagu, *Lead and Lead Poisoning in Antiquity*, Environmental Science and Technology Series, ed. Robert L. Metcalf and Werner Stumm (New York: John Wiley & Sons, 1983).

5. Studies of almost any subject in modern history will deal at least tangentially with some aspects of these changes. Medical and scientific scholars and environmental historians have made the most explicit statements about the complex issues that shape the definition, assessment, and regulation of risks; see the forum on risk in *Dædalus* 119 (1990). Social historians and economists are making direct contributions as well; see Viviana A. Zelizer, *Pricing the Priceless Child: The Changing Social Value of Children* (New York: Basic Books, 1985); and Suellen Hoy, *Chasing Dirt: The American Pursuit of Cleanliness* (New York: Oxford University Press, 1995).

6. Lloyd G. Stevenson's 1949 doctoral dissertation, "A History of Lead Poisoning" (The Johns Hopkins University), deals with the subject from antiquity to "modern times," but only skims the surface of the American experience and ignores pediatric lead poisoning. Richard P. Wedeen's *Poison in the Pot: The Legacy of Lead* (Carbondale, Ill.: Southern Illinois University Press, 1984) covers similar chronological acreage, but his focus on one aspect of lead poisoning—its effect on the renal system—gives his work depth and analytical bite. Wedeen, a kidney specialist with a longstanding interest in lead poisoning, was instrumental in pushing the Occupational Safety and Health Administration (OSHA) to reduce the amount of lead permitted in workplaces. A series of articles by industrial physician Carey P. McCord forms the outline for a history of lead poisoning in the United States before 1900; see, e.g., "Lead and Lead Poisoning in Early America: Lead Compounds," *Industrial Medicine and Surgery* 23 (Feb. 1954): 75–80.

7. Patricia Vawter Klein, "From Knowledge to Policy: The Use and Abuse of Expert Opinion and Scientific Research in the Foundation of Exclusionary Policies toward Women in Lead Industries, 1900–1925 and 1965–1980" (Ph.D. diss., State University of New York, Buffalo, 1989); Christopher Clare Sellers, *Hazards of the Job: From Industrial Disease to Environmental Health Science* (Chapel Hill: University of North Carolina Press, 1997); quote, Christopher Clare Sellers, "Manufacturing Disease: Experts and the Ailing American Worker" (Ph.D. diss., Yale University, 1992), 15.

8. William L. Roper (administrator, U.S. Department of Health and Human Services), cover letter accompanying educational pamphlets for distribution to pediatricians, 15 Jan. 1991.

9. A number of medical books on lead poisoning appeared after 1970, inevitably including a historical overview; see, e.g., Jane Lin-Fu, "Lead Poisoning and Undue Lead Exposure in Children: History and Current Status," in *Low Level Lead Exposure: The Clinical Implications of Current Research*, ed. Herbert Needleman (New York: Raven Press, 1980), 23-44; Jane Lin-Fu, "Modern History of Lead Poisoning: A Century of Discovery and Rediscovery," in *Human Lead Exposure*, ed. Herbert Needleman (Boca Raton, Fla.: CRC Press, 1992). Publications by the Environmental Defense Fund best characterize the surveys from the perspective of environmental advocacy; see Karen L. Florini, George Krumbaar, and Ellen K. Silbergeld, *Legacy of Lead: America's Continuing Epidemic of Childhood Lead Poisoning* (Washington: Environmental Defense Fund, 1990); Pete Reich, *The Hour of Lead: A Brief History of Lead Poisoning in the United States over the Past Century and of Efforts by the Lead Industry to Delay Regulation* (Washington: Environmental Defense Fund, 1992); see also Richard M. Stapleton, *Lead Is a Silent Hazard* (New York: Walker, 1994).

10. Richard Rabin, "Warnings Unheeded: A History of Child Lead Poisoning," *AJPH* 79 (Dec. 1989): 1668-74; Barbara Berney, "Round and Round It Goes: The Epidemiology of Childhood Lead Poisoning, 1950-1990," *Milbank Quarterly* 71 (1993): 3-39; Elizabeth Fee, "Public Health in Practice: An Early Confrontation with the 'Silent Epidemic' of Childhood Lead Paint Poisoning," *Journal of the History of Medicine* 45 (1990): 570-606.

11. For example, see Shep Melnick, *Regulation and the Courts: The Case of the Clean Air Act* (Washington: Brookings Institution, 1983), 265-76.

12. David Rosner and Gerald Markowitz, "'A Gift of God'?: The Public Health Controversy over Leaded Gasoline during the 1920s," in *Dying for Work: Workers' Safety and Health in the Twentieth Century*, ed. David Rosner and Gerald Markowitz, Interdisciplinary Studies in History, ser. ed. Harvey Graff (Bloomington: Indiana University Press, 1989), 121-39; William J. Kovarik, "The Ethyl Controversy: The News Media and the Public Health Debate over Leaded Gasoline, 1924-1926" (Ph.D. diss., University of Maryland, 1993).

13. For example, Charles Rosenberg in *The Cholera Years: The United States in 1832, 1849 and 1866, with a New Afterword* (Chicago: University of Chicago Press, 1987) analyzed the shifts in public understanding and the public health community's response to cholera in the nineteenth century while simultaneously instructing the reader on the value of an empowered secular civic system of public health regulation. Similarly, Allan Brandt's *No Magic Bullet: A Social History of Venereal Disease in the United States since 1880* (New York: Oxford University Press, 1985) is a searing cautionary tale of the dangers of an overconfident and uncritical scientific community working in political and social isolation. In both books, the disease as suffered by the community is only one of several historical

subjects considered. This list of citations could be extended almost indefinitely, but the only beneficiary would be my argument for the notable absence of scholarship on lead poisoning.

14. This is not the same thing as positing that lead poisoning is monocausal. Many of the controversies about lead originate in the fact that the dose-response to exposure is so wildly affected by such factors as diet, health, age, climate, and (it has been argued) ethnicity. Nor is there one source of lead poisoning: it comes from lead in the air, dust, food, and nonfood sources. What is not disputed is that the necessary agent for lead poisoning—lead somewhere in the environment—is man made.

15. Charles Dickens, *The Uncommercial Traveller* (1861; London: Mandarin Paperbacks, 1991), 338–39; William Ryan, *Blaming the Victim* (New York: Pantheon, 1971), 23.

16. For fifty years the lead industry convinced the scientific and medical communities that lead exposure was natural, that "the natural environment of this planet is such that the occurrence of lead in the tissues, body fluids and excreta of its human inhabitants is inevitable"; see Robert Kehoe, "The Harben Lectures, 1960," *Journal of the Royal Institute of Public Health and Hygiene* 24, nos. 4–8 (Apr.-Aug. 1961): passim; quote, Lecture 3, "Present Hygienic Problems Relating to the Absorption of Lead" (Aug. 1961): 177.

17. Alice Hamilton, *Exploring the Dangerous Trades: The Autobiography of Alice Hamilton, M.D.* (Boston: Little, Brown, 1943); Samuel B. Hays, "The Role of Values in Science and Policy: The Case of Lead," in *Human Lead Exposure,* ed. Needleman, 267–86.

18. Benjamin Franklin to Benjamin Vaughan, 31 July 1786; quoted in Carey P. McCord, "Lead and Lead Poisoning in Early America: Benjamin Franklin and Lead Poisoning," *Industrial Medicine and Surgery* 22 (Sept. 1958): 393–99. The "Subject" Franklin discussed involved Massachusetts rum makers in the late seventeenth century, who inadvertently contaminated their product by using lead in several components of their stills.

19. *Comprehensive and Workable Plan for the Abatement of Lead-Based Paint in Privately Owned Housing: Report to Congress,* a U.S. Department of Housing and Urban Development (HUD) study (Washington: GPO, 1990), concluded that 57 million homes in the United States contained lead-based paint and that 3.8 million units that housed children under the age of 7 contained such paint in poor condition or high levels of lead in house dust (xvii–xix).

20. "Lead Poisoning in Bridge Demolition Workers—Georgia, 1992," *MMWR* 42 (28 May 1993): 388–90; John Rekus, "Lead Poisoning: OSHA's Existing Lead Regulations Do Not Provide Construction Employees the Same Level of Protection Afforded Workers in General Industry," *Occupational Health and Safety* 61 (Aug. 1992): 14.

21. W. Schweisheimer, "Newly Painted Walls in Relation to Health," *The Painter and Decorator* 51 (June 1937): 15; "Homes Built with Assistance of

U.S.D.A. Extension Service Painted with Pure White Lead and Oil," *Lead* 9 (Nov. 1939): 9. The latter article continues by stressing the long-term savings involved in using the higher priced lead paint: "It is interesting to note the consistency with which builders of low-cost homes rely on white lead."

22. See the exchange between Harold B. Finger, HUD's assistant secretary for research and technology, and Kennedy, in Senate Committee on Labor and Public Welfare, *Lead-Based Paint Poisoning Amendments of 1972: Hearings before the Subcommittee on Health of the Committee on Labor and Public Welfare,* 91st Cong., 2d sess., 6, 9, and 10 March 1972, 17–18. Kennedy favored the explanation that "the inexperience perhaps of hospitals in rural communities" left them unable to diagnose the illness.

23. David Hackett Fischer, "The Braided Narrative: Substance and Form in Social History," in *The Literature of Fact,* ed. Angus Fletcher (New York: Columbia University Press, 1976), 109–33. William Cronon ("A Place for Stories: Nature, History, and Narrative," *Journal of American History* 78 [Mar. 1992]: 1347–76) cautions historians to detect the narrative in others' work as well as their own, to remain keenly aware of the "hidden agendas" within every plot-line, agendas "that influence what the narrative includes and excludes" to the point of wresting control from the "historian as author" who employs them.

24. Alice Hamilton, "Lead Poisoning in American Industry," *JIH* 1 (May 1919): 21.

25. This notion of trusting the body to "balance" its intake and output came to be associated with Robert Kehoe of the Kettering Laboratory in Cincinnati, whose clinical experiments in the 1930s established the basic facts of human lead metabolism; see the four-part series of articles by Robert Kehoe, Frederick Thamann, and Jacob Cholak, "On the Normal Absorption and Excretion of Lead," *JIH* 15 (Sept. 1933): 257–305. But a British article from fifty years before Kehoe et al.'s publication both articulated the "balance" theory and warned of its limits. The French researcher M. Armand Gauthier of the Paris Academy of Medicine (quoted in "An Ubiquitous Poison," *British Medical Journal* [26 Nov. 1881]: 862–63) established that "so long as the quantity of lead which remains unassimilated remains equal to that daily absorbed, lead poisoning, properly so-called, does not make its appearance. The lead circulates slowly, becoming assimilated and eliminated in nearly equal quantities, until the day when an increase in the dose of the poison, an arrest in the elimination, an affection of the kidney, a weakening of the vital reactions, an exhaustion of the tolerance of the animal economy, allow the phenomena of acute or chronic lead-poisoning suddenly to declare themselves." One example of this is reported in William McNally, "Lead Poisoning Caused by a Bullet Embedded for Twenty-seven Years," *Industrial Medicine* 18 (Feb. 1949): 77–78. A 66-year-old woman for years carried a bullet lodged in her knee. Suddenly, she developed acute symptoms and died from lead encephalopathy. The lead levels in her liver and heart were three times those associated with serious plumbism. For a more recent discussion, see Wilfred C. Peh and William R.

Reinus, "Lead Arthropathy: A Cause of Delayed Onset Lead Poisoning," *Skeletal Radiology* 24 (July 1995): 357–60.

26. Historians of the "silent epidemic" have not ignored this cyclical pattern, as the title of Berney's "Round and Round It Goes" suggests. The cycles I find in the public health community's responses to lead poisoning closely resemble the cyclical nature of treating an individual with "chelation" therapy. Chelates (from the Greek *khele,* or "claw") bond with ions of lead, "clawing" them from the soft tissues or blood to form a nontoxic compound that can be safely excreted in the urine. Doctors treat severe lead poisoning until the symptoms abate. But they cannot remove all of the lead; much remains in the bones and soft tissues, and eventually the patient's blood-lead level rises again—perhaps not as high as before, perhaps not as obvious. Only physicians sensitive to the more subtle symptoms of chronic low-level lead exposure will know to treat again. America's response to lead-poisoned children has been chelation therapy writ large.

27. John M. Keating, *Cyclopædia of the Diseases of Children* (1890), 4: 616.

28. Herbert Needleman and Richard J. Jackson, "Lead Toxicity in the 21st Century: Will We Still Be Treating It?" *Pediatrics* 89 (Apr. 1992): 679.

CHAPTER 1: PLUMBING THE DEPTHS

1. Carey P. McCord, "Lead and Lead Poisoning in Early America: Clinical Lead Poisoning in the Colonies," *Industrial Medicine and Surgery* 23 (Mar. 1954): 125; Nriagu, *Lead Poisoning in Antiquity,* 379 (see intro., n. 4).

2. Robert A. Goyer, "Lead Toxicity," in *Lead Absorption in Children: Management, Clinical and Environmental Aspects,* ed. J. Julian Chisolm and David M. O'Hara (Baltimore: Urban & Schwarzenberg, 1982), 22.

3. CDC, *Preventing Lead Poisoning in Young Children* (Atlanta: CDC, 1985), 14.

4. Wedeen, *Poison in the Pot,* 2 (see intro., n. 6). J. Julian Chisolm ("Treatment of Lead Poisoning," *Modern Treatment* 8 [Aug. 1971]: 593–611) was confident of this: "There is no known significant toxicity associated with the portion [of body lead] that has been well incorporated into the matrix of the bone."

5. Goyer, "Lead Toxicity," 23–24.

6. Ibid., 30–31; Sergio Piomelli et al., "Management of Childhood Lead Poisoning," Journal of Pediatrics 105 (Oct. 1984): 523–32.

7. Goyer, "Lead Toxicity," 26–28; Piomelli et al., "Management of Childhood Lead Poisoning."

8. See Ernest L. Abel, comp., *Lead and Reproduction: A Comprehensive Bibliography* (Westport, Conn.: Greenwood Press, 1984), xi–xiv. For a recent historical analysis of this century-long debate, see Allison L. Hepler, "Protecting Mothers and Fetuses: Lead Poisoning of Women in Industry" (paper presented at the meeting of the American Society for Environmental History, March 1997). On the

continuing consequences of parents' occupational lead exposure, see Elizabeth A. Whelan et al., "Elevated Blood Lead Levels in Children of Construction Workers," *AJPH* 87 (Aug. 1997): 1352–55.

9. CDC, *Preventing Lead Poisoning*, 16; John W. Graef, "Clinical Outpatient Management of Childhood Lead Poisoning," in *Lead Absorption in Children*, ed. Chisolm and O'Hara, 161–62.

10. Chisolm, "Treatment of Lead Poisoning."

11. David Purves, *Trace Element Contamination of the Environment*, Fundamental Aspects of Pollution Control and Environmental Science series (Amsterdam: Elsevier Science Publishers, 1977; rev. ed. 1985), 7: 4.

12. Nriagu, *Lead Poisoning in Antiquity*, 274; P. C. Srivastava and S. Varadi, letter, *British Medical Journal* 1 (2 Mar. 1968): 578.

13. Noel H. Gale and Zofia Stos-Gale, "Lead and Silver in the Ancient Aegean," *Scientific American* 244 (June 1981): 181–83; Nriagu, *Lead Poisoning in Antiquity*, 87.

14. Gale and Stos-Gale, "Lead and Silver in the Ancient Aegean," 184; Clair C. Patterson, "An Alternative Perspective—Lead Poisoning in the Human Environment: Origin, Extent, and Significance," in *Lead in the Human Environment*, ed. National Research Council, Committee on Lead in the Human Environment (Washington: National Academy Press, 1980), 281.

15. Both Gale and Stos-Gale ("Lead and Silver in the Ancient Aegean," 181) and Nriagu (*Lead Poisoning in Antiquity*, 92–98) give detailed sketches of the process of cupellation; see also Georgius Agricola, *De Re Metallica* (1556; trans. Herbert Clark Hoover and Lou Henry Hoover, 1912; reprint, New York: Dover Publications, 1950), 475.

16. Pliny, *Natural History*, 33.4, quoted in Nriagu, *Lead Poisoning in Antiquity*, 66.

17. Patterson, "Lead Poisoning in the Human Environment," 281; Nriagu, *Lead Poisoning in Antiquity*, 200–201.

18. In Britain, *plumber* referred to roofers as well—sheet lead has been used in roofing since ancient Egypt; Nriagu, *Lead Poisoning in Antiquity*, 237, 240–41. A roof was called "a lead" even if it was made from something else.

19. In *Lead Poisoning in Antiquity*, Nriagu describes in great detail the sources and distribution for each major aqueduct (322–26) and discusses Roman methods for testing pH and sediments (327). Roman baths were lined with sheets of lead; the ones at Bath, England, were slightly more than half an inch thick and weighed 332 pounds per square yard (239).

20. Ibid., 245–48.

21. Thomas T. Read, *Our Mineral Civilization* (Baltimore: Williams and Wilkins, 1932), 61. In 1979 the EPA (*Lead* [Washington: GPO], 3) estimated that 180 metric tons of lead entered the environment each year as waste products of the printing trades. Lead's long relationship with typesetting survives—at least linguistically—in the age of computers, laser printing, and desktop publishing. Even

today the amount of space between lines of type is termed "leading," a reference to the former typesetting practice of inserting strips of lead of varying widths between lines of type to add space and improve readability or appearance of the printed page.

22. Nriagu, *Lead Poisoning in Antiquity* 13, 259. Some of those whose professions required weighing of goods learned to use lead's mass to unscrupulous advantage. Tradesmen in Rome added minium to pepper in order to increase its weight and medieval pigment sellers likewise substituted minium for the costlier cinnabar (mercuric sulphide); see Matt Clark, "Here's Lead in Your Wine," *Newsweek*, 28 Mar. 1983, 53. In 1986, a New York art conservator developed plumbism while restoring paintings; see Berton Rouche, "Annals of Medicine: Cinnabar," *New Yorker* 62 (8 Dec. 1986): 94–102. She had been using authenticated old pigment labelled "cinnabar," which tests revealed to be heavily tainted with lowly minium.

23. Nriagu, *Lead Poisoning in Antiquity*, 249, 12; Emma Lila Fundaburk and Mary Douglass Fundaburk Foreman, eds., *Sun Circles and Human Hands: The Southeastern Indians Art and Industries* (Montgomery, Ala.: Paragon Press, 1957), 37.

24. Nriagu, *Lead Poisoning in Antiquity*, 34–41, 50. Industrial hygiene pioneer Alice Hamilton's favorite professor at the Johns Hopkins University, William Osler, infused his lectures on pathology with references to antiquity; see her *Exploring the Dangerous Trades*, 53 (see intro., n. 17). In addition to saturnism, he noted other scientific expressions dating to antiquity that were still in use in the early twentieth century: silver nitrate was called "lunar caustic" and treating rubber with sulphur was (and continues to be) known as "vulcanizing."

25. Nriagu, *Lead Poisoning in Antiquity*, 294.

26. Nicander, *Alexipharmaca* 75–85, 88–114, in *Nicander: The Poems and Poetical Fragments,* ed. and trans. A. S. Gow and A. F. Scholfield (London: Cambridge University Press, 1953).

27. Vannoccio Biringuccio, *The Pirotechnia* (1540; trans. and with introduction and notes by Cyril Stanley Smith and Martha Teach Gnudi, 1942; reprint, Cambridge, Mass.: MIT Press, 1966), 55.

28. Xenophon, *Oeconomics* 10.7, quoted in McCord, "Early America: Lead Compounds," 75–76 (see intro., n. 6); Nriagu, *Lead Poisoning in Antiquity*, 26.

29. Wedeen, *Poison in the Pot,* 50. In 1924 Manchurian statistics showed that plumbism was the fourth highest killer of infants; see "Lead in Cosmetic Kills Many Infants," *New York Times,* 31 Dec. 1924, 6. In Japan, lead meningitis was first noted in 1784 and its source discovered in 1923, and lead powders were discouraged from 1901; see Katsuji Kato, "Lead Meningitis in Infants: Résumé of Japanese Contributions on the Diagnosis of Lead Poisoning in Nurslings," *AJDC* 44 (Sept. 1932): 571. In 1933, Kato (letter, *JAMA* 101 [7 Oct. 1933]: 1135) decried lead powder's continuing use.

30. Nriagu, *Lead Poisoning in Antiquity,* 233–36.

31. Ibid., 378–79. In 1980 the U.S. Food and Drug Administration banned sales of all ceramic kitchenware with lead solubility levels higher than 2.7 μg/ml, but enforcement is nearly impossible, say Russel Flegal, Donald R. Smith, and Robert W. Elias in "Lead Contamination in Food," in *Food Contamination from Environmental Sources,* ed. Jerome Nriagu and Milagros S. Simmons, Wiley Series in Advances in Environmental Science and Technology (New York: John Wiley and Sons, 1990), 109. For a discussion of Mexican pottery, see Gustavo Olaíz et al., "Risk Factors for High Levels of Lead in Blood of Schoolchildren in Mexico City," *AEH* 51 (Mar.–Apr. 1996): 122–26.

32. Nriagu, *Lead Poisoning in Antiquity,* 223–32. Janet Raloff ("Beverages Intoxicated by Lead in Crystal," *Science News* 139 [26 Jan. 1991]: 54) reported the findings of two New York City researchers, who concluded that "long-term storage of *anything* in lead crystal is to be avoided." Of special concern is acidic baby food such as apple juice or sauce.

33. Bernardino Ramazzini, *Diseases of Workers* ([*De Morbis Artificum*], 1713), History of Medicine Series no. 23, trans. Wilmer Cave Wright (New York: Hafner Publishing Company, 1964), 55.

34. O. Harn, *Lead, the Precious Metal,* quoted in Nriagu, *Lead Poisoning in Antiquity,* 262.

35. Nriagu, *Lead Poisoning in Antiquity,* 352–78 (see particularly the table summarizing ancient uses of lead compounds, 357–62); N. Sivin, *Chinese Alchemy: Preliminary Studies* (Cambridge, Mass.: Harvard University Press, 1968), 175, quoted in Nriagu, *Lead Poisoning in Antiquity,* 373; McCord, "Early America: Lead Compounds," 76–77; Wedeen, *Poison in the Pot,* 52.

36. Nriagu, *Lead Poisoning in Antiquity,* 353; Wedeen, *Poison in the Pot,* 55; Ramazzini, *Diseases of Workers,* 41.

37. Wedeen, *Poison in the Pot,* 54–60; Read, *Our Mineral Civilization,* 67; McCord, "Early America: Lead Compounds," 77; "Wood Sees Results of Lead in Cancer," *New York Times,* 19 Jan. 1926, 3.

38. Aristotle, *Historia animalium* (*Works* 4.583a), quoted in Nriagu, *Lead Poisoning in Antiquity,* 367 (Nriagu also cites Soranos of Ephesus [A.D. 98–138], Oribasios, and Aetios); Thomas Oliver, ed., *Dangerous Trades: The Historical, Social, and Legal Aspects of Industrial Occupations as Affecting Health, By a Number of Experts* (1902); Abel, *Lead and Reproduction,* xi. In 1977 women in Malaysia were still resorting to lead paste to induce abortions; see Malcolm Potts, Peter Diggory, and John Peel, *Abortion* (London: Cambridge University Press, 1977), 257.

39. Nriagu, *Lead Poisoning in Antiquity,* 338–44.

40. Wedeen, *Poison in the Pot,* 15–20.

41. Joseph Eisinger, "Lead and Wine: Eberhard Gockel and the Colica Pictonum," *Medical History* 26 (1982): 296. Wedeen (*Poison in the Pot,* 20–22) credits Samuel Stockhausen with making a connection, in 1656, between occupational plumbism among miners and sufferers of wine colic and gout.

42. Wedeen, *Poison in the Pot,* 26–28; Sidney Mintz, *Sweetness and Power: The Place of Sugar in Modern History* (New York: Viking Penguin, 1985), 137.

43. George Baker, "An Examination of Several Means by which the Poison of Lead May Be Supposed Frequently to Gain Admittance into the Human Body, Unobserved and Unsuspected" (1768), quoted in McCord, "Early America: Benjamin Franklin" (see intro., n. 18). In his essay, first of a seven-part series, McCord argues (with more patriotic fervor than documentation) that Franklin, who was widely known in the European scientific community, may have inspired much of the eighteenth century's progress in discovering the etiology of the various "dry gripes," colics, and dropsies associated with lead poisoning.

44. Franklin to Vaughan (see intro., n. 18); Massachusetts Bay Colony, *An Act for Preventing Abuses in Distilling of Rum and Other Strong Liquors, with Leaden Heads or Pipes* (3 Sept. 1723), quoted in McCord, "Early America: Benjamin Franklin," 397–99.

45. Fredrick Accum, *A Treatise on Adulteration of Food, and Culinary Poisons* (1820); Wedeen, *Poison in the Pot,* 172, 181.

46. John F. Moffitt, "Painters 'Born under Saturn': The Physiological Explanation," *Art History* 11 (June 1988): 195.

47. Ibid., 197; Rouche, "Cinnabar," 96; Lloyd C. Taylor, Jr., *The Medical Profession and Social Reform, 1885–1945* (New York: St. Martin's Press, 1974), 57; Ivan Amato, "Singing the Cadmium Blues," *Science News* 138 (15 Sept.1990), 168; "Possible Ban on Some Artists' Materials," *American Artist* 53 (Nov. 1989): 22.

48. "Canned Food Sealed Icemen's Fate," *History Today* 37 (Oct.1987): 3; Owen Beattie and John Geiger, *Frozen in Time: Unlocking the Secrets of the Franklin Expedition* (New York: E. P. Dutton, 1988), 161–62; Irving Shapiro, Philippe Grandjean, and Ole Vagn Nielsen, "Lead Levels in Bones and Teeth of Children in Ancient Nubia: Evidence of Both Minimal Lead Exposure and Lead Poisoning," in *Low Level Lead Exposure,* 35–42 (see intro., n. 9); H. A. Waldron, "Postmortem Absorption of Lead by the Skeleton," *American Journal of Physical Anthropology* 55 (July 1981): 395–98.

49. Arthur C. Aufderheide et al., "Lead in Bone II: Skeletal-Lead Content as an Indicator of Lifetime Lead Ingestion and the Social Correlates in an Archaeological Population," *American Journal of Physical Anthropology* 55 (July 1981): 285–91; Arthur C. Aufderheide et al., "Anthropological Applications of Skeletal Lead Analysis," *American Anthropologist* 90 (1988): 931–36.

CHAPTER 2: CHILDHOOD LEAD POISONING BEFORE 1930

1. Henry N. Thomas and Kenneth D. Blackfan, "Recurrent Meningitis, Due to Lead, in a Child of Five Years," *AJDC* 8 (Nov. 1914): 377–80.

2. See, e.g., Jack Newfield, "Silent Epidemic in the Slums," *Village Voice,* 18 Sept. 1969, 3.

3. The on-line catalog *Readers' Abstracts* listed 274 articles on lead poisoning published in popular magazines from 1983 to 1998; in addition, indexes to medical and social science journals brim with citations to hundreds of reports, with Medline, e.g., showing 341 titles in U.S. publications for 1998 alone.

4. Initial values were set by extensive animal and epidemiological research conducted by Robert Kehoe, who for over fifty years ran the Kettering Laboratory in the Department of Preventive Medicine and Industrial Health at the University of Cincinnati, through a program funded in large part by the lead industry. The clearest presentation of Kehoe's methodology and findings is given in his "Harben Lectures" (see intro., n. 16); in Lecture 3 he states unequivocally that the 80 μg/dL limit is "the critical concentration of lead in the blood of *child or adult* [italics added], below which, in our experience . . . no case of even the mildest type of poisoning has been induced by the absorption of inorganic compounds of lead" (180).

5. *Baltimore City Health Department Annual Report for the Year Ending 1952* (Baltimore: City Health Department, 1953), 286. Since the 1970s, the federal government, through the National Institute for Occupational Safety and Health (NIOSH) and the CDC, has largely taken over setting these levels and sponsoring most lead research, but the falling threshold values still often seem to reflect what is deemed manageable—"acceptable" is closely linked to "treatable." On the gradual reduction in threshold values, see Florini, Krumbaar, and Silbergeld, *Legacy of Lead,* 11–12 (see intro., n. 9); CDC, *Strategic Plan for the Elimination of Childhood Lead Poisoning* (intro., n. 2).

6. Pittsburgh psychiatrist Herbert Needleman, for example, has studied the long-term effects of low-level lead exposure among working- and middle-class children. His work has aroused bitter debate in public health circles; the associations he finds between even low lead levels and reduced intellectual capabilities are frequently disputed, perhaps not least because of the costs that would be involved in reducing children's environmental exposures to the levels Needleman's findings suggest are necessary. In any case, defining such low exposures as "poisonings" is surely not as clear-cut as when doctors were dealing with palsies and painful lead colics.

7. Two fairly complete narratives of the growth of lead-poisoning awareness are Rabin, "Warnings Unheeded" (see intro. n. 10) and Berney, "Round and Round It Goes" (intro., n. 10). Lin-Fu has also written several short histories; see, e.g., her "Modern History of Lead Poisoning" (intro., n. 9). Fee's "Public Health in Practice " (intro., n. 10), although tightly focused on the Baltimore experience, includes an excellent analysis of some of the reasons health departments were slow to act on lead poisoning.

8. Australian physicians began writing about childhood lead-paint poisoning

as early as 1892 (see J. Lockhart Gibson et al., "Notes on Lead Poisoning as Observed among Children in Brisbane," *Transactions of the Intercolonial Medical Congress of Australasia* 3 [Sept. 1892]: 76), but Gibson's definitive article, "A Plea for Painted Railings and Painted Walls of Rooms as the Source of Lead Poisoning among Queensland Children," appeared in 1904 (*Australian Medical Gazette* 23: 149–53).

9. One American physician who did take to heart Gibson and Turner's findings was Baltimore pediatrician Kenneth Blackfan, who discussed the Australians' reports in his two most important papers on childhood lead poisoning, "Lead Poisoning in Children with Especial Reference to Lead as a Cause of Convulsions" (*American Journal of the Medical Sciences* 153 [June 1917]: 877–87) and Thomas and Blackfan, "Recurrent Meningitis, Due to Lead."

10. Bruno Latour, *The Pasteurization of France* (Cambridge, Mass.: Harvard University Press, 1988), 148. For a similar expression of this subjective relationship to evidence as it relates to lead research, see Hays, "The Role of Values" (see intro., n. 17).

11. As late as 1920, Robert Strong, clinical professor of pediatrics at Tulane University, reported that lead poisoning "does not seem to be common in children"; see his "Meningitis, Caused by Lead Poisoning, in a Child of Nineteen Months," *Archives of Pediatrics* 37 (Aug. 1920): 532.

12. John C. Ruddock, "Lead Poisoning in Children with Special Reference to Pica," *JAMA* 82 (24 May 1924): 1682.

13. L. Emmett Holt, Jr., "Lead Poisoning in Infancy," *AJDC* 25 (1923): 231.

14. EPA, *Air Quality Criteria for Lead* (Research Triangle Park, N.C.: EPA, Environmental Criteria and Assessment Office, 1986); Melnick, *Regulation and the Courts,* 265–76 (see intro., n. 11). Evidence from the beginning of this story — of a sudden increase in baseline blood-lead levels as leaded gasoline first appeared in the 1920s — is not readily available.

15. In Mexico City, where leaded gasoline is still burned, average blood-lead levels remain above the limits set by the CDC ($>$15 μg/dL). Some studies of Mexico City children have found strong correlations between lead levels and the use of lead-glazed cooking pots, but one of the strongest associations is the amount of traffic on the road the child lives on; see Isabelle Romieu et al., "Vehicular Traffic as a Determinant of Blood-Lead Levels in Children: A Pilot Study in Mexico City," *AEH* 47 (July–Aug. 1992): 246–49.

16. The only lead-poisoning case found for this period was described in Boston Children's Hospital Archives (no. 92–25), Department of Diseases of the Nervous Systems, book UU, case 130112, 176. A 9-year-old boy was hit in the head with a bat and taken to City Hospital, where he received four stitches. Neurological problems ensued, as well as stomach pains. After twelve days he suffered convulsions and blindness. No explanation is given for why lead was suspected, but City Hospital analyzed his urine for lead, with positive results. Hence, he was diagnosed with "lead neuropathy." It is significant that City Hospital, as a facility

for the general population, was more likely to see occupational lead poisoning and to test for it in cases with both abdominal and neurological symptoms.

17. The increased reporting in medical journals is reflected in an increase in fatal childhood lead poisoning noted in census mortality records: from 1911 to 1920 eight fatalities among children under the age of 5 were reported; from 1921 to 1930, the number rose to thirty-one; see Bureau of the Census, *Mortality Statistics 1911–1930* (Washington: GPO, 1912–31).

18. For Baltimore's early struggles with lead-paint poisoning, see Fee, "Public Health in Practice."

19. J. Schirmer, H. A. Anderson, and L. A. Saryan, "Fatal Pediatric Poisoning from Leaded Paint—Wisconsin, 1990," *MMWR* 40 (29 Mar. 1991): 193–95.

20. The death statistics come from a variety of published reports. For Philadelphia 1956–60, see *AEH* 3 (Nov. 1961): 90; New York City 1954–64, *Scientist and Citizen* (Apr. 1968): 56; Baltimore 1940–49, *Baltimore Health News* 26 (Apr. 1949), and 1950–53, *Baltimore City Health Department Annual Report* (for the years ending 1931–67) and Baltimore Public Health Department, "Lead Paint Poisoning in Children" (1968); Chicago 1957–65, *Scientist and Citizen* (Apr. 1968): 56. Population figures are the average of the 1950 and 1960 city populations of children under 5 years reported in the 1960 federal census. No effort was made to use population estimates of inner cities, as data for the four cities named incorporate inner-city and lower density urban areas.

21. Historians are often warned about carelessly arguing from silence—one can easily see what fallacies it led early lead researchers to—but here it is justified, as one of my chief intentions is to argue *about* that silence.

22. Scholars of lead poisoning in the ancient world have been suspected of just such willy-nilly attribution. See H. A. Waldron's essay on the "lead in the fall of Rome" thesis, "Lead Poisoning in the Ancient World," *Medical History* 17 (1973): 391–99; and his review of Nriagu's *Lead and Lead Poisoning in Antiquity* in *Medical History* 29 (1984): 107–108. An even harsher review of the thesis can be found in John Scarborough's review of Nriagu's book in *Journal of the History of Medicine* 39 (1984): 469–75.

23. See, e.g., Boston Children's Hospital Archives no. 92–25, Book L, 168, 188; ibid., Book M, 150.

24. Ibid., log book 18 April 1906, 74.

25. New York City, Department of Health, "Special Investigation of Infantile Paralysis," box "Cases A–Be," manuscript collection no. 33, American Philosophical Society. In a similar case, a 2-year-old developed a facial paralysis after being "kept on premises and in their apartment since this epidemic, (three weeks)." As in this and many other reported cases, the apartment was being painted at the time.

26. For modern diagnosis and pathology, consult Malcolm Parsons, *Tuberculosis Meningitis: A Handbook for Clinicians* (Oxford, UK: Oxford University Press, 1979); "versatile" quote, 1. Boston pediatrician Charles F. McKhann ("Lead Poisoning in Children, with Notes of Therapy," *AJDC* 32 [1926]: 387) remarked

that in severe cases of lead poisoning, "the signs are variable, and may change rapidly, simulating tuberculosis meningitis from which it is often only with difficulty distinguished." The confusion had not stopped even two decades later. For example, three of the fourteen cases of lead poisoning treated at a Brooklyn hospital in 1951 were initially diagnosed as tubercular meningitis; see Rudolph Giannattasio, Andrew Bedo, and Michael Pirozzi, "Lead Poisoning, Observations in Fourteen Cases," *AJDC* 84 (Sept. 1952): 316–21.

27. John M. Hunter, "The Summer Disease: An Integrative Model of the Seasonality Aspects of Childhood Lead Poisoning," *Social Science and Medicine* 11 (1977): 691–703.

28. In Boston from 1904 to 1913, it was found that the peak months for TB meningitis were June and January, though only the January rise was seen in the years 1909–13; see Alfred Meyers, "A Study of 105 Cases of Tuberculous Meningitis," *AJDC* 9 (May 1915): 427–45. A New York pediatrician reported in 1940 that TB meningitis in that city peaked in March; see L. Emmett Holt, Jr., *The Diseases of Infancy and Childhood,* 11th ed. (New York: Appleton, 1940), 1328. In 1922, a German pediatrician found seasonal peaks in March, July, and November; see Emma Stelling, "Untersuchungen über Meningitis Tuberculosa," *Archiv für Kinderheilkunde* 70 (1922): 196.

29. I examined complete years for 1901–12, 1914, and 1917–19. I ignored 1913 because the admission logs for several months of that year contain almost no information about patients. One log book, for January 1915 to June 1916, was lost, and so I excluded all data for both those years.

30. Meyers, "105 Cases of Tuberculous Meningitis."

31. In 1890, Keating (*Diseases of Children,* 634–39 [see intro., n. 27]) tabulated the cases of pediatric lead poisoning he had encountered in his practice and in the journals. Of thirty cases, ten were attributed to contaminated foods and six to water, five were nursing infants poisoned by their nurses' breast ointment or cosmetics, three were poisoned by fresh paint vapors, two were traced to medicines, another two to the child's occupation in lead trades, and one to a parent's occupation. Keating found only one case where the victim had eaten a lead-painted object, and this was not house paint.

32. David Denison Stewart, "A Clinical Analysis of Sixty-four Cases of Poisoning by Lead Chromate (Chrome Yellow), Used as a Cake Dye" (paper read before the Philadelphia County Medical Society, 14 Sept. 1887; College of Physicians Library, Philadelphia). Stewart wrote several articles over the course of his investigation; see, e.g., "Lead Convulsions, a Study of Sixteen Cases," *American Journal of the Medical Sciences* 109 (Mar. 1895): 288–306. One of the bakers, a Mr. Palmer, had lost his first wife and six children to convulsions over the course of a year and a half, and a journeyman baker living with the Palmer family contracted severe abdominal cramps and constipation and quit, suspecting something in the house was the cause of his illness.

33. Early in the twentieth century, most paints sold in the United States con-

tained some lead, either as the chief component in suspension in oil or as a pigment. The most expensive paint, prized for its durability and covering power, was pure white lead (lead chromate) in linseed oil.

34. L. Emmett Holt, Jr., *The Diseases of Infancy and Childhood*, 6th ed. (New York: Appleton, 1911), 400.

35. Blackfan, "Lead Poisoning in Children"; Strong, "Meningitis, Caused by Lead Poisoning," 532; McKhann, "Lead Poisoning in Children, with Notes on Therapy."

36. McKhann, "Lead Poisoning in Children, with Notes on Therapy," 392; "pica," Latin for magpie, a bird famous for its catholic appetite, refers to any craving for unnatural food but is usually associated with dirt eating. This eating "disorder" (which is merely an exaggeration of the hand-to-mouth and teething activity normal in very young children) was used to brand children who ate paint (and their supposedly inattentive parents) as at least partly to blame for their illness.

37. To see how forceful the impact of foreign experience can be when the time is right, contrast the speed with which research about phossy jaw, a disfiguring disease of phosphorus workers, was transmitted from England and France to the United States in the first years of the Progressive Era occupational hygiene campaigns; see David Moss, "Kindling a Flame under Federalism: Progressive Reformers, Corporate Elites, and the Phosphorous Match Campaign of 1909–1912," *Business History Review* 68 (Summer 1994): 244–75.

38. On the ubiquity of TB infection in the nonsymptomatic public, see Paul Starr, *The Social Transformation of American Medicine* (New York: Basic Books, 1982), 191. My research at Children's Hospital began, and almost ended, in the Pathology Department. I was nearly convinced that my pursuit would prove fruitless because almost every TB victim's autopsy revealed either a form of visible tubercle (whether or not located in the meninges) or microscopic evidence of TB bacilli. But only a handful of TB meningitis cases were autopsied, and the microscopic findings do not eliminate lead as an additional suspect.

39. On the reasons for the delay of the decline in infant mortality in the nineteenth century, see Richard A. Meckel, *Save the Babies: American Public Health Reform and the Prevention of Infant Mortality, 1850–1929* (Baltimore: Johns Hopkins University Press, 1990).

40. In addition to these dominant competing theories, Sylvia Noble Tesh identifies a "personal behavior theory," similar to today's "lifestyle theory" as well as to premodern supernatural theories; see her *Hidden Arguments: Political Ideology and Disease Prevention Policy* (New Brunswick, N.J.: Rutgers University Press, 1988), 8–9.

41. For how contagionism, germ theory, and miasmism worked together, see James H. Cassedy, "The Flamboyant Colonel Waring: An Anticontagionist Holds the American Stage in the Age of Pasteur and Koch," *BHM* 36 (1962): 163–76. The New Public Health, which arose in the Progressive Era, stressed scientism and professional involvement and emphasized personal, industrial, and mental "hy-

giene" (no value-neutral term like *sanitation*). On the New Public Health, see Barbara Rosenkrantz, *Public Health and the State: Changing Views in Massachusetts, 1842–1936* (Cambridge, Mass.: Harvard University Press, 1972), 131; Starr, *Social Transformation of American Medicine*, 190–92.

42. Bacteriology changes public health, according to Elizabeth Fee in *Disease and Discovery: A History of the Johns Hopkins School of Hygiene and Public Health, 1916–1939* (Baltimore: Johns Hopkins University Press, 1987): it became "an ideological marker, sharply differentiating the 'old' public health, the province of untrained amateurs, from the 'new' public health, which belonged to scientifically trained professionals" (19); see also Starr, *Social Transformation of American Medicine*, 189.

43. The Bartholdt bill, drafted in 1910, would have declared lead paint a poison and restricted its manufacture and use, but it never made it out of committee.

44. Zelizer, *Pricing the Priceless Child*, 5 (see intro., n. 5).

45. See Meckel's "Cities as Infant Abattoirs," a dark and troubling chapter on changing attitudes toward infant mortality in his *Save the Babies*. On the need to be cautious when relying on mortality figures from the nineteenth century, see Maris A. Vinovskis, "Mortality Rates and Trends in Massachusetts before 1860," *Journal of Economic History* 32 (1972): 184–213.

46. Murdina M. Desmond, "A Review of Newborn Medicine in America: European Past and Guiding Ideology," *American Journal of Perinatology* 8 (Sept. 1991): 308–22; Leon Eisenberg, "From Circumstance to Mechanism in Pediatrics during the Hopkins Century," *Pediatrics* 85 (Jan. 1990): 42–49; Meckel, *Save the Babies*, ch. 2.

47. Thomas McKeown in *The Modern Rise of Population* (New York: Oxford University Press, 1976) studied the decline of mortality rates in the nineteenth century, and suggested that real progress in mortality was made independent of antibiotics and vaccines: it was the improved standard of living, not medicine, that made the difference. Nowhere is the McKeown argument about improvements in diet and water supplies stronger than in explaining the reduction in infant mortality.

48. Of course, the environmental assaults in many workplaces obviated the need for more sophisticated empirical methods. The role of these occupational health specialists in the formation of the environmental movement has recently been enlarged by historical studies; see, e.g., Christopher Sellers, "Factory as Environment: Industrial Hygiene, Professional Collaboration and the Modern Sciences of Pollution," *Environmental History Review* 18 (Spring 1994): 55–83; Robert Gottlieb, *Forcing the Spring: The Transformation of the American Environmental Movement* (Washington: Island Press, 1993).

49. As Holt wrote in "Lead Poisoning in Infancy" in 1923, "As far as clinical descriptions of the disease are concerned, the early writings of Tanquerel have yet to be improved upon" (231), and he cites Tanquerel's *Lead Diseases* (1848). On Kehoe's domination of lead research, see William Graebner, "Hegemony through

Science: Information, Engineering and Lead Toxicology, 1925–1965," in *Dying for Work*, ed. Rosner and Markowitz, 140–59 (see intro., n. 12).

50. Australian infant mortality in 1901 was 103.6 per thousand live births, the lowest of all industrial nations. By 1910 the rate had fallen to 74.81, less than half that of the average for all of Europe; *Transactions of the Fifteenth International Congress on Hygiene and Demography* 6: sect. 9, "Demographics" (Washington: GPO, 1913), 161–62.

51. In the years before the discovery of the Broken Hill lode in New South Wales, Queensland produced almost 30% of Australia's silver and lead ore.

52. Ruddock, "Lead Poisoning in Children," 1682.

CHAPTER 3: TOXIC PURITY

1. House Committee on Interstate and Foreign Commerce, *Hearings on H. R. 21901, Manufacture, Sales, Etc., of Adulterated or Mislabeled White Lead and Mixed Paint*, 61st Cong., 2d sess., 31 May 1910, 14.

2. Actually, one aspect of his testimony would have been remarkable: his emphasis on dust as a possible source of lead poisoning. Most of the initial concern over lead-paint poisoning focused on preventing children from eating paint chips. The problems of paint dust were not addressed until much later.

3. Advertisement, *Painter and Decorator* 24 (1910), unpaged.

4. William Pulsifer, *Notes for a History of Lead and an Inquiry into the Development of the Manufacture of White Lead and Lead Oxides* (1888), 314–16; McCord, "Early America: Lead Compounds" (see intro., n. 6).

5. The only major corroders who retained independent ownership were Eagle of Cincinnati and Wetherill of Philadelphia; George B. Heckel, *The Paint Industry: Reminiscences and Comments* (St. Louis: American Paint Journal Company, 1928), 257–65; quote, 257–58. According to Heckel, Harn did not invent the Heinz 57 varieties slogan—Heinz himself did.

6. For the century's first decade, I have found consumption figures only for 1907–1909. In 1907, white lead manufacturers in the United States consumed 115,000 short tons, or 29.72% of the total; in 1908, 118,000 tons, or 35.12%; and in 1909, 134,000 tons, or 36.31%; U.S. Geological Survey, "Part 1: Metals," in *Mineral Resources of the United States* 1909 (Washington: GPO, 1911). Figures for lead in batteries from U.S. Department of the Interior, Bureau of Mines, "Lead," in *Minerals Yearbook* 1920–88 (Washington: GPO, 1922–90).

7. H. V. Kent, president of Kent and Purdy Paint Company, to James R. Mann, 18 June 1910, in House Committee on Interstate and Foreign Commerce, "Papers Accompanying Specific Bills and Resolutions," file H. R. 61A-D6, H. R. 21832–25825, RG 233, National Archives.

8. On the French role in developing zinc oxide, see Stevenson, "History of Lead Poisoning" (see intro., n. 6), 381–82; Heckel, *Paint Industry*, 280–81; quote,

291. Heckel (*Paint Industry,* 4) describes how he entered the paint business in his youth. His father, like William Dean Howells's *Silas Lapham,* got caught up in a "mineral paint craze" raging through western Pennsylvania after the Civil War. He hired a man to dig holes about his farm. When one yielded iron oxide on a base of aluminum silicate, he built a paint factory.

9. From 1900 to 1910 annual zinc production in the United States increased in value from $11 million to $27 million and 435% by weight, while lead increased from $23 million to $33 million and only 147% by weight; Bureau of the Census, *Historical Statistics of the United States, Colonial Times to 1970,* Part 1 (Washington: GPO, 1975), 583–84, 603–604.

10. The value of paint and varnish produced in 1899 was about $69,560,000 and in 1909 it was $124,889,000 (both amounts converted to constant 1958 dollars), for an increase of about 79.5%; U.S. Department of Commerce, *Statistical Abstract of the United States, 1915* (Washington: GPO, 1916), 196. The value of goods and services in paints rose by 45.7% from $299,800,000 in 1899 to $436,670,000 in 1909—an impressive rate of growth but well short of that of the nation's gross national product, at 56.1%; Bureau of the Census, *Historical Statistics of the United States,* 224.

11. Bureau of the Census, *Historical Statistics of the United States,* 224.

12. McCord, "Early America: Lead Compounds," 79–80; Hamilton, *Exploring the Dangerous Trades* (see intro., n. 17), 131.

13. This narrative could have begun at the mine or the primary smelter, but that would introduce too many variables in what is meant to be a brief discussion of paint manufacture. For detailed descriptions of particular factories, see Alice Hamilton, "White-Lead Industry in the United States, with an Appendix on the Lead-Oxide Industry," *Bulletin of the Bureau of Labor* 95 (July 1911): 189–259; Edward E. Pratt, *Occupational Diseases: A Preliminary Report on Lead Poisoning in the City of New York, with an Appendix on Arsenical Poisoning* (Albany, N.Y., 1912).

14. Pratt, *Occupational Diseases,* 454–55.

15. Ibid., 478–80.

16. In a lecture at Harvard Medical School in 1919, Hamilton declared, "Our information on the effect of lead on women in America is fortunately very scanty, for we have never employed women in any of the more dangerous lead processes with the exception of litho transfer work and the finishing of glazed pottery and tiles. Industrial lead poisoning in women is still a rarity in the United States." Hamilton's lecture summarized the research on lead poisoning from the preceding decade, and was published as "Lead Poisoning in American Industry" (see intro., n. 24). For a detailed discussion of women in the potteries, see Marc Jeffrey Stern, *The Pottery Industry of Trenton: A Skilled Trade in Transition, 1850–1929* (New Brunswick, N.J.: Rutgers University Press, 1994); for printers and potters, see Klein, "From Knowledge to Policy" (intro., n. 7).

17. It was important to choose your manure wisely! That of dogs, pigs, or any meat-eating animal was to be avoided, as its gas would produce black sulphide of lead; horses made the best manure, but American factories employed tannin almost exclusively—preferring it for the whiter whites it produced. See Thomas Oliver, "Industrial Lead Poisoning, with Descriptions of Lead Processes in Certain Industries in Great Britain and the Western States of Europe," in *Bulletin of the Bureau of Labor* 95 (July 1911): 11.

18. In the plants Pratt investigated (see *Occupational Diseases*, 451), the same workers who built the blue beds also stripped the white beds. Alex P——, a Polish farmer who emigrated to the United States at age 21, had worked for fourteen years as a stacker and stripper in a New York City white lead factory. Although he had only recently contracted lead poisoning in the form of colic and paralysis, four of his eight children had been stillborn or had died in their first week. Excellent descriptions of the stack process appear in Gordon Thayer, "The Lead Menace," *Everybody's Magazine*, 1 Mar. 1913, 327, which focuses rather emotionally on U.S. production; and in Oliver, *Dangerous Trades* (see ch. 1, n. 38), 288, which focuses on Britain and the continent; see also Harrison Brothers and Company, "The Chemistry of Paints" (1890), 10.

19. Pratt (*Occupational Diseases*, 397) observed of a mechanical separator in a particular plant that although its mouth was "protected by a hood, this is one of the most dangerous processes in the factory. . . . The worker stands constantly over a rising cloud of white lead dust." Typically, the separator was located "in a poorly ventilated, dark basement." See also Oliver, *Dangerous Trades*, 288.

20. Pratt, *Occupational Diseases*, 489, 450.

21. For conditions in the Pullman plant, see Alice Hamilton, "What One Stockholder Did," *Survey* 28 (June 1912): 387–89; Hamilton, *Exploring the Dangerous Trades*, 145–48, 156–59.

22. Pratt, *Occupational Diseases*, 505–506.

23. Senate Committee on Manufactures, *Adulterated or Mislabeled Paint, Turpentine, or Linseed Oil: Hearings before the Committee on Manufactures*, 61st Cong., 2d sess., 17–18 Feb. 1910; reprinted in *Painters Magazine* 37 (Mar. 1910): 259.

24. Which is not to say that the lead companies did not experiment—they seemed always on the lookout for newer, faster methods of producing their ancient product. Paint manuals and historical accounts are filled with references to the Kremnitz process, the Carter quick process, the Acme process, and so on. But some variant of the "Old Dutch Process," with its premodern employment of ceramic pots, spent tanbark, and cheap, unskilled labor, remained the standard for most white lead production well into the twentieth century.

25. By 1911, labeling laws were on the books in Kansas, Minnesota, Nevada, and Texas. On the Nebraska law, see Heckel, *Paint Industry*, 322; for the other state laws, see issues of *Paint, Oil and Drug Review* 12 (1911): for Kansas, 5 July,

10; Minnesota, 19 July, 12; Nevada and Texas, 6 Sept., 16; see also *Heath and Milligan Manufacturing Company v Worst,* 207 US 338, 52 L. Ed. 236 (28 S. Cir 114, 1907).

26. The association paid $10 per year to informants in each state whose sole duty was to report the introduction of labeling laws. "What it cost the proponents of these measures to carry on the fight I do not know," recalled Heckel (*Paint Industry,* 324), "but we all had our unkind suspicions at the time—probably entirely unjustified."

27. Ladd, who trained as a chemist in Maine, moved to North Dakota in 1890 to run the experiment station at the state's new agricultural college. From 1899, he published the *North Dakota Farmer and Sanitary Home* and pushed for the state's Food and Drugs Act, passed in 1901. When he was sued by food processors in 1904, he gained a national forum for exposing blatant frauds in food processing and labeling. In 1920, North Dakota elected him to the U.S. Senate, where he sponsored yet another paint labeling law. See Elwyn B. Robinson, *History of North Dakota* (Lincoln: University of Nebraska Press, 1966), 261–62. James Harvey Young (*Pure Food: Securing the Federal Food and Drugs Act of 1906* [Princeton, N.J.: Princeton University Press, 1989], 281–82) sees Ladd as representative of reformers, "neither Populists nor Progressives" but convinced that the Midwest was the "colonial territory" of eastern manufacturers.

28. Heckel (see *Paint Industry,* 326) met Ladd in Fargo "on a bleak fall day" in 1906 to try to convince him of the meaninglessness of "pure" paint and to dissuade him from publishing damning evidence of mislabeling, but to no avail.

29. *Heath and Milligan v Worst.*

30. Edwin F. Ladd, "Paints and Their Composition," bulletin no. 70, North Dakota Agricultural College, Government Agricultural Experiment Station of North Dakota; reprinted in Senate Committee, *Adulterated or Mislabeled Paint,* 15.

31. These were the years "when a dealer who also sold anything else besides paint was insulted if you addressed him as a paint dealer"; St. Louis paint manufacturer Allen W. Clark, quoted in Heckel, *Paint Industry,* 343.

32. The draft legislation was known as the "Joint Conference Bill," after the conference held by the Paint Manufacturers' Association, the Paint Oil and Varnish Association, and the International Association of Master House Painters and Decorators; John Dewar to John Esch, 2 May 1908. The alliance was short lived because opposition ran deep among the ranks of the mixed-paint manufacturers.

33. Heyburn helped frame Idaho's constitution in 1899, and he served in the U.S. Senate from 1903 until his death in 1912; see U.S. Congress, *Biographical Directory of the United States Congress, 1774–1989, the Continental Congress, September 5, 1774, to October 21, 1788, and the Congress of the United States, from the First through the One Hundredth Congresses, March 4, 1789, to January 3, 1989, Inclusive* (Washington: GPO, 1989), 1181. With Porter McCumber of North Dakota, he cosponsored the bill that became the Pure Food and Drugs Act of 1906

and pushed it through the Senate; see Young, *Pure Food,* 204–10. The paint labeling bill was first presented in the House by Rep. Thomas Frank Marshall of North Dakota as H.R. 17824, then combined to form Heyburn's bill; since only Heyburn reintroduced the legislation in the next session, I refer to all three bills as "the Heyburn bill." Hence, some of the correspondence regarding it is written to representatives, not senators. A number of these letters can be found in committee correspondence, and many were reproduced in published committee hearings; see, e.g., testimony of Massachussets painter P. J. Imberger in Senate Committee, *Adulterated or Mislabeled Paint,* 123; see also Heckel, *Paint Industry,* 331–40.

34. Retail Hardware Association of the Carolinas to Rep. Wyatt Aiken, 6 Apr. 1908, House Committee on Interstate and Foreign Commerce, "Naval Stores and Paint," file H. R. 60A-H16.12, RG 233, National Archives.

35. Under state laws, turpentine and linseed oil adulterers were prosecuted more frequently than paint manufacturers, perhaps because standards of purity were more clear cut for these chemicals.

36. Senate Committee, *Adulterated or Mislabeled Paint,* 257.

37. Senate Committee, *Adulterated or Mislabeled Paint,* 15.

38. Paint Manufacturers' Association of the United States to Rep. Irving Wagner, House Committee, "Naval Stores and Paint," 4 Mar. 1908; testimony of Henry Wood of Massachusetts in Senate Committee, *Adulterated or Mislabeled Paint,* 261.

39. A. Burdsal Company to Senate Committee on Manufactures, 18 May 1908, file Sen 60A J81, RG 46, National Archives.

40. Dewar named at least six paint manufacturers who had recently begun labeling their paints, including Acme Paint & Colorworks of Detroit, Sherwin-Williams, and the NLC; Senate Committee, *Adulterated or Mislabeled Paint,* 134.

41. James Patton, president of Patton Paint Company, quoted in testimony of Dewar in Senate Committee, *Adulterated or Mislabeled Paint,* 134; testimony of William A. Buddecke in House Committee, *Hearings on H. R. 21901,* 40.

42. Bade Brothers to Rep. William Calder, House Committee, "Naval Stores and Paint," 11 Apr. 1908; D. T. Wier to Ladd, quoted in testimony of Dewar, Senate Committee, *Adulterated or Mislabeled Paint,* 15.

43. A. Burdsal Company to Senate Committee on Manufactures.

44. House Committee, *Hearings on H. R. 21901,* 3. Bartholdt, from St. Louis, served in the House from 1893 to 1915. He was born in Germany, migrated to the United States at age 19, and before entering politics was a publisher, rising to the post of editor in chief of the *St. Louis Tribune;* see *Biographical Directory of the U.S. Congress,* 584.

45. Hearings on the Heyburn bill were held on 17–18 Feb. 1910; see Senate Committee, *Adulterated or Mislabeled Paint,* 174. On 26 Feb., the Bartholdt bill was referred to the Committee on Interstate and Foreign Commerce; see *Congressional Record,* 61–62, 2480–81; House Committee, *Hearings on H. R. 21901,* 24–26.

46. *Painters Magazine* 37 (Mar. 1910): 266.

47. Ibid.

48. L. Matern, "The Prevention of Lead Poisoning," *Painter and Decorator* 24 (May 1910): 360–61.

49. The same German laws were adapted by the American Association for Labor Legislation when drafting its model state legislation, "Standard Bill for the Prevention of Occupational Diseases with Special Reference to Lead Poisoning"; *American Labor Legislation Review* 4 (Dec. 1914): 541–46.

50. France's stance was due in part to the fact that in the early 1840s Jean-Edne Leclaire, an influential chemist and philanthropic industrialist, developed a commercial-quality zinc oxide paint. Between 1849 and 1891, the French government made at least four separate proclamations banning lead for interior paints in government buildings. On various European laws, see Stevenson, "History of Lead Poisoning," 333–90; for the French law of 1909, see transcript of the Meeting of the Chamber of Deputies, 10 July 1909, reprinted in House Committee, *Hearings on H. R. 21901*, 30; on the 1918 reinstatement, see Stevenson, "History of Lead Poisoning," 382.

51. House Committee, *Hearings on H. R. 21901*, 26–28. Rhodes had been a member of the House of Representatives in the 59th Congress, was a former mayor of Potosi, Missouri, and from 1919 to 1923 would serve again in the House, chairing the Committee on Mining; *Biographical Directory of the U.S. Congress*, 1710.

52. House Committee, *Hearings on H. R. 21901*, 14–15.

53. Ibid., 4, 18.

54. No painters' unions testified at the hearings, although the St. Louis Painters' District Council and the Los Angeles District Council of Painters sent resolutions in support, as did the Indianapolis District Council of the Brotherhood of Painters, Decorators, and Paperhangers of America; ibid., 8. But the voices of these small groups carried little weight compared to that of Dewar's national association.

55. Testimony of Kent in House Committee, *Hearings on H. R. 21901*, 7; Kent to Mann, 18 June 1910, House Committee, "Papers Accompanying Specific Bills and Resolutions."

56. Wiley also dismissed all the bill's restrictions on manufacture and provisions for worker protection, except for the one requiring that "workmen shall not drink liquor in the place they work"; text of Bartholdt bill, House Committee, *Hearings on H. R. 21901*, 20–21. On Wiley and the Pure Food and Drugs Act, see James Harvey Young, *The Toadstool Millionaires: A Social History of Patent Medicines in America before Federal Regulation* (Princeton, N.J.: Princeton University Press, 1961); for a more cynical assessment of Wiley's role in regulation, see Peter Temin, *Taking Your Medicine: Drug Regulation in the United States* (Cambridge, Mass.: Harvard University Press, 1980). For Wiley's conservative, tradi-

tional interpretations of purity and artificiality, see Jack High and Clayton A. Coppin, "Wiley and the Whiskey Industry: Strategic Behavior in the Passage of the Pure Food Act," *Business History Review* 62 (Summer 1988): 286–309.

57. Kent to Wiley, 17 June 1910, House Committee, "Papers Accompanying Specific Bills and Resolutions."

58. House Committee, *Hearings on H. R. 21901*, 17.

59. Testimony of Philbin in House Committee, *Hearings on H. R. 21901*, 44–51.

60. Philbin to Mann, 7 June 1910, House Committee, "Papers Accompanying Specific Bills and Resolutions."

61. In 1955, paint manufacturers voluntarily adopted a 1% limit on lead in paints for interior use, though in practice the limit was widely ignored—and not only by small companies whom, it might be argued, such restrictions damaged; Senate Committee on Labor and Human Resources, *Lead-Based Paint Poisoning: Hearing before the Subcommittee on Health*, 91st Congr., 2d sess., 23 Nov. 1970, 229. Much of the debate in hearings on the early 1970s legislation regarding lead poisoning concerned the amount of lead to be allowed in paints. Child health advocates wanted the cap at 0.06%, the limits of analysis, while the paint industry wanted the limit at 0.5%, which would allow for small amounts of lead "adulterants" as well as for lead hardening agents; see Senate Committee, *Lead-Based Paint Poisoning Amendments of 1972* (see intro., n. 22).

62. House Committee, *Hearings on H. R. 21901*, 12.

63. Heckel, *Paint Industry*, 353.

64. Ibid., 609. Earlier lead-industry attacks on "cheap substitutes" tended to paint zinc, barytes, and whiting with the same brush. The NLC's 1900 pamphlet "Uncle Sam's Experience with Paints" is typical.

65. Heckel, *Paint Industry*, 266. Harn and Heckel's cooperative venture culminated in the "Save the Surface" campaign of the war years; ibid., 401. Lead-industry advertisements continued to feature exposés of fraudulent salesmen hawking adulterated "dope lead" (see, e.g., *Carter Times,* June 1918, 13), but for the most part, they stopped castigating legitimate (and accurately labeled) competitive products.

66. William Lucas to Wagner, House Committee, "Naval Stores and Paint," 4 Mar. 1908.

67. Hamilton, *Exploring the Dangerous Trades,* 254.

68. "Deplores Our Absence," *New York Times,* 26 Oct. 1921, 1; "Sets a White Lead Limit," *New York Times,* 19 Nov. 1921; "International Labor Conference at Geneva," *MLR* 14 (Jan. 1922): 51–56. The resolution banning lead house paint gave the signatories six years to adopt their own regulations. Interior paint was to contain no measurable amount of lead; exterior paint could not exceed 2% lead by weight. The only exceptions were for public buildings and railroads. By 1926, the *MLR*'s pages were peppered with reports of countries in Europe "coming on

line"; e.g., for France and Spain, see *MLR* 22 (Apr. 1926): 91–92; for Belgium and the United Kingdom, 23 (July 1926): 63; for Sweden, 23 (Dec. 1926): 71. The examples of Cuba and Czechoslovakia appear in International Labor Office (ILO), "Report of the International Labor Office upon the Working of the Convention Concerning the Use of White Lead in Painting" (Geneva: ILO, 1933), 8–9.

69. Hamilton, *Exploring the Dangerous Trades*, 254. In 1930, Emery Hayhurst and Leonard Greenburg printed observations from the British industrial hygienist Thomas Legge on safety of lead workers in various nations since World War I; see their "Industrial Hygiene," *AJPH* 20 (1930): 436–38.

70. ILO, "White Lead: Data Collected by the International Labor Office in Regard to the Use of White Lead in the Painting Industry," Studies and Reports, Series F (Industrial Hygiene), no. 11 (Geneva: ILO, 1927), 30, 35.

71. The Lead-Based Paint Poisoning Prevention Act of 1970 marked the beginning of decades of legislative and bureaucratic mudwrestling to eliminate lead paint from store shelves and tenement walls; in 1978, Congress finally banned the manufacture of paints with more than 0.06% lead for any residential purposes, but procuring funding for abatement and adequate screening and treatment continues to be a problem.

72. I have found no source on production of lead pigments between 1910 and 1920, so values for these years have been interpolated. In addition to white lead, another four thousand tons of red lead and other lead pigments were consumed between 1920 and 1976; see U.S. Geological Survey, "Metals," in *Mineral Resources;* Bureau of Mines, "Lead," in *Minerals Yearbook* 1920–76.

73. On the origins of the OSHA legislation, see Carl Gersuny, *Work Hazards and Industrial Conflict* (Hanover, N.H.: University Press of New England, 1981), 111–15; for a critique of OSHA, see David Rosner and Gerald Markowitz, "Research or Advocacy: Federal Occupational Safety and Health Policies during the New Deal," in *Dying for Work,* 83–102 (see intro., n. 12); on the hearings to establish occupational standards for lead under OSHA, see "Conference on Standards for Occupational Lead Exposure," *Journal of Occupational Medicine* 17 (Feb. 1975).

74. Lead Industries Association, *Useful Information about Lead* (New York: n.p., 1934), 51.

CHAPTER 4: OCCUPATIONAL LEAD POISONING IN THE PROGRESSIVE ERA

1. Occupational medicine was a second career for Hamilton; see Barbara Sicherman, *Alice Hamilton: A Life in Letters* (Cambridge, Mass.: Harvard University Press, 1984), 358. She conducted her first industrial investigation in 1910 at the age of 40 (and completed her last, of the rayon industry, 30 years later); prior to that she had established herself as a bacteriologist and lecturer in pathology at

Northwestern University. In addition, her unpaid calling as reformer and activist predated her stint as occupational investigator and persisted well beyond her retirement from paid service and nearly to the end of her hundred-year life.

2. Hamilton, *Exploring the Dangerous Trades,* 127–28 (see intro., n. 17). This encounter has been described in a number of studies of Hamilton; see, e.g., Angela Nugent Young, "Interpreting the Dangerous Trades: Workers' Health in America and the Career of Alice Hamilton, 1910–1935," (Ph.D. diss., Brown University, 1983), 1; Sicherman, *Alice Hamilton,* 159.

3. Quote, Carey P. McCord, *A Blind Hog's Acorns: Vignettes of the Maladies of Workers* (Chicago: Cloud, 1945), 21. For other internalist histories, see Hamilton herself in *Exploring the Dangerous Trades,* 145–48; Joseph C. Aub and Ruth K. Hapgood, *Pioneer in Modern Medicine: David Linn Edsall of Harvard* (Cambridge, Mass.: Harvard Medical Alumni Association, 1970).

4. John Burnham provides a powerful indictment of the Progressives' "social control" in "Medical Specialists and Movements Toward Social Control in the Progressive Era: Three Examples," in *Building the Organizational Society,* ed. Jerry Israel (New York: Free Press, 1972), 19–30. Medical historians have tended to balance the issues of science and class dynamics, although they are quicker to allow middle-class reformers their "social control" so long as they were saving lives; see Rosenkrantz, *Public Health and the State* (see ch. 2, n. 41); Brandt, *No Magic Bullet* (intro., n. 13); Meckel, *Save the Babies* (ch. 2, n. 39).

5. Some scholars, who argue that scientism eventually excised moral imperative from scientific inquiry, frame their narratives in terms of a rise and fall, in which Hamilton's cohort serves as a foil to the succeeding science-jaded generation. In *Hazards of the Job,* Sellers (see intro., n. 7) sees the Progressive hygienists as having formulated a "public and constructive knowledge" about occupational disease that tragically contained the seeds of its own destruction in its reliance upon science; Graebner discusses what those seeds brought forth in "Hegemony through Science" (ch. 2, n. 49). This chapter acknowledges the enormous power of business and cultural interests to define scientific "truth" but does not succumb to despair over whether "objective truth" exists. Tesh's *Hidden Arguments* (ch. 2, n. 40) has influenced my thoughts on this subject.

6. I am not arguing here that nineteenth-century physicians—let alone paint-factory foremen—understood the complex physiological issues of the body's metabolic response to lead or the subtle chronic hazards of low-level exposure. But a foreman could probably point out a case of lead poisoning more readily than an American-trained physician—despite the fact that, if he chose to look, the latter had access to a growing body of medical research on diagnosing plumbism. By the 1940s, however, the company doctor was able to "prove," by technical medical reasoning, that what the foreman knew was lead poisoning could not be. On nineteenth-century physicians' awareness of the physiology of lead poisoning, see Wedeen, *Poison in the Pot,* 159 (see intro., n. 6).

7. American women could be found in many lead-using industries, but

mostly in those where lead was employed in processes related to trades in which women were already accepted, such as printing, decoration, and pottery.

8. David Montgomery brilliantly narrates the decline of workers' control in *Workers' Control in America: Studies in the History of Work, Technology, and Labor Struggles* (Cambridge, UK: Cambridge University Press, 1980). While I acknowledge the loss in many areas of work life, I will argue that as plant physicians and "rational" plant hygiene undermined workers' control, workers lived longer and healthier lives.

9. Alice Hamilton, "Industry Is Health Conscious," *Medical Woman's Journal* (Oct. 1948): 33.

10. For more on lead poisoning in the colonial period, see McCord, "Early America: Clinical Lead Poisoning," 120–25 (see ch. 1, n. 1).

11. Carey P. McCord, "Lead and Lead Poisoning in Early America: The Pewter Era," *Industrial Medicine and Surgery* 22 (Dec. 1953): 577. In recent years, historians and anthropologists have put a bit more flesh on McCord's allusory skeleton. Wedeen's *Poison in the Pot* is especially strong on gouts and other renal effects of lead poisoning in the eighteenth, nineteenth, and twentieth centuries; see also Aufderheide et al., "Lead in Bone II" (see ch. 1, n. 49); Aufderheide et al., "Anthropological Applications" (ch. 1, n. 49).

12. McCord, "Early America: Clinical Lead Poisoning," 123.

13. Carey P. McCord, "Occupational Health Publications in the United States prior to 1900," *Industrial Medicine and Surgery* 24 (Aug. 1955): 363–68. I did not attempt to record or classify all the articles about occupational lead poisoning in the nineteenth century that I found in such sources as the *Index and Catalogue of the Library of the Surgeon General's Office* (Washington, 1886), but many do not appear in McCord's bibliography, which was in fact largely drawn from the "Proceedings of the Second National Conference on Industrial Diseases, Atlantic City Sessions of 1912," *American Labor Legislation Review* 2 (1912).

14. "Colica Pictonum, Cured by Vinegar," *Boston Medical Intelligencer,* 21 Oct. 1826, quoted in McCord, "Early America: Clinical Lead Poisoning," 121.

15. Benjamin McCready, "On the Influence of Trades, Professions, and Occupations in the United States, in the Production of Disease," in *Transactions/State Medical Society of New York* 3 (1837), quoted in McCord, "Early America: Clinical Lead Poisoning," 122. McCord argues that much of McCready's essay was lifted directly from C. T. Thackrah, presumably his "The Effects of Arts, Trades and Professions . . . on Health and Longevity" (1832).

16. The bibliography on nineteenth-century urban public health continues to burgeon. A useful starting point is Joel A. Tarr, *The Search for the Ultimate Sink: Urban Pollution in Historical Perspective* (Akron, Ohio: University of Akron Press, 1996); see also Martin Melosi, ed., *Pollution and Reform in American Cities, 1870–1930* (Austin: University of Texas Press, 1980); Rosenkrantz, *Public Health and the State;* Rosenberg, *Cholera Years* (see intro., n. 13). William J. Novak's *The People's Welfare: Law and Regulation in Nineteenth-Century America*

(Chapel Hill: University of North Carolina Press, 1996) places city and state public health measures within a context of an expansive movement to regulate all manner of health, safety, and moral issues, a movement that began in earnest long before the conventional post-bellum reform period.

17. Citizens' Association of New York, Council of Hygiene and Public Health, *Sanitary Conditions of the City* (1865; reprint, New York: Arno Press, 1970), 341; New York City Board of Health, *Third Annual Report* (1873), 312, discussed in Sellers, "Manufacturing Disease," 1 (see intro., n. 7); Massachusetts Bureau of Statistics, *Annual Reports,* 1870–1931; Caroll Wright, *The Working Girls of Boston,* in Massachusetts Bureau of Statistics of Labor, *Annual Report* (1884; printed separately 1889; reprint, New York: Arno Press, 1969), 73–74.

18. Carey P. McCord, "Lead and Lead Poisoning in Early America: The Lead Pipe Period," *Industrial Medicine and Surgery* 23 (Jan. 1954): 29; Young, "Interpreting the Dangerous Trades," 4.

19. Germany, Britain, and France are the obvious choices for comparison: after the United States, these nations produced the largest amounts of lead paints and products during the early twentieth century; Christopher J. Schmitz, *World Non-Ferrous Metal Production and Prices, 1700–1976* (Totowa, N.J.: Frank Cass, 1979). In addition, their hygiene programs were sufficiently different from one another to allow comparison, but each had a profound impact upon the strategies of American hygienists, reformers, and governments.

20. For a succinct overview of the historiography of welfare-state formation, see Seth Koven and Sonya Michel, "Womanly Duties: Maternalist Politics and the Origins of Welfare States in France, Germany, Great Britain, and the United States, 1880–1920," *American Historical Review* 95 (1990): 1080–81.

21. On the development of professional medicine in western Europe, see Paul Weindling, "Medicine and Modernization: The Social History of German Health and Medicine," *History of Science* 24 (1986): 277–301; for a brief comparison of hospital building, see Charles E. Rosenberg, *The Care of Strangers: The Rise of America's Hospital System* (New York: Basic Books, 1987), 169–71; for a comparison of medical training in this period, see Thomas Neville Bonner, *Becoming a Physician: Medical Education in Britain, France, Germany and the United States, 1750–1945* (New York: Oxford University Press, 1995).

22. L. Tanquerel des Planches, *Traité des Maladies de Plomb ou Saturnines,* 2 vols. (1839). Forty percent of the plumbism victims at La Charité during the time of Tanquerel's investigation worked in lead-compound businesses, and 32% were painters who used white lead. Subsequent studies have validated many of Tanquerel's conclusions about the origins of neurological plumbism, and industrial toxicologists still refer to his work. Benjamin Franklin visited La Charité in 1767 and collected and analyzed its treatment records; McCord, "Early America: Benjamin Franklin" (see intro., n. 18).

23. On early French manufacturers of zinc, see Oliver, "Industrial Lead Poisoning" (see ch. 3, n. 17); for more on nineteenth-century efforts in France, see

Stevenson, "A History of Lead Poisoning," 333–90 (intro., n. 6). The French government's moves to replace lead with zinc are discussed in Oliver, *Dangerous Trades,* 293 (ch. 1, n. 38), and Wedeen, *Poison in the Pot,* 71.

24. In 1900, British physician Thomas Oliver (*Dangerous Trades,* 290) surveyed the French lead industries for the Home Office's White Lead Committee; he cited seven reasons for Besançon et Cie's success. The push to expel women from the "dangerous trades" was on, both in Britain and on the continent. Industrial physicians were convinced that women were more susceptible to plumbism; Britain banned them from most lead processes in 1898. Whatever statistical reasons medical men may have cited, sexism certainly played its part. Oliver (ibid., 300) wrote that women workers were more casual, "ignorant and careless."

25. On Germany's early hygiene laws, see Josef Rambousek, *Über die Verhütung der Bleigefahr* (Vienna, 1908), 3–5; on the impact of German trade associations in promoting the economic benefits of hygiene, see Lee K. Frankel, "Industrial Insurance the Basis of Industrial Hygiene," in *Transactions of the Fifteenth International Congress on Hygiene and Demography* (Washington: GPO, 1913), 3: 893; on early laws requiring medical inspections, see Ludwig Teleky, "Diagnosis and Supervision of Industrial Diseases in Germany," *JIH* 4 (Sept. 1922): 214; for workmen's compensation and compulsory insurance laws in Germany, see Theda Skocpol, *Protecting Soldiers and Mothers* (Cambridge, Mass.: Harvard University Press, 1992), 172. See also Starr, *Social Transformation of American Medicine,* 237.

26. Rambousek, *Verhütung der Bleigefahr,* 4. German industry's fight against lead poisoning was intense enough to support a professional journal, *Bleivergiftungen,* which appeared for at least four years and highlighted engineers' and doctors' strategies for industrial hygiene and plans for model factories. On Germany's impact on British efforts, see A. M. Anderson, "Historical Sketch of Development of Legislation for Injurious and Dangerous Industries in England," in Oliver, *Dangerous Trades,* 58–59.

27. Dickens, *The Uncommercial Traveller* (see intro., n. 15). The "poor craythur" was likely dying from the last stages of lead encephalopathy. For more on the British response to sex-specific protection in the lead industry, see Barbara Harrison, "'Some of Them Gets Lead Poisoned': Occupational Lead Exposure in Women, 1880–1914," *Social History of Medicine* 2 (Aug. 1989): 171–95. For more on Dickens's visit on "the borders of Ratcliff and Stepney, eastward of London," see Lloyd G. Stevenson, "All According to the Constitooshun: Charles Dickens and Lead Poisoning," in *Healing and History: Essays for George Rosen,* ed. Charles Rosenberg (New York: Science History Publications, 1979), 137–48.

28. The Workshops' Regulation Act of 1867 turned out to be ineffective—as did the Consolidating Act eleven years later—because enforcement fell to nearly powerless local authorities. Even the White Lead Act of 1883, despite its use of licensing power to set regulations for personal hygiene in white lead factories, proved impotent. See Thomas Oliver, "The Rise and Progress of Factory Legisla-

tion," *Journal of State Medicine* 22 (Aug. 1914): 477–79; see also Harrison, "'Some of Them Gets Lead Poisoned,'" 171–78.

29. Oliver, "Rise and Progress of Factory Legislation," 481; Thomas Oliver, "Lead Poisoning, Past and Present," *Journal of the Royal Institute of Public Health and Hygiene* 4 (Apr. 1941): 96.

30. Thomas Legge, "Twenty Years' Experience of the Notification of Industrial Diseases," *JIH* 1 (Apr. 1920): 590.

31. Taylor, *The Medical Profession,* 49.

32. Gottlieb (*Forcing the Spring,* 10–11 [see ch. 2, n. 51]) sees Hamilton leading a generation of reformers who invented urban environmentalism, hailing her as "the mother of American occupational and community health. . . . clearly as much an environmentalist as John Muir"; Nancy Cott, who criticizes Hamilton's opposition to the Equal Rights Amendment, introduces her as "Alice Hamilton, pioneer of industrial medicine," in *The Grounding of American Feminism* (New Haven, Conn.: Yale University Press, 1987), 127; in Sellers's detailed analysis of her career she embodies a "fusion of medical and state authority" (*Hazards of the Job,* 70 [see intro., n. 7]). I agree with Sicherman's judicious appraisal (*Alice Hamilton,* 1): "If she did not single-handedly found the field, as has sometimes been claimed, she was its foremost practitioner in the early years of this century."

33. Alice was not the only Hamilton to make a name for herself. The most famous Hamilton was Edith, a noted classicist whose *The Greek Way* is one of the best-selling surveys of ancient Greek culture. Readers interested in Alice Hamilton's life have a number of places to turn, starting with her autobiography, *Exploring the Dangerous Trades,* or Sicherman's annotated edition of her letters, *Alice Hamilton;* Wilma Ruth Slaight's "Alice Hamilton: First Lady of Industrial Medicine" (Ph.D. diss., Case Western Reserve University, 1974) is a much more conventional biography; Taylor (*The Medical Profession,* 49–52) offers a concise biographic essay. Because of her long association with Hull House, Hamilton wanders in and out of a number of studies of Jane Addams, Hull House, women reformers, and the Chicago milieu. Hamilton's pivotal place at the intersection of gender history and medical history in the Progressive Era has encouraged a number of theoretical analyses of her career, including Young, "Interpreting the Dangerous Trades," and, more recently, Sellers, *Hazards of the Job.*

34. Elizabeth Sargeant, "Alice Hamilton, M.D.," *Harper's Magazine* 152 (May 1926): 766, quoted in Taylor, *The Medical Profession,* 50.

35. Hamilton, *Exploring the Dangerous Trades,* 53.

36. Sicherman, *Alice Hamilton,* 111–15. For the wide sweep of Chicago's reform elite, see Ray Ginger, *Altgeld America: The Lincoln Ideal versus Changing Realities* (New York: New Viewpoints, 1973); on reformers in Chicago who traveled within Jane Addams's orbit from the late nineteenth century, see Kathryn Kish Sklar, "Hull House in the 1890s: A Community of Women Reformers," *Signs: Journal of Women in Culture and Society* 10 (Summer 1985): 658–77. Histo-

rian Ellen Fitzpatrick analyzes the careers of four University of Chicago graduates, including Breckinridge and Abbott, in *Endless Crusade: Women Social Scientists and Progressive Reform* (New York: Oxford University Press, 1990).

37. Sicherman (*Alice Hamilton*, 153) concludes that 1907 was probably the year Hamilton received a copy of Oliver's book; Hamilton, *Exploring the Dangerous Trades*, 115.

38. William Hard, "The Law of the Killed and Wounded," *Everybody's Magazine*, 19 Sept. 1908, 371, quoted in Gersuny, *Work Hazards and Industrial Conflict*, 53–54, 100 (see ch. 3, n. 73).

39. Robert Asher, "Failure and Fulfillment: Agitation for Employers' Liability and the Origins of Workmen's Compensation in New York State, 1876–1910," *Labor History* 24 (Spring 1983): 198–222; see also Robert Asher, "The Limits of Big Business Paternalism: Relief for Injured Workers in the Years before Workmen's Compensation," in *Dying for Work*, 19–33 (see intro., n. 12).

40. Frank W. Lewis, "Employers' Liability," *Atlantic Monthly*, Jan. 1909, 57–65.

41. On railroad workers' unions as key players in early Progressive Era liability reform, see Asher, "Failure and Fulfillment."

42. For sources on Progressive Era employers' preference for workmen's compensation laws over "a radical employers' liability system," see Asher, "Failure and Fulfillment," 22, n. 73. If it is necessary to dispel the notion that injured workers universally benefited from reformed liability laws, Crystal Eastman (*Work Accidents and the Law*, The Pittsburgh Survey, ed. Paul Underwood Kellogg [1910; reprint, New York: Arno, 1969], 271) reported that even in 1907–1909, fewer than one-quarter of the dependents of married New York rail workers killed on the job received even a year's salary in compensation; see also Anthony Bale, "America's First Compensation Crisis: Conflict over the Value and Meaning of Workplace Injuries under the Employers' Liability System," in *Dying for Work*, 34–52.

43. "Organization," *American Labor Legislation Review* 4 (Dec. 1914): 511. On the AALL's founding, see Sellers, *Hazards of the Job*, 50–81; Skocpol, *Protecting Soldiers and Mothers*, 176–79.

44. On the influence of European thought, especially the German historical school, on Richard Ely, the AALL's first president, see Sellers, "Manufacturing Disease," 143.

45. See "Phosphorus Poisoning in the Match Industry in the United States," in U.S. Department of Commerce and Labor, *Bulletin of the Bureau of Labor* 86 (Jan. 1910): 67–85.

46. For the development of nonpoisonous matches, see Moss, "Kindling a Flame under Federalism," 252–53 (see ch. 2, n. 37).

47. The first known case was in Massachusetts in 1851; Hamilton, *Exploring the Dangerous Trades*, 116–18.

48. For an analysis of the methodological innovations in the AALL's investiga-

tion, see Sellers, "Manufacturing Disease," 160–61; see also Sicherman, *Alice Hamilton,* 155–56; Hamilton, *Exploring the Dangerous Trades,* 117–18; editor's note in Thayer, "The Lead Menace," 325 (see ch. 3, n. 18); Moss, "Kindling a Flame under Federalism," 244–75.

49. Connecticut, Minnesota, New Jersey, Ohio, and Wisconsin also created commissions in 1909 to investigate the prospects of workmen's compensation; see Asher, "Failure and Fulfillment," 221. Quote, Illinois Commission on Workingmen's Insurance, "Report to Governor," in Charles Henderson, "Illinois Commission of Occupational Diseases," in AALL, *First National Conference on Industrial Diseases,* (New York, 1910), 19. See also Hamilton, *Exploring the Dangerous Trades,* 118–19.

50. Hamilton, *Exploring the Dangerous Trades,* 120; Sicherman, *Alice Hamilton,* 157.

51. Pratt, *Occupational Diseases* (see ch. 3, n. 13); "Survey of Lead Poisoning in St. Louis and Vicinity," *Survey* 31 (17 Jan. 1914): 470; Aub and Hapgood, *Pioneer in Modern Medicine,* 182–184. Some of Hamilton's studies include "White-Lead Industry in the United States" (ch. 3, n. 13); "Hygiene of the Painters' Trade," *Bulletin of the Bureau of Labor Statistics* 120 (Washington: GPO, 1913), and "Women in the Lead Industries," *Bulletin of the Bureau of Labor Statistics* 253 (Washington: GPO, 1919).

52. Thayer, "The Lead Menace," 332.

53. For the Illinois survey's recommendations, see "Work Poisons: The Report of the Illinois Commission on Occupational Diseases," *Survey* 25 (18 Feb. 1911): 842–45.

54. The six states that passed occupational disease bills in 1911 were California, Connecticut, Illinois, Michigan, New York, and Wisconsin; "Laws Enacted during 1911 Requiring the Report of Occupational Diseases," in U.S. Department of Commerce and Labor, *Bulletin of the Bureau of Labor* 95 (July 1911): 283–88.

55. Taylor, *The Medical Profession,* 57; Sicherman, *Alice Hamilton,* 158–59; Thayer, "The Lead Menace," 325–34.

56. Massachusetts passed the first workmen's compensation law in 1910; see Asher, "Failure and Fulfillment," 220. Mississippi was the last state to do so, in 1948; see Gersuny, *Work Hazards and Industrial Conflict,* 99. Morton Keller provides a concise analysis of the political and social forces behind the broad support for workmen's compensation in *Regulating a New Society: Public Policy and Social Change in America, 1900–1933* (Cambridge, Mass.: Harvard University Press, 1994), 198–202.

57. Asher, "Failure and Fulfillment," 220–21.

58. John B. Andrews, "Compensation for Occupational Diseases," *Survey* 30 (5 Apr. 1913): 18.

59. See Andrews' fiery appeal for compensation-law reform in "Compensation for Occupational Diseases," 15–19. In 1929, Hamilton ("Nineteen Years in the

Poisonous Trades," *Harpers* 159 [Oct. 1929]: 591) reported that only five states compensated for all workplace accidents and illness, and another five for specific schedules of diseases. The situation in the mid-1940s is outlined in Ludwig Teleky, "Compensation for Occupational Diseases," *Compensation Medicine* 1 (Feb. 1946): 8–15. On the courts and compensation for occupational diseases, see David Rosner and Gerald Markowitz, *Deadly Dust: Silicosis and the Politics of Occupational Disease in Twentieth-Century America* (Princeton, N.J.: Princeton University Press, 1991).

60. "Accidental" poisonings, likewise, occur at specific times, with finite durations. Quote, Judge J. Stone's opinion in *Adams v Acme White Lead & Color Works*, 182 MI 157, 148 NW 485 (1914).

61. Ibid.

62. Ibid. Compensation boards could be very strict about the element of time in distinguishing between accident and disease; see, e.g., *Industrial Commission of Ohio v Roth et al.*, 98 OH 34, 120 NE 172 (1918). Edwin S. Roth was an 18-year-old laborer. He was painting a building, but in the cold November air his paint would not flow from his brush. His foreman told him to take his can of paint into a small shed and heat it over a fire, a process Roth repeated over the course of two days. Three weeks later, he died. Although he was exposed to the noxious fumes for only several hours, the Ohio Industrial Commission defined Roth's condition as a disease, not an accident. After two and half years, the Ohio Supreme Court overturned the commission's ruling.

63. Labanoski brought his suit two months after being committed to the State Hospital for the Insane at Alton; *Labanoski v Hoyt Metal Company* 292 IL 218, 126 N. E. 548 (1920). After July 1921, an amendment to the Occupational Disease Act of 1911 limited what poisoned workers could recover for occupational diseases to that provided by workmen's compensation.

64. *Maxwell Motor Corporation v Winter*, 118 OH 622, 163 N. E. 198 (1924).

65. *Gerald's Case*, 247 MA 229, 141 N. E. 862 (1924).

66. A similar case, in which Alice Hamilton testified, involved the death of a print compositor for International Harvester. The outcome hinged on whether she had died of pernicious anemia or lead poisoning—a symptom of which is pernicious anemia. See *Wilcox v International Harvester Co. of America*, 278 IL 465, 116 N. E. 151 (1917).

67. May R. Mayers, "Lead Poisoning and Compensation," *Compensation Medicine* 1 (Jan. 1946): 13–22.

68. Alice Hamilton to C. H. Verill of the U.S. Department of Commerce and Labor, 12 Feb. 1913, folder 29: "Lead" (1), A-22, Alice Hamilton Papers, Schlesinger Library.

69. Alice Hamilton, "The Hygiene of the Lead Industry" (address presented at the Meeting of Superintendents of the National Lead Company, Chicago, 7 Dec. 1910), folder 29: "Lead" (1), Alice Hamilton Papers.

70. Hamilton, *Exploring the Dangerous Trades,* 127–28.

71. Oliver, "Industrial Lead Poisoning," 25.

72. Hamilton was referring specifically to her confrontation with smelter operators in Joplin, Missouri, who hastily orchestrated a town-wide clean-up in anticipation of the survey. For this anecdote, see Hamilton, "Nineteen Years in the Poisonous Trades," 584. The same story, with a few added details and a discouraging follow-up on the same village twenty years later, appears in Hamilton, *Exploring the Dangerous Trades,* 145–48.

CHAPTER 5: PROTECTING WORKERS AND PROFITS IN THE LEAD
INDUSTRIES

1. Hamilton, *Exploring the Dangerous Trades,* 131–32 (see intro., n. 17).

2. Winthrop Talbot, "Some Economic Aspects of Factory Hygiene," *AJPH* 2 (Oct. 1912): 773–75.

3. Frederick L. Hoffman, "The Decline in Lead Poisoning" (paper presented at the Health Congress of the Royal Institute of Public Health, Ghent, Belgium, 1927), 5, 8. Hoffman's speech was intended to prove the "needlessness of prohibition, while fully supporting the urgency of rational regulations" of white lead paint, which the ILO was considering.

4. According to census mortality records, the average was 130 lead-poisoning deaths per year from 1910 to 1940. The five-year average death rate for the years 1911–15 was 2.38 per million; in 1936–40 it had fallen 68% to 0.77. I chose 1941 as my end date for the deaths per ton (d/T) rate because, with the drastic drop in production during the Depression, the same hygienic conditions produced far more d/T; note that the lowest rate prior to 1941 was 0.14 d/T in 1930, when auto production was still in full swing. The peak d/T rate around 1910 results partly from an enlargement of the registry area. After 1910, more deaths were reported because the Census Bureau accepted mortality data from a larger area of the country but production data continued to cover the same regions. It is also probable that increased awareness of occupational lead poisoning resulted in more frequent reports. Data for deaths by lead poisoning, Bureau of the Census, *Mortality Statistics* 1901–46 (see ch. 2, n. 17); for lead production and consumption, Bureau of Mines, "Lead," in *Minerals Yearbook* (ch. 3, n. 6).

5. Of course, pediatric cases in the same period seemed to be on the rise. From 1911 to 1930, physicians reported an average of only two lead-poisoning deaths per year among children under 6 years old, while from 1931 to 1940, they reported almost thirty per year; Bureau of the Census, *Mortality Statistics.* By 1940, pediatric cases accounted for over one-quarter of all reported lead poisonings.

6. E. F. Smith, in discussion at the Section on Preventive and Industrial

Medicine and Public Health of the Seventy-eighth Annual Session of the American Medical Association, 18 May 1927, quoted in J. P. Leake, "Lead Hazards," *JAMA* 89 (1 Oct. 1927): 1113.

7. On the state of treatment in the 1930s, see Irving Gray, "Recent Progress in the Treatment of Plumbism," *JAMA* 104 (19 Jan. 1935): 200–205.

8. Hamilton, "White-Lead Industry in the United States" (see ch. 3, n. 13), 218, 216; Alice Hamilton, "Lead Poisoning in Illinois," in AALL, *First National Conference on Industrial Diseases*, 32 (ch. 4, n. 49).

9. Hamilton, "White-Lead Industry in the United States," 219.

10. Hamilton, *Exploring the Dangerous Trades*, 117; on phossy jaw, see Moss, "Kindling a Flame under Federalism" (see ch. 2, n. 37).

11. Hamilton describes phossy jaw victims in "Healthy, Wealthy—If Wise—Industry," *American Scholar* 7 (Winter 1938): 13; for her generous appraisal of Diamond's actions, see *Exploring the Dangerous Trades*, 117–18. Phossy jaw was eliminated from the match industry, but workers in fireworks and insecticide plants continued to be disfigured; Jean Spencer Felton, "Classical Syndromes in Occupational Medicine: Phosphorus Necrosis—A Classical Occupational Disease," *American Journal of Industrial Medicine* 3 (1982): 77–120.

12. Rosner and Markowitz (*Deadly Dust*, 5 [see ch. 4, n. 59]) argue that the "king of occupational diseases," silicosis, "represented a new kind of disease, unlike those, such as lead or phosphorous poisoning, that had posed major threats in earlier times." Silicosis came to the public's attention with the disaster at Gauley Bridge, West Virginia, where hundreds of workers died while digging a tunnel through a mountain for Union Carbide; see Martin Chernisck, *The Hawk's Nest Incident, America's Worst Industrial Disaster* (New Haven, Conn.: Yale University Press, 1986).

13. Eighteenth-century physician and historian of occupational disease Bernardino Ramazzini (*Diseases of Workers*, 27 [see ch. 1, n. 33]) gave lead the distinction as the oldest reported occupational disease, citing Leyden, *Collectanea Chymica*, on the prevalence of occupational lead poisoning.

14. See Hamilton, *Exploring the Dangerous Trades*, 122. Both David Edsall, who founded Harvard's Division of Industrial Hygiene, and Frederick Hoffman saw lead poisoning as the natural primary subject of their studies.

15. Read, *Our Mineral Civilization*, 66 (see ch. 1, n. 21).

16. Heckel, *The Paint Industry*, 611–13 (see ch. 3, n. 5). Robert H. Wiebe (*The Search for Order, 1877–1920* [New York: Hill and Wang, 1967], 175) noted that progressive industrialists had plenty of opportunity to hear "the woman's way," because "women of good families such as Jane Addams and Florence Kelley were learning how to shame their contemporaries with surprising results. Rant though they might, men in authority simply could not seal themselves from these voices as they had a generation before from a Terence Powderly or a Henry George."

17. Hamilton, *Exploring the Dangerous Trades*, 269.

18. Geier, "General Discussion," 247. Despite his disclaimer, the "cry of hu-

manity" seemed to ring loudly in Otto Philip Geier's ears. The son of a German émigré, he began practicing medicine in 1900 and quickly became deeply involved in public health, conducting milk campaigns and directing Cincinnati's Anti-Tuberculosis League and its welfare department; *National Cyclopædia of American Biography* (Ann Arbor, Mich.: 1967), vol. 49, 271–72. According to Sellers ("Manufacturing Disease," 334 [see intro., n. 7]), Geier founded "the only other year-long program in industrial medicine outside that at Harvard, at the Cincinnati Medical College," but Hamilton disparaged this program as "a purely consulting and commercial affair" (ibid., 363–65).

19. Hamilton, "Lead Poisoning in Illinois," 33.

20. Heckel, *The Paint Industry,* 613, 615.

21. Wiebe (*Search for Order,* 177) describes the urban progressives as "generally younger men with a passion for the future." Both Edward Cornish and Frank Hammar, whom Alice Hamilton described as pioneers in corporate support for hygiene, were of Hamilton's generation; see *Exploring the Dangerous Trades,* 8–10, 137. It seems difficult to imagine that Felix Wormser, who was born one generation later and whose career is discussed below, would have been as cooperative. Compare also the careers of Hamilton and Geier with that of Carey McCord, discussed in the next chapter.

22. The proud plant owner appears in Hamilton's "Nineteen Years in the Poisonous Trades," 588–89 (see ch. 4, n. 59), and *Exploring the Dangerous Trades,* 135–36; for the story of the owner's daughter, see *Exploring the Dangerous Trades,* 8–9.

23. Historians of the Progressive Era seem to be moving toward the mature position of admitting that many of the factors they and their opponents wrangled over could apply simultaneously. On Taylorism, see Robert Kanigel, *The One Best Way: Frederick Winslow Taylor and the Enigma of Efficiency* (New York: Viking, 1997); Samuel Haber, *Efficiency and Uplift: Scientific Management in the Progressive Era, 1890–1920* (Chicago: University of Chicago Press, 1964). On trade associations, mergers, and the transformation of American businesses from the inside, see Alfred Chandler, *The Visible Hand: The Managerial Revolution in American Business* (Cambridge, Mass.: Belknap Press, 1977); Martin Sklar, *The Corporate Reconstruction of American Capitalism, 1890–1916* (New York: Cambridge University Press, 1988). For an appraisal that eschews polemics in favor of an argument based on a plurality of interests and dynamics, see Morton Keller, *Regulating a New Economy: Public Policy and Economic Change in America, 1900–1933* (Cambridge, Mass.: Harvard University Press, 1990). On conservation for use, see Samuel P. Hays, *Conservation and the Gospel of Efficiency: The Progressive Conservation Movement, 1890–1920* (Cambridge, Mass.: Harvard University Press, 1959); Gottlieb, *Forcing the Spring* (see ch. 2, n. 48).

24. Talbot, "Some Economic Aspects of Factory Hygiene," 774; Frederick Hoffman, "Industrial Accidents and Industrial Diseases," *Publications of the American Statistical Association* 11 (Dec. 1909): 567. The National Safety Council,

established by industry in 1912, reflected the progressive industrial ideals of voluntarism, association, and reliance on experts, and was largely composed of physicians and engineers involved in plant hygiene; see Harry E. Mock, "Industrial Medicine and Surgery—A Résumé of Its Development and Scope," *JIH* 1 (May 1919): 4–5. Alice Hamilton served on NSC committees and praised the council for 1920s' work on spray painting and the use of benzol; see *Exploring the Dangerous Trades*, 142, 296.

25. Nor, it must be added in support of the argument that moralism dictated hygienic improvements, did the destruction of natural resources stir the employer's conscience in the same way that the waste of human life did.

26. Heckel, *The Paint Industry*, 613–14. Uncritical "social control" historians would no doubt bristle at lead manufacturers' blatant exercise of their power to prescribe social solutions to economic problems. Undoubtedly, as Heckel's tone makes clear, native-born, middle-class managers trusted they were improving their charges' souls as well as enhancing their chances of survival. In fact, both of this factory's requirements were probably highly effective. There is no doubt of the association between adequate diet and absorption of lead, and the efficacy of milk for reducing the amount absorbed through the digestive tract was established before 1913; see A. J. Carlson and A. Woelfel, "The Solubility of White Lead in Human Gastric Juice, and Its Bearing on the Hygiene of the Lead Industries," *AJPH* 3 (Aug. 1913): 755–69. As to sending workers to the showers, recent studies of plumbism among the children of lead workers show a clear relationship between occupational exposure, bathing facilities (and policies), the amount of lead in the home environment, and children's blood-lead levels. Regardless of employers' moralism, making lead workers bathe and change clothes at the end of the work day made perfect sense.

27. Industrial welfarism encompassed genuine improvements in working conditions, embellished by inexpensive amenities such as company libraries, profit sharing, entertainment facilities, club tours, and parties; see Stuart Brandes, *American Welfare Capitalism, 1880–1940* (Chicago: University of Chicago Press, 1976), 136–37; Lizabeth Cohen, *Making a New Deal: Industrial Workers in Chicago, 1919–1939* (New York: Cambridge University Press, 1990), ch. 4; Bruno Ramirez, *When Workers Fight: The Politics of Industrial Relations in the Progressive Era, 1898–1916*, Contributions in Labor History, no. 2 (Westport, Conn.: Greenwood Press, 1978).

28. Sellers, "'A Prejudice Which May Cloud the Mentality': An Overview of the Birth of the Modern Science of Occupational Disease," in *Toxic Circles: Environmental Hazards from the Workplace into the Community*, ed. Helen Sheehan and Richard P. Wedeen (New Brunswick, N.J.: Rutgers University Press, 1993), 236–37.

29. See, e.g., *Maxwell Motor Corporation v Winter* (see ch. 4, n. 64). After failing to gain compensation under workmen's compensation, Winter sued his old employer before a jury, which awarded him $4000.

30. J. W. Schereschewsky, "A Plan for Education in Industrial Hygiene and the Avoidance of Occupational Complaints," *AJPH* 6 (Oct. 1916): 1032. Historians continue to disagree about which prediction was more accurate. Recently, they have looked past the obvious shifts in mood and the crests and falls in reformers' optimism to see important lines of continuity across the century's first and third decades; see, e.g., Keller, *Regulating a New Society;* and Martin Sklar, *The United States as a Developing Country: Studies in U.S. History in the Progressive Era and the 1920s* (New York: Cambridge University Press, 1992).

31. Hamilton ("Nineteen Years in the Poisonous Trades," 584) also spoke of "the arrogance of the manufacturers, the indifference of those higher up, and the contempt of the trades unions for non-union labor."

32. On the PHS's enlarged role during World War I, see Christopher Sellers, "The Public Health Service's Office of Industrial Hygiene and the Transformation of Industrial Medicine," *BHM* 65 (1991): 42–73, esp. 61–63; and Young, "Interpreting the Dangerous Trades," 68–106 (see ch. 4, n. 2).

33. Allen F. Davis, "Welfare, Reform and World War I," *American Quarterly* 19 (Fall 1967): 516–33.

34. For the impact of World War I on the efforts of Frances Kellor and Katherine Davis, two Chicago-based reformers who cast their lot with the growing federal government, see Fitzpatrick, *Endless Crusade,* 160–62, 201–202 (see ch. 4, n. 36).

35. On the Hull House delegation to the Hague, see Hamilton, *Exploring the Dangerous Trades,* 161–82; and Jane Addams, *Women at the Hague: The International Congress of Women and Its Results* (New York: Macmillan, 1915). Quote, Hamilton, "Nineteen Years in the Poisonous Trades," 584. For Hamilton's most dramatic description of wartime industrial conditions, see *Exploring the Dangerous Trades,* 183–99. As for the War Labor Board, she recalled that "only in the Spring of 1919 did Samuel Gompers [whose AFL she suggested was "quite indifferent to the lot of the unskilled men"] triumphantly produce a code, less stringent than the English, lacking any compulsory feature and adopted months after the Armistice"; see Hamilton, "Healthy, Wealthy—If Wise—Industry," 16–17.

36. Bureau of the Census, *Historical Statistics of the United States,* 603–604 (see ch. 3, n. 9); and Carey P. McCord, Dorothy K. Minster, and Robert Kehoe, "Lead Poisoning in the United States" (abridged report of the American Public Health Association's Committee on Lead Poisoning, 1928), *AJPH* 19 (June 1929): 632. The numbers cited do not agree exactly with those published, but neither source is specific as to what lead production is included. The 1929 numbers are perhaps incomplete, though the general trend they portray is conclusive.

37. This quote referred to her first postwar survey, of steel mills still reeling from the 1919 steel strike; see Hamilton, "Nineteen Years in the Poisonous Trades," 585.

38. For a brief introduction to the origins of professional occupational medi-

cine, see Angela Nugent, "Fit for Work: The Introduction of Physical Examinations in Industry," *BHM* 57 (Winter 1983): 578–95.

39. Mock, "Industrial Medicine and Surgery"; Emery R. Hayhurst, "The Industrial Hygiene Section, 1914–1934," *AJPH* 24 (Oct. 1934): 1039–44. The National Council for Industrial Safety's directorate for the next twenty years reads like a *Who's Who* of industrial medicine: its chairs included Alice Hamilton, Emery Hayhurst, J. W. Schereschewsky, Carey P. McCord, A. J. Lanza, and George M. Price.

40. "Proceedings of Conference on Industrial Hygiene," 1919, folder 1, Victor Heiser Papers, American Philosophical Society Archives. The Rockefeller Foundation held this conference to help its officers determine the most beneficial course to follow in promoting industrial health.

41. Mock, "Industrial Medicine and Surgery"; Hayhurst, "The Industrial Hygiene Section," 1039–44; Sellers, "The Public Health Service's Office of Industrial Hygiene," 63–64.

42. Hamilton (*Exploring the Dangerous Trades,* 152–53, 157) noted that in addition to overwork, the typical company doctor had poor training, questionable allegiances, and poor standing: in a Pullman car factory she visited, the first person called in after an accident was not the physician but the company lawyer, who noted the injured parts and summoned the doctor. If the injuries were too serious for the physician—an ex–Civil War surgeon—an ambulance was summoned from eight to ten miles away. See also Taylor, *The Medical Profession,* 52 (see ch. 1, n. 47).

43. Harry E. Mock, *Industrial Medicine and Surgery* (Philadelphia, 1917), 91, quoted in Gersuny, *Work Hazards and Industrial Conflict,* 23 (see ch. 3, n. 73).

44. For example, Sellers ("The Public Health Service's Office of Industrial Hygiene," 50) asserts that "beginning around 1910, the new state workers' compensation laws created an added incentive for companies to hire physicians, to defend against false claims"; see also Anthony Bale, "Compensation Crisis: The Value and Meaning of Work-Related Injuries and Illnesses in the United States, 1842–1932" (Ph.D. diss., Brandeis University, 1986). On physical exams, see Nugent, "Fit for Work."

45. Although never given tenure, Hamilton stayed at Harvard until 1935, when the new president, James Bryant Conant, enforcing his "uniform retirement policy," asked for her resignation. She returned to government work, taking a job as medical consultant to the Division of Labor Standards in the Department of Labor, under Secretary Frances Perkins; see Hamilton, *Exploring the Dangerous Trades,* 252–53; Sicherman, *Alice Hamilton,* 357 (see ch. 4, n. 1). Neither source says whether she was allowed in the Harvard Club, but she did not sit on the platform!

46. Joseph S. Aub et al., *Lead Poisoning* (Baltimore: Williams & Wilkins, 1926). The book includes a chapter by Alice Hamilton on the prevalence of industrial lead poisoning in the United States.

47. Taylor, *The Medical Profession,* 18.

48. David Linn Edsall (1869–1945) was the son of a New Jersey merchant and state legislator. He studied biology at Princeton and received his medical degree in 1893 from the University of Pennsylvania. His interest in occupational medicine developed early and was the subject of his research at the University of Pennsylvania's Pepper Laboratory, where he studied from 1897 to 1910. From 1909 to 1912 he was instrumental in instituting reforms in the medical program at St. Louis's Washington University. He was frustrated in his attempts to institute similar reforms at the University of Pennsylvania and left for Harvard, where he rose to become dean of the Medical School in 1918. In 1922, he accepted a second deanship, of Harvard's new School of Public Health, and resigned from medical practice. He remained dean until his retirement in 1935. See Aub and Hapgood, *Pioneer in Modern Medicine* (see ch. 4, n. 3).

49. Sicherman, *Alice Hamilton,* 209–10; Taylor, *The Medical Profession,* 39–40, 182–84.

50. Taylor, *The Medical Profession,* 183, 40; Wade Wright, "An Industrial Clinic," *MLR* 5 (Dec. 1917): 189–90.

51. Aub and Hapgood, *Pioneer in Modern Medicine,* 186.

52. Technically, the division was a part of the joint Harvard–Massachusetts Institute of Technology School for Health Officers. For more on Edsall, see Sellers, "Manufacturing Disease," 95–163.

53. Edsall to Frank Hammar, 7 May 1920, folder "Lead Fund, 1900–1928," Box 2, Dean's Files, 1922–49, Harvard School of Public Health Archives, Countway Library, Harvard Medical School (henceforth "Lead Fund, 1900–28").

54. Hammar to Edsall, 3 June 1920, "Lead Fund, 1900–28"; Edsall to Hammar, 18 Dec. 1920, ibid.

55. Edsall to Hammar, 18 Dec. 1920; Hammar to Edsall, 9 Feb. 1921, "Lead Fund, 1900–28."

56. Aub to G. W. Thompson, 29 Aug. 1924, "Lead Fund, 1900–28"; quote, Aub to A. L. Endicott, Harvard University bursar, 10 May 1927, folder "Lead Fund, Dr. Joseph Aub, 1921–34," Box 2, Dean's Files, 1922–49, Harvard School of Public Health Archives, Countway Library, Harvard Medical School (henceforth "Lead Fund, 1921–34"). By 1929, the LIA was the formal liaison between manufacturers and Harvard, and Secretary-Treasurer Felix Wormser assured Aub of the association's continued support; Wormser to Aub, 14 Nov. 1929, "Lead Fund, 1921–34."

57. David Edsall, "The Activities in Industrial Hygiene at the Harvard Medical School since 1918 and at the Harvard School of Public Health since 1921," file, "Committee on Industrial Hygiene," Box 2, Dean's Files, 1922–49, Harvard School of Public Health, Countway Library, Harvard Medical School.

58. [Joseph C. Aub], "Third Report on Investigation of Lead Poisoning," "Lead Fund, 1900–28"; Aub to Thompson, 29 Aug. 1924, ibid.

59. Cornish to Edsall, 12 May 1921, "Lead Fund, 1900–28."

60. Cornish to Edsall, 29 Mar. 1923, "Lead Fund, 1900–28." Cornish did not succeed in convincing American painters, who continued to suffer the nation's highest rates of lead poisoning. According to Hamilton, Cornish was exceptional. She recalled (*Exploring the Dangerous Trades,* 11) that after she confronted him with the conditions in his plants, he undertook to correct their shortcomings and successfully campaigned for other LIA affiliates to do the same.

61. Thomas Brown, Eagle Picher Lead Co., to Edsall, 12 Nov. 1921, "Lead Fund, 1900–28."

62. Most of the other principal investigators maintained a public front of disinterest. Phillip Drinker briefly protested the introduction of tetraethyl lead but later, it would seem, saw the error of his ways, confessing privately to an editor of *Consumer's Reports* that "we were proved to be wrong and we did not enjoy the result any more than anyone else who guesses badly"; Drinker to F. J. Schlink, 19 Aug. 1948, folder "Bowditch, Manfred [3], 1948–1960" (henceforth "Bowditch folder"), Box 2: Benison-Bowman, Aub, Joseph C., Office Files (henceforth Aub files), Countway Library, Harvard Medical School. Only Alice Hamilton—whose friendly relations with the lead magnates she met during her early surveys were critical in forging the LIA-Harvard link but whose "unscientific" methods kept her somewhat estranged in the clinically oriented Harvard program—remained a staunch critic of lead for the rest of her long life. Aub would play a critical role in downplaying the controversy arising from a Children's Hospital neurologist's finding that lead caused permanent damage even among patients declared "cured."

63. For a sharply critical appraisal of the lead industries' aggressive experimental programs, see Graebner, "Hegemony through Science," 140–59 (see ch. 2, n. 49).

64. For Hamilton's observations on the 1925 conference, see Sicherman, *Alice Hamilton,* 239.

CHAPTER 6: COMPANY DOCTORS ON THE JOB

1. Much of the following discussion draws on the confidential proceedings of one such conference, held on 6 Apr. 1937 at the Palmer House in Chicago, sponsored by the Lead Industries Association; LIA, "Lead Poisoning: Report of Conference, Physicians and Surgeons of Member Companies," Countway Library, Harvard Medical School (henceforth LIA, "Conference").

2. Elston L. Belknap, "Control of Lead Poisoning in the Worker," *JAMA* 104 (19 Jan. 1935): 205–10. Preemployment exams began before the 1930s; for a detailed history of such exams and the controversies they spawned, see Nugent, "Fit for Work" (see ch. 5, n. 38). Alice Hamilton recommended them to avoid endangering workers who were easily leaded. By the 1930s, they had become the norm

in large lead industries, where they served more than the purpose their opponents usually claimed—screening "undesirable" workers at the whim of management.

3. K. A. Koerber, in LIA, "Conference," 82; Belknap, "Control of Lead Poisoning in the Worker," 205–10.

4. G. H. Gehrmann, "Prevention of Lead Poisoning in Industry," *AJPH* 23 (July 1933): 687–92.

5. Belknap, "Control of Lead Poisoning in the Worker," 206.

6. Koerber, in LIA, "Conference," 69.

7. G. E. Johnson, International Smelting and Refinery Company, in LIA, "Conference," 71.

8. In many industries, rules were habitually ignored—often at the insistence of managers who pushed for higher productivity, turned up the speed of the assembly line, or demanded longer hours than the rules allowed; Gersuny, *Work Hazards and Industrial Conflict,* 27 (see ch. 3, n. 73).

9. McCord developed the basophilic aggregation test (BAT) in the early 1920s; see Carey P. McCord, F. R. Holden, and Jan Johnson, "Basophilic Aggregation Test in the Lead Poisoning Epidemic of 1934–1935," *AJPH* 25 (Oct. 1935): 1089–98. Essentially, it measured the presence of the compounds in blood cells that would eventually cause stippling and other signs of damage. These "basophilic substances," when stained, formed "basophilic aggregations" that were readily visible under the microscope. See Roy R. Jones, "Symptoms in Early States of Industrial Plumbism," *JAMA* 104 (19 Jan. 1935): 195–200; for McCord's humorous recollections on "The Birth of the B.A.," see his *A Blind Hog's Acorns,* 213–19 (see ch. 4, n. 3).

10. It is customary to divide lead poisoning itself into three categories, based on the systems affected. Belknap ("The Actual Control of Poisons in Industry," *Painter and Decorator* 54 [Oct. 1940]: 32) distinguished among mild, or gastrointestinal, plumbism; neurological plumbism, exhibiting palsies and limb drops; and lead encephalopathy, or brain involvement.

11. Koerber, in LIA, "Conference," 74.

12. Quote, "They taste lead . . . ," G. E. Brockway, NLC New York, in LIA, "Conference," 82; "if you are going to . . . ," Koerber, in LIA, "Conference," 74.

13. Koerber, in LIA, "Conference," 87.

14. Lawrence Fairhall, in LIA, "Conference," 87; quote, "in the last analysis . . . ," is Jones's summary of Aub's position presented in *Lead Poisoning* (see ch. 5, n. 46), from "Symptoms in Early States of Industrial Plumbism."

15. Aub, in LIA, "Conference," 8–9, 17. In a symposium three years earlier, Aub ("The Biochemical Behavior of Lead in the Body," *JAMA* 104 [12 Jan. 1935]: 87–90) beamed, "Nothing more dramatic in treatment can be desired than the rapid subsidence of lead colic following [calcium injections]."

16. Gray, "Recent Progress in the Treatment of Plumbism" (see ch. 5, n. 7).

17. Gesswein, in LIA, "Conference," 68.

18. Brockway, Koerber, in LIA, "Conference," 52.

19. Brockway, D. R. Johns, in LIA, "Conference," 73, 69.

20. Clark, Koerber, in LIA, "Conference," 36, 27.

21. See, e.g., the exchange between Belknap and Kehoe during an Industrial Medicine and Public Health session at the AMA's Cleveland conference, 13 June 1934; "Abstract of Discussion," *JAMA* 104 (19 Jan. 1935): 210–11. Belknap defended routine deleading as a way "to keep pace with the small amount that [the worker] absorbs from day to day." Kehoe—somewhat removed from daily plant operation and a firm believer that the role of industrial hygiene was prevention, not treatment—questioned risking increased liberated blood lead "for what seems the slight and evanescent advantage" of excreting small amounts. Kehoe's opposition to deleading would last into the era of safe chelaters; see his "Misuse of Edathamil Calcium-Disodium for Prophylaxis of Lead Poisoning," *JAMA* 157 (22 Jan. 1955): 341–42.

22. L. G. Reichhard, in LIA, "Conference," 88.

23. McCord, *A Blind Hog's Acorns*, 220.

24. Upon retirement from Chrysler, he returned to Ann Arbor, where the university retained him as a lecturer and consultant in environmental health and he increased his editorial role for the journal. The title of his engaging autobiography, *A Blind Hog's Acorns*, refers to the words of a phrenologist, who sadly informed Carey's parents that the structure of their young son's skull predicted that he would do poorly in school. But not to worry too much, the quack added, "Even a blind hog gets an acorn once in a while." McCord died in 1979, at age 93, having gathered his share of acorns; for additional biographic details, see his obituary by Doris Flornoy, *Journal of Occupational Medicine* 21 (Dec. 1979): 841. His history of lead poisoning in America appeared in vols. 22 and 23 (1953–54) of *Industrial Medicine and Surgery;* see, e.g., "Early America: Benjamin Franklin" (see intro., n. 18).

25. Presumably, these events (and those of the second case described) occurred between 1920 and 1935, when McCord was most active in consulting work. This case history appears as "A Minor Medical Miracle," in *A Blind Hog's Acorns*, 208–12; McCord does not say so, but one assumes that "Marshall Grant" is a pseudonym.

26. McCord does not identify the other doctor, other than to call him "a young physician, now famous as an investigator of lead poisoning, who then was just starting on a brilliant career"; ibid., 209. Given this detail and the fact that McCord practiced in the Cincinnati area, it is likely that he was Robert Kehoe.

27. Ibid., 199–202.

28. McCord tells another story, straight out of Perry Mason, in which a painter, suffering (or pretending to suffer, McCord suspected) all the classic symptoms of acute lead colic, sued his employer; ibid., 220–25. Things were looking pretty grim for the defense, until, right in the middle of McCord's testimony, the

painter fell from his chair, doubled over with what his doctor suddenly recognized as acute appendicitis.

29. Benjamin Cohen, the owner, operator, and sole employee of State Battery Repair in Roxbury, Mass., provides an extreme case of hazardous small plants; see Dr. Dart to Manfred Bowditch, 16 May 1942, folder "Bowditch, Manfred [2], 1941–1946," Box 2: Benison-Bowman, Aub files. In 1942, investigators at the state Department of Labor reported that Cohen, age 38, was suffering with wrist drop, cramps, and marked constipation. He had been sick to some extent the entire five years that he had operated his battery repair and recycling business, where he often melted lead in a pot with no hood. For about fifteen years prior to opening his own shop, he had worked for a battery company.

30. A Dr. McGee recalled this story at an LIA-sponsored hygiene conference held in Chicago, 6–7 Nov. 1958; see discussion transcript following John H. Foulger, "Precautionary Labeling of Lead Products," *Industrial Medicine and Surgery* 28 (Mar. 1959): 125. McGee set up a small laboratory at the hospital, and "with Aub's publications as [their] bible," conducted studies of all the men who worked at the smelter, finding several with serious levels of absorption.

31. Carey P. McCord, "Industrial Health Promotion in Small Plants," *AJPH* 15 (Apr. 1925): 299–302.

32. Alice Hamilton, "The Prevalence and Distribution of Industrial Lead Poisoning," in *Lead Poisoning,* 224. White lead production for 1920–30 fluctuated between 117,000 and 150,000 tons, but after 1930 there was a steady decline as substitutes became more widely used; U.S. Department of the Interior, Bureau of Mines, *Mineral Resources of the United States* 1928 (Washington: GPO, 1931), 639; Bureau of Mines, "Lead," in *Minerals Yearbook* 1936, 146 (see ch. 3, n. 6).

33. "Clinical Study of Frequency of Lead, Turpentine and Benzene Poisoning in Painters" (review of a study by Louis I. Harris, director of the preventable diseases unit of the New York City Department of Health), *Archives of Internal Medicine* (Aug. 1918): 129–56; Frederick Hoffman, "Deaths from Lead Poisoning, 1925–1927," *Bulletin of the Bureau of Labor Statistics* 488 (June 1929): 1–7; "Safety and Health: Lead Poisoning in the United States," *MLR* 46 (Feb. 1938): 420–33; Frederick Hoffman, "Lead Poisoning in 1943 and Earlier Years," *MLR* 59 (Nov. 1944): 976–78; Hamilton, "Hygiene of the Painters' Trade," 48 (see ch. 4, n. 51).

34. Hamilton ("Hygiene of the Painters' Trade," 5) cited the 1910 census figures of 277,541 painters, glaziers, and varnishers, and claimed that about 72,500 of these were organized. In 1911, the Chicago District Council of the Brotherhood of Painters, Decorators and Paper Hangers estimated the seasonal employment of its nearly 1,400 painter members at eight weeks, two days; ibid., 6.

35. Ibid., 63. Hamilton associated these demands with a survey—"the first instance of a study of industrial hygiene made by the industry itself"—conducted earlier that year by J. A. Runnberg, the union's statistician; ibid., 49.

36. "Painters' Health Campaign," *New York Times,* 8 Jan. 1922, 21; "Trade's

Tragic Toll," *New York Times*, 1 Apr. 1923, section VII, 13; David Rosner and Gerald Markowitz, "Safety and Health as a Class Issue: The Workers' Health Bureau of America during the 1920s," in *Dying for Work*, 57 (see intro., n. 12).

37. According to Rosner and Markowitz ("Safety and Health," 53–64), union support waned for three reasons: the bureau supported laborers more than "labor," and from the unions' standpoint, this ambivalence weakened their footing; it stressed job safety instead of the union's primary interests in wages and hours; and it was run by women.

38. In *Painter and Decorator* 46 (May 1932), for example, three of twelve advertisements promoted lead products exclusively. These ads amounted to 2.5 pages, or 53% of the issue's 4.75 advertising pages.

39. Compare W. Schweisheimer, "Lead Poisoning among Painters: 'Painters Colic'," *Painter and Decorator* 52 (Apr. 1938): 18–19, with his "Heart Trouble among Painters," *Painter and Decorator* 53 (Aug. 1939): 38. Other *Painter and Decorator* articles have titles that suggest they might deal with public health aspects of lead poisoning, though they do not. Schweisheimer's "Newly Painted Walls in Relation to Health" (51 [June 1937]: 15) discusses getting rid of germs in dust and dirt; Matthew Woll's "Safeguarding Your Children" (50 [Feb. 1936]: 19) turns out not to be a plea about protecting children from lead paint but a pitch for union life insurance. (Woll was president of Union Labor Life Insurance Company and a vice-president of the American Federation of Labor.) One of the magazine's most remarkable articles on health was Belknap's "The Actual Control of Poisons in Industry," published without comment in 1940. Belknap was an effective company physician and able activist for industrial hygiene, and he was anything but pro-labor.

40. William Absalon, "Cultural Lag in Painting," *Painter and Decorator* 53 (Apr. 1939): 38.

CHAPTER 7: INTRODUCING LEADED GASOLINE

The epigraph to this chapter comes from "Report of the Investigations Made by Drs. Thompson and Schoenleber at the General Motors Research Laboratories, Dayton, Ohio, and the United States Bureau of Mines Laboratories, Pittsbugh, PA" (henceforth "Report of Thompson and Schoenleber"), 17–18 May 1924, folder "Material from Carpenter Relative to Ethyl Corp. at Time of Federal Suit re. Restraint of Trade Petroleum Conference" (Carpenter folder), Box 19, Robert A. Kehoe Archives (Kehoe Archives), Cincinnati Medical Heritage Center.

1. Consumption increased from 434 tons in 1919 to 949 tons in 1929, the peak prewar year for automobile production; see Bureau of Mines, "Lead," in *Minerals Yearbook* 1920–30 (see ch. 3, n. 6).

2. In 1993, U.S. battery manufacturers accounted for 1,119,858 of the 1,357,127 metric tons of metallic lead and lead oxides consumed; see U.S. Depart-

ment of the Interior, Bureau of Mines, "Lead," in *Minerals Yearbook Statistical Compendium* 1993 (Washington: GPO, 1993), 532.

3. In 1986, the EPA reported in *Air Quality Criteria for Lead* (see ch. 2, n. 14; cited in Morton Lippmann, "Lead and Human Health: Background and Recent Findings" [1989 Alice Hamilton Lecture], *Environmental Research* 51 [1990]: 6) that U.S. consumption of lead in gasoline was over 6.6 million metric tons for the period 1929–83; see also Clair C. Patterson, "Contaminated and Natural Lead Environments of Man," *AEH* 11 (Sept. 1965): 350.

4. Average blood-lead levels in 1965 are estimated to have been around 25 µg/dL; Patterson, "Contaminated and Natural Lead Environments," 355. The NIOSH defines as "elevated" blood levels at or above 25 µg/dL, which falls within Class III of the CDC's 1991 treatment protocols for childhood plumbism calling for "environmental evaluation and remediation" and a medical evaluation for "pharmacological treatment"; "Adult Blood Lead Epidemiology and Surveillance: United States, Fourth Quarter 1994," *MMWR* 44 (14 Apr. 1995): 286–87; CDC, *Preventing Lead Poisoning in Young Children* (Atlanta: CDC, 1991), 3.

5. Quote "gift of God," Frank Howard, cited in Rosner and Markowitz, "'A Gift of God'?," 130 (see intro., n. 12); quote "allowed high performance vehicles," Donald R. Lynam, G. B. Meyers, and Gary L. Ter Haar, letter to the editor, *AJPH* 75 (Dec. 1985): 1452.

6. Recent scholarship has revealed just how contingent the development of tetraethyl lead, the twentieth century's answer to these technological "requirements," was upon industry alliances, forecasts of gasoline shortages, and predictions about what Americans were going to want automotive culture to look like; on this broader question, see Alan Loeb, "Steam versus Electric versus Internal Combustion: Choosing the Vehicle Technology at the Start of the Automotive Age" (typescript). On the many alternatives to tetraethyl lead considered during the search for a gasoline additive, see Kovarik, "The Ethyl Controversy," esp. ch. 3 and 5 (see intro., n. 12).

7. Mary Ross, "The Standard Oil's Death Factory," *Nation*, 26 Nov. 1924, 561–62; "Odd Gas Kills One, Makes Four Insane," *New York Times*, 27 Oct. 1924, 1. For a detailed analysis of contemporary press coverage of these events, see Kovarik, "The Ethyl Controversy," ch. 4; Charles Norris and Alexander Gettler, "Poisoning by Tetra-Ethyl Lead: Postmortem and Chemical Findings," *JAMA* 85 (12 Sept. 1925): 818–20.

8. "Progress Report No. VI of the Committee upon the Medical Aspects of Tetraethyl Lead Poisoning" (typescript), 1 Apr. 1925, Box 23, Kehoe Archives.

9. George Sweet Gibb and Evelyn H. Knowlton, *History of Standard Oil Company (New Jersey): The Resurgent Years 1911–1927* (New York: Harper & Brothers, 1956), 543; Kovarik, "The Ethyl Controversy," 62–63.

10. Kovarik, "The Ethyl Controversy," 62.

11. Gibb and Knowlton, *History of Standard Oil Company*, 540. Ownership and management of GM and DuPont were tightly interlocked. In 1923, Pierre S.

du Pont was president of GM, and ten of seventeen top managers at GM were also DuPont executives; see *United States v E. I. Du Pont de Nemours and Company, General Motors Corporation, et al.*, 126 F. Supp. 235, 1954 U.S. Dist. Lexis 2471; 1954 Trade Cas. (CCH) P67,905.

12. "Ethyl fluid"—three parts tetraethyl lead in two parts solvent—was mixed with gasoline at the pumps in the ratio of 5 cubic centimeters (cc) per gallon; J. H. Schrader, "Tetra-ethyl Lead and the Public Health," *AJPH* 15 (Mar. 1925): 213. The wine coloring apparently originated with Midgley and Kettering; see Rosamond McPherson Young, *Boss Ket: A Life of Charles F. Kettering* (New York: Longmans, Green & Co., 1961), 135–36, 155. According to Kettering family history, Midgley suggested that "ethyl" be given "a distinctive color to make it stand out from other gasolines." They chose red, because when Kettering first put Midgley on the fuel-additive project, he had half-seriously suggested looking at red dyes, assuming that their color might change the vaporization rate.

13. "Progress Report No. VI." It is unknown how many men died at the Wilmington plant—spokesmen simply admitted that "some" had; "Denies Lead Gas Killed 9," *New York Times*, 3 Nov. 1924, 6. On the initial deaths at Deepwater, see David Hounshell and John Smith, *Science and Corporate Strategy: Du Pont R&D, 1902–1980* (New York: Cambridge University Press, 1988), 151–52.

14. Tanquerel, *Traité des Maladies de Plomb ou Saturnines* (see ch. 4, n. 22).

15. Schrader, "Tetra-ethyl Lead and the Public Health," 213.

16. Hounshell and Smith, *Science and Corporate Strategy*, 151; Gibb and Knowlton, *History of Standard Oil*, 540–41.

17. The DuPont process used ethyl bromide, but bromine was difficult to acquire; in the early stages, Kettering built a small processing plant on a ship to extract bromine from sea water. Standard Oil hired Charles Kraus of Clark University, who found that ethyl chloride was as effective; Gibb and Knowlton, *History of Standard Oil*, 541; Hounshell and Smith, *Science and Corporate Strategy*, 151–53.

18. Hounshell and Smith, *Science and Corporate Strategy*, 152; "United States v du Pont, General Motors, et al.—Memorandum of Conference with Dr. Robert A. Kehoe" (carbon copy of typed draft), 24 Jan. 1952, folder "Material from Carpenter" (henceforth "Carpenter folder"), Box 19, Kehoe Archives; William F. Ashe, "Robert Arthur Kehoe, MD," *AEH* 13 (July 1966): 138.

19. Hounshell and Smith, *Science and Corporate Strategy*, 152–53; "Progress Report No. VI."

20. *New York Times*, 22 June 1925, 3, cited in Rosner and Markowitz, "'A Gift of God'?," 128; Silas Bent, "Deepwater Runs Still," *Nation*, 8 July 1925, 62–64.

21. "Report of Thompson and Schoenleber."

22. "Progress Report No. VI."

23. Henderson reported the exchange with Midgley in a cover letter, dated 1 Mar. 1924, to a letter he was sending to the *New York Times*. A typed copy of the cover letter and letter proposed for publication appear as exhibits A and B in

"Report of Thompson and Schoenleber." For Henderson's work for the Bureau of Mines, see Sellers, "Manufacturing Disease," 397 (see intro., n. 7); for Midgley's request for a study, see U.S. Treasury Department, PHS, *Proceedings of a Conference to Determine Whether or Not There Is a Public Health Question in the Manufacture, Distribution, or Use of Tetraethyl Lead Gasoline,* bulletin 158 (Washington: GPO, 1925): 108, quoted in William Graebner, "Private Power, Private Knowledge, and Public Health: Science, Engineering, and Lead Poisoning, 1900–1970," in *The Health and Safety of Workers: Case Studies in the Politics of Professional Responsibility,* ed. Ronald Bayer (New York: Oxford University Press, 1988), 30.

24. On the Bureau of Mines study, see Rosner and Markowitz, "'A Gift of God'?"

25. Kehoe's animal experiments involved shaving rabbits' bellies and applying 0.1 cc tetraethyl lead to the shaved skin, twice a week; see handwritten notebook begun ca. 1923, entries for July 1924 through early 1925, Box 32, Kehoe Archives. Kehoe's rabbits died.

26. In May 1924 the medical committee recommended that "in view of the present lack of scientific experimental data" on the intermediate risk of direct contact with leaded gasoline, the company should arrange a second Bureau of Mines study to weigh the public health risks of such exposure; "Report of Thompson and Schoenleber," 8–9.

27. Ibid., 9.

28. Schrader, "Tetra-ethyl Lead and the Public Health," 213–16.

29. Midgley to H. S. Cumming, U.S. Surgeon General, 30 Dec. 1922, file "Tetraethyl Lead," Folder T, General Files 0425, Records of the National Institutes of Health, RG 443, National Archives (henceforth "NIH tetraethyl lead file").

30. "Another Man Dies from Insanity Gas," *New York Times,* 20 Nov. 1924, edition M, cited in Graebner, "Private Power, Private Knowledge," 31; "Report on the Tetraethyl Lead Plant at Bayway. Oct. 27, 1924," Carpenter folder.

31. The Standard Oil representative's comment appeared in every report I read about the disaster, but it appeared first in "Odd Gas Kills One, Makes Four Insane." For an excellent narrative of the entire tetraethyl lead scandal, see Rosner and Markowitz, "'A Gift of God'?," 121–39. Blaming the workers began almost immediately, but this rationalization appeared after investigations into the rash of lead-poisoning incidents were underway; see "Report Condemns Making of Lead Gas," *New York Times,* 27 Nov. 1924, 14.

32. Ross, "The Standard Oil's Death Factory," 562; "Nine of Du Pont Plant Died," *New York Times,* 2 Nov. 1924, 22; "Denies Lead Gas Killed 9," 6.

33. *New York Times,* 22 June 1925, 3, cited in Rosner and Markowitz, "'A Gift of God'?," 128; Hounshell and Smith, *Science and Corporate Strategy,* 154; Bent, "Deepwater Runs Still," 62–64.

34. Rosner and Markowitz, "'A Gift of God'?," 123–24; *New York Times,* 1 Nov. 1924, 1.

35. Ethyl Corp. advertisement, quoted in [Emery Hayhurst], "Ethyl Gasoline" (editorial), *AJPH* 15 (Mar. 1925): 239–40.

36. For scientific objections to the bureau report, see Alice Hamilton, Paul Reznikoff, and Grace Burnham, "Tetra-Ethyl Lead," *JAMA* 84 (1925): 1481–86; see also Kovarik, "The Ethyl Controversy," 128–29.

37. John E. Mitchell, "Will Ethyl Gasoline Poison All of Us? Scientists Disagree," *New York World*, 3 May 1925, clipping in NIH tetraethyl lead file.

38. Ross, "The Standard Oil's Death Factory," 562; Rosner and Markowitz, "'A Gift of God'?," 121; "Results of Studies of Hazards Connected with Use of Tetraethyl Lead Gasoline," *MLR* 22 (Mar. 1926): 126; "The Washington Conference on the Ethyl Gasoline Hazard," *AJPH* 15 (July 1925): 632–34. The Washington Conference has attracted considerable historical attention; see, e.g., Sellers, "Manufacturing Disease," 484–86; Young, "Interpreting the Dangerous Trades," 173–76 (see ch. 4, n. 2); Graebner, "Private Power, Private Knowledge," 33–38.

39. Rosner and Markowitz, "'A Gift of God'?," 129; "Menace of Tetraethyl Lead to Garage Workers," *MLR* 20 (May 1925): 174–75.

40. Gehrmann, "Prevention of Lead Poisoning in Industry," 687–92 (see ch. 6, n. 4); Stuart W. Leslie, *Boss Kettering* (New York: Columbia University Press, 1983), 165. Accounts of Midgley's lead poisoning appear in a number of additional sources; see, e.g., Rosner and Markowitz, "'A Gift of God'?," 130–31; Kovarik, "The Ethyl Controversy," 85–86.

41. Rosner and Markowitz, "'A Gift of God'?," 131–32.

42. "Results of Studies of Hazards Connected with Use of Tetraethyl Lead Gasoline," report to U.S. Surgeon General, 17 Jan. 1926, NIH tetraethyl lead file, 126.

43. Ross, "The Standard Oil's Death Factory," 562.

44. In 1970, the year of greatest consumption, tetraethyl lead accounted for 20.5% of all lead consumed for the year; see Bureau of Mines, "Lead," in *Minerals Yearbook 1970*, 656.

45. Young, *Boss Ket*, 165; Hounshell and Smith, *Science and Corporate Strategy*, 155.

46. Ethyl's ad men did not shy away from associating their product with sexual power. Many Ethyl ads in the late 1920s featured provocatively dressed women who extolled the virtues of leaded gasoline.

47. Gibb and Knowlton, *History of Standard Oil*, 543.

48. For a summary of the Kettering Institute's control over lead research, see Graebner, "Hegemony through Science," 140–59 (see ch. 2, n. 49).

49. Ashe, "Robert Arthur Kehoe, MD," 138.

50. Ibid., 139.

51. Ibid., 139; Irene R. Campbell, "The House That Robert A. Kehoe Built," *AEH* 13 (July 1966): 143–51.

52. The classic statement of the "natural lead balance" paradigm is Kehoe's "The Harben Lectures, 1960" (see intro., n. 16). For early balance studies, see

Robert Kehoe, Frederick Thamann, and Jacob Cholak, "Lead Absorption and Excretion in Relation to Diagnosis of Lead Poisoning," *JIH* 13 (1933): 320–39; see also Robert Kehoe, "Industrial Lead Poisoning," in *Industrial Hygiene and Toxicology*, ed. F. A. Patty (2d rev. ed., New York: Interscience Publishers, 1963), 2: 941–85.

53. Donald R. Lynam, "Lead Industries Association Position," *Journal of Occupational Medicine* 17 (Feb. 1975): 84–90. Kehoe repeated in 1965 that 80 µg/dL was the lowest concentration the Kettering lab had seen in association with clinical plumbism; U.S. Department of Health, Education and Welfare, PHS, *Symposium on Environmental Lead Contamination* (Washington: GPO, 1966), 55. Of course, as diagnostics improved and more of the long-range effects of low-level lead exposure came to light, the threshold was gradually lowered.

54. In a report about a DuPont tetraethyl lead plant that had adopted Kehoe's strategies for lowering plumbism rates, Gehrmann ("Prevention of Lead Poisoning in Industry," 689) spelled out this reduction of emphasis on environment in favor of monitoring workers: it is "the moral and legal obligation of the employer to exert every possible human effort to maintain and protect [the worker's] health by (1) careful and frequent medical examinations, (2) rigid rules of conduct during working hours, and (3) installation and maintenance of proper mechanical protective and operative devices."

55. Peter T. Kilborn, "Who Decides Who Works at Jobs Imperiling Fetuses," *New York Times*, 2 Sept. 1990, section I, 28.

56. In his 1926 study of lead poisoning, Aub's researchers had argued against lead's being considered a "natural" ingredient in the body, but gradually Kehoe's explanation took hold; Patterson, "Contaminated and Natural Lead Environments."

57. For an early history and criticism of the threshold concept, see Herbert Stockinger, "Concepts of Thresholds in Standards Setting," *AEH* 25 (Sept. 1972): 153–57.

58. Kehoe's comments appear in the discussion section following Charles F. McKhann and Edward C. Vogt, "Lead Poisoning in Children," *JAMA* 101 (7 Oct. 1933): 1131–35.

59. Harold Schmeck, "Senate Pollution Hearing Told Leaded Gasoline Is Not a Peril," *New York Times*, 10 June 1966, 90; Harry A. Waldron, "The Blood Lead Threshold," *AEH* 29 (Nov. 1974): 273.

CHAPTER 8: DEFINING CHILDHOOD LEAD POISONING AS A DISEASE OF POVERTY

1. The booklets are described in "Winning the Children," *Dutch Boy Painter* 16 (Sept. 1923): 125, in Warshaw Collection of Business Americana, Archives Center, National Museum of American History, Smithsonian Institution, s.v. "paint";

"When You Were a Kid," *Dutch Boy Painter* 20 (Aug. 1927): 114, in Warshaw Collection.

2. In 1928, e.g., *Scientific American* suggested that silk, especially silk underwear, that had been weighted with lead could pose a hazard, noting that "it is a question whether lead in silk on coming into contact with the various secretions of the skin might not be absorbed to some extent"; D. H. Killeffer, "Industries from Atoms," *Scientific American* 138 (Jan. 1928): 70. The article does not mention tetraethyl lead—in fact, Ethyl was a regular advertiser in *Scientific American* throughout the late 1920s—but it does reflect a growing skepticism about possible dangers from seemingly innocuous lead applications.

3. Louis Dublin, form letter to pediatricians, 11 Sept. 1930; Dublin to Ella Oppenheimer, 14 Sept. 1933; both in "Disease Due to Poisons," file 4-5-17, Central Files, 1933–36, Records of the U.S. Children's Bureau, RG 102, National Archives (henceforth Children's Bureau, "Disease Due to Poisons").

4. Oppenheimer to Dublin, 11 Sept. 1933; quote, "either directly or by inference" (handwritten postscript initialed by Dublin), Dublin to Oppenheimer, 14 Sept. 1933; Oppenheimer to Dublin, 27 Sept. 1933; all in Children's Bureau, "Disease Due to Poisons." Oppenheimer did not examine the files for at least eighteen months.

5. McCord, "Early America: Clinical Lead Poisoning," 125 (see ch. 1, n. 1).

6. Charles McKhann of Boston's Children's Hospital, who published the most important papers of this period, made clear distinctions between pediatric and adult plumbism; see McKhann, "Lead Poisoning in Children, with Notes on Therapy," 386–92 (see ch. 2, n. 26); McKhann and Vogt, "Lead Poisoning in Children," 1131–35 (ch. 7, n. 58); Charles McKhann, "Lead Poisoning in Children: The Cerebral Manifestations," *Archives of Neurology and Psychiatry* 27 (1932): 294–304.

7. For cases of lead poisoning before the toddler years, see Murray H. Bass and Sidney Blumenthal, "Fatal Lead Poisoning in a Nursing Infant Due to Prolonged Use of Lead Nipple Shields," *Journal of Pediatrics* 15 (1939): 724–32. Older children, like adults, generally require much higher exposures to lead before exhibiting symptoms. Apart from Gibson's report ("A Plea for Painted Railings" [see ch. 2, n. 8]) of older children exposed to white lead paint, the most famous cases of lead poisoning in this age group were caused by massive exposure to lead fumes from burning battery cases; see Huntington Williams et al., "Lead Poisoning from the Burning of Battery Casings," *JAMA* 100 (13 May 1933): 1485–89.

8. The bureau's use of pamphlets and radio addresses fit its straitened resources; see Kriste A. Lindenmeyer, *"A Right to Childhood": A History of the U.S. Children's Bureau and Child Welfare, 1912–46* (Urbana: University of Illinois Press, 1997).

9. Mrs. Michael V. Sacharoff to U.S. Children's Bureau, 31 Aug. 1933, and

reply, 12 Sept. 1933; Fletcher Dodge, Toy Manufacturers of the U.S.A., to Oppenheimer, 23 Sept. 1931; all in Children's Bureau, "Disease Due to Poisons."

10. It is unclear when Macy's initiated this policy. One correspondent suggested it began about 1927. In 1931, the A. Schoenhut Company sold Macy's a large quantity of colored play blocks. Macy's specified lead-free pigments, and Shoenhut's paint supplier asserted that its pigments did not contain the mineral. Soon after Christmas, however, after selling 75% of the blocks, Macy's returned the remainder, complaining that they did not meet the specification. Only then did Schoenhut hire a chemist, who confirmed that the green and yellow blocks did indeed contain as much as 11% lead. After many manufacturers told him it was impossible, Schoenhut secured a contract with a paint manufacturer who guaranteed his paints would have only trace lead—and agreed to carry liability insurance to protect Schoenhut should the paints be proven to contain toxic pigments. See A. F. Schoenhut to Oppenheimer, 17 Apr. 1935, in Children's Bureau, "Disease Due to Poisons."

11. E.g., Herbert Klopper, president of Newark Varnish Works, wrote of reformulating his paints for toy companies, "Without themselves passing judgment on the question of health hazard, our customer nevertheless directed us to use thereafter only coloring pigments that were absolutely free from lead"; Klopper to Oppenheimer, 25 Apr. 1935, in Children's Bureau, "Disease Due to Poisons"; see also Charles Pajean, The Toy Tinkers, Inc., to Oppenheimer, 29 Apr. 1935, ibid.

12. Felix Wormser (1894–1981) received a master's degree in engineering from Columbia University and from 1917 to 1919 worked for the Bureau of Mines; for biographical details, see *Who's Who in America,* Vol. 33 (1964–65); Senate Committee on Public Works, *Air Pollution—1966: Hearings before a Subcommittee on Air and Water Pollution,* 89th Cong., 2d sess., 7–15 June 1966, 234. For most of his career he was a consulting engineer to mining companies. He helped establish the LIA in 1928 and was its secretary-treasurer until 1947, when he became president. That year he also became vice-president of St. Joseph Lead Company, one of the largest lead producers. In 1953, he was appointed Assistant Secretary of the Interior for Mineral Resources.

13. The text of the LIA survey and twelve responses appear in Frederick L. Hoffman, *Lead Poisoning Legislation and Statistics* (Newark, N.J.: Prudential Insurance Company, 1933), 19.

14. Alice Hamilton to Martha Eliot, 19 Nov. 1935, and reply, 23 Nov. 1935, in Children's Bureau, "Disease Due to Poisons."

15. McKhann and Vogt, "Lead Poisoning in Children," 1131, n. 5; Hoffman, *Lead Poisoning Legislation and Statistics,* 20.

16. Hoffman, *Lead Poisoning Legislation and Statistics,* 19; H. Eliott, Halsam Products, to Oppenheimer, 13 Apr. 1935, in Children's Bureau, "Disease Due to Poisons."

17. Eliot passed Hamilton's letter to Oppenheimer, promising Hamilton "a

brief resume of the information we have in the Children's Bureau at the present time"; Hamilton to Eliot, 19 Nov. 1935, and reply, 23 Nov. 1935, in Children's Bureau, "Disease Due to Poisons." In the 1940s the Children's Bureau maintained a list of toxic and nontoxic pigments, but conducted very little—if any—testing.

18. Wormser added "that the medical profession will do the lead industry a great favor if it will report to us the presence of lead paint on any crib, or toy. . . . so we may take remedial measure"; see "Facts and Fallacies of Lead Exposure" (paper presented at a meeting of LIA and AMA members, Oct. 1946, subsequently published as "Facts and Fallacies Concerning Exposure to Lead," *Occupational Medicine* 3 [Feb. 1947]: 135–44), 15.

19. Fee, "Public Health in Practice," 570–606 (see intro., n. 10); Williams et al., "Lead Poisoning from the Burning of Battery Casings," 1485–89.

20. Although Brailey belittled Easter—"Melrose had had a year or so at Tuskegee Institute and . . . had had ideas of studying medicine . . . but the circulation of the blood had proved too much for him"—she identified him as "the most important informant in the investigation." Brailey's report appears in Williams et al., "Lead Poisoning from the Burning of Battery Casings," 1485; Fee's description in "Public Health in Practice," 579, gives Easter his due.

21. Williams et. al ("Lead Poisoning from the Burning of Battery Casings," 1485) acknowledged that efforts had only stopped the local problem: "Reports from Philadelphia, Long Island and Detroit would indicate . . . that it may be looked for in widely separated parts of the country." On other cities, see Abraham Levinson and Mary Zeldes, "Lead Intoxication in Children: A Study of 26 Cases," *Archives of Pediatrics* 56 (Nov. 1939): 738–48; George Cooper, "An Epidemic of Inhalation Lead Poisoning with Characteristic Skeletal Changes in the Children Involved," *American Journal of Roentgenology and Radium Therapy* 58 (Aug. 1947): 129–41; "Death at the Hearth," *Time,* 21 Feb. 1955, 38.

22. I have not located the disputed report, apparently entitled "Lead Toys—Lead Paint—Lead Poisoning," which Wormser ("Facts and Fallacies of Lead Exposure," 12) said had appeared "a few years ago" in a journal of the "National Safety Education." It is probable that to counter it, Wormser identified Levinson and Zeldes's "Lead Intoxication in Children," which discusses twenty-six cases of lead poisoning from battery burning in the years 1935–38. Of these, five—all children—died.

23. "Lead Poisoning and the Eighteenth Amendment" (editorial), *American Journal of Surgery* 10 (Oct. 1930): 32–34; Robert Kehoe, "Under What Circumstances Is Ingestion of Lead Dangerous?" in PHS, *Symposium on Environmental Lead Contamination,* 52 (see ch. 7, n. 53).

24. Kato, "Lead Meningitis in Infants," 569–91 (see ch. 1, n. 29); Gibson, "A Plea for Painted Railings."

25. Levinson and Zeldes, "Lead Intoxication in Children," 738; Hoffman, *Lead Poisoning Legislation and Statistics,* 18.

26. Anne Mathews to Frances Perkins, 12 Apr. 1933, in Children's Bureau, "Disease Due to Poisons."

27. Ibid.

28. Mathews to President Roosevelt, 8 Nov. 1933, in Children's Bureau, "Disease Due to Poisons."

29. Clara Beyer to Mrs. J. W. Moore, 11 Dec. 1933, in Children's Bureau, "Disease Due to Poisons."

30. The last letter in this correspondence was written in June 1935; prior to that, Indiana did not include occupational diseases in its workmen's compensation laws. Moore and her associates in Indiana had been lobbying for an enlarged compensation law. She reported that "there was a bill introduced this last term . . . but it was not drawn up properly—it was too cumberson [sic] and contained too many features to handle effectively"; Moore to Ella Merritt, 5 June 1935, in Children's Bureau, "Disease Due to Poisons."

31. According to Anne Mathews, a year after Frank got sick the company installed some ventilation equipment; Mathews to Roosevelt, 8 Nov. 1933.

32. A Medline search found no studies from 1966 to 1998 in which spina bifida was associated with parental lead exposure. For summaries of early research in occupational lead poisoning, see Oliver, *Dangerous Trades* (see ch. 1, n. 38). See also Klein, "From Knowledge to Policy" (intro., n. 7).

33. Ironically, fetal risks played little part in the Supreme Court's unanimous 1991 decision in *Automobile Workers v Johnson Controls* (*Supreme Court Bulletin* 89–1215 [Washington: GPO, 1991], B1163) that found fetal-protection policies illegal under Title VII, especially as amended by the 1978 Pregnancy Discrimination Act. The court focused instead on whether the ability to bear children could be used by a company as a legitimate reason for disqualifying a job candidate.

34. See, e.g., Gibson, "A Plea for Painted Railings"; Blackfan, "Lead as a Cause of Convulsions" (see ch. 2, n. 9); Ruddock, "Lead Poisoning in Children with Special Reference to Pica," 1682 (ch. 2, n. 12); McKhann and Vogt, "Lead Poisoning in Children," 1131–35. Although these doctors might agree that pica could be considered a "disorder," it was very common in teething children. Even industrial hygienist Robert Kehoe had a good sense of the correct etiology in most cases of childhood lead poisoning: "most cases of lead poisoning in infants and children come from chewing objects coated with metallic lead and lead pigments"; Kehoe to Frederick Hoffman, 26 June 1937, "Correspondence," Box 100, Kehoe Archives (see ch. 7, n. to epigraph).

35. Levinson and Zeldes, "Lead Intoxication in Children," 748; Nell Conway, "Lead Poisoning—from Unusual Causes," *Industrial Medicine* 9 (Sept. 1940): 471–77. Conway was employed at the Kettering Laboratory, which meant she worked for Robert Kehoe—a fact that is nearly impossible to square with Kehoe's 1937 letter to Hoffman, mentioned in the preceding note. Kehoe had a reputation for monitoring everything coming out of Kettering that had to do with public

relations, so it is difficult to imagine that he did not know of Conway's article. One possible explanation is Kehoe's aversion to involving himself in public debate over pediatric lead poisoning; he confined his harsh criticisms of the lead-paint industry to private correspondence. This would perhaps also explain Conway's curious omission of any discussion of house paints.

36. Randolph K. Byers and Elizabeth E. Lord, "Late Effects of Lead Poisoning on Mental Development," *AJDC* 66 (Nov. 1943): 471–94. Byers (1896–1988) attended Harvard Medical School and interned at MGH at the time Joseph Aub and his team were conducting their lead studies. He completed his internship at Children's Hospital, where he continued practice. He was appointed chief of the Neurologic Service in 1951, and stayed at Children's until his retirement in 1979. Biographical information drawn from Byers, "Memoirs of Randolph K. Byers" (1986, in possession of Byers's family, photocopy); see also Randolph Byers, "Introduction," in *Low Level Lead Exposure*, 1–4 (see intro., n. 9).

37. Another student was described as working close to his age level, despite "lively knee jerks, ankle clonus," restlessness, and inattentiveness; Byers and Lord, "Late Effects of Lead Poisoning," 476–77.

38. "Paint Eaters," *Time*, 20 Dec. 1943, 49.

39. Wormser to Kehoe, 19 Jan. 1944, folder "Lead Industries Association" (henceforth "LIA folder"), Box 90, Kehoe Archives.

40. Kehoe to Wormser, 7 Feb. 1944, "LIA folder."

41. Ibid.

42. Wormser to Kehoe, 21 Feb. 1944, "LIA folder."

43. Byers, "Memoirs," 78a. After this meeting, Byers met with lead executives once a year to update them on his progress, "and in spite of the three to five martinis for each of them and one for me," he recalled, "they gave me a grant of several thousand dollars annually"; ibid., 78a–79a. Byers published the results of his study in 1954; see Randolph Byers and Clarence Maloof, "Edathamil Calcium-Disodium (Versenate) in Treatment of Lead Poisoning in Children," *AJDC* 87 (May 1954): 559–69.

44. The chief proponent of this view is Graebner, "Hegemony through Science" (see ch. 2, n. 49). To measure the effects of this imposed technocracy, compare Byers and Lord's study in "Late Effects of Lead Poisoning" with Byers and Maloof, "Edathamil Calcium-Disodium (Versenate)." The science may have improved and a measure of fire may have been drained from the prose, but the later paper clearly advocates aggressive case finding and abatement and in no obvious way favors its sponsors.

45. Byers and Lord, "Late Effects of Lead Poisoning," 476. This simple truth persists today in regional differences in case finding, which suggest, against all logic, that lead poisoning hardly exists in southern cities.

CHAPTER 9: URBAN PHYSICIANS DISCOVER THE SILENT EPIDEMIC

1. *Better Homes and Gardens* (Mar. 1944): 97. The ad shows the NLC's trademark character cheerily straining to hand an artillery shell to a G.I. manning a piece of heavy artillery. Shells, labeled "Save Fats," "Save Food," "Save Metals," and so on, are being passed to the Dutch Boy by a number of hands, two pairs of which appear to be women's.

2. Fred P. Peters, "Lead Carries Its Weight," *Scientific American* 170 (June 1944): 250–52. Although there was less lead in each bullet than in earlier years, over two hundred tons of lead went into ammunition during the war; see Bureau of Mines, *Minerals Yearbook* 1939–45 (see ch. 3, n. 6). Despite the demand for lead batteries in trucks, jeeps, and submarines, consumption of lead for that purpose dropped during the war, only to surge ahead of all other uses in the postwar automobile craze. *Lead* declared that "all gasoline being used by the air forces of the United States and its Allies contains tetraethyl lead," and that Ethyl was "now producing for military use more than was used in all automobiles and airplanes in the United States and allied nations in 1940"; "Leaded Gasoline Helping to Win Battles," *Lead* 14 (Sept./Oct. 1944): 3. See also "National Lead Company in Wartime," *Dutch Boy Quarterly* 21 (1943): 17–24; "Lead Foil in Packaging Saves Weight and Space while Releasing Other Metals for Armament Use," *Lead* 13 (Mar. 1943): 3.

3. Lead consumption rose from 667 tons in 1939 to 1,119 tons in 1944, an increase of 67.8%; Bureau of Mines, *Minerals Yearbook* 1939–45. This increase was in keeping with the GNP, which rose by 72.5% over the same years; Bureau of the Census, *Historical Statistics of the United States*, Table F1–5, 224 (see ch. 3, n. 9).

4. "National Lead Company in Wartime," 24. The use of lead pigments was restricted for one month in 1942; see "White Lead in Wartime," *Dutch Boy Quarterly* 20 (1942): 9. This article included a photo of war workers' homes in Canton, Ohio, and boasted that "Dutch Boy white lead was used for both exteriors and interiors thruout [sic] the project" (10). The wartime restrictions that really affected the lead industry were those on oils, hastening the development of water-based paints that homeowners could apply themselves.

5. Ernest T. Trigg, "'After War—What?'," *Paint Industry Magazine* 59 (Nov. 1944): 368–70.

6. Ibid.; Bureau of the Census, *Historical Statistics of the United States*, Table N156–69, 639.

7. For suburban housing in the postwar years, see Kenneth T. Jackson, *Crabgrass Frontier: The Suburbanization of the United States* (New York: Oxford University Press, 1985): 231–45. See also "Lead Fittings for Levitt's Towns," *Lead* 19 (1953): 6; "Levittown and Lead," *Lead* 24 (1960): 6. Tetraethyl lead use rose with

the nation's automotive horsepower. From 1946 to 1962, the amount of lead consumed annually in gasoline rose threefold, from 57 tons to 169 tons; Bureau of Mines, *Minerals Yearbook* 1945-62.

8. "Women do not paint" (advertisement for NLC paints), *Country Life in America* (date unknown, but between 1906 and 1917): 531, Warshaw Collection (see ch. 8, n. 1). Contrast with an ad in *Paint, Oil & Chemical Review* (5 Apr. 1956) for Titanox paint, an NLC subsidiary, featuring a close-up of a paint-spattered but determined woman, roller in hand. "Modern Living: Everyone a Painter" (*Time*, 17 Dec. 1951, 91) featured a photo of a woman, with hand on hip and *sans* hat or shoes, standing on a desk with a paint roller in her hand, painting the ceiling. The shift in paint companies' policies can be seen in Sherwin-Williams's wartime promotion of "Kem-Tone," one of its first water-based paints. See "The Paint Industry Goes to War," *Business Week*, 17 Oct. 1942, 23.

9. Ironically, ancient Romans wore lead plates to produce just the opposite effect; see Niriagu, *Lead Poisoning in Antiquity*, 262 (see intro., n. 4); Wormser, "Facts and Fallacies of Lead Exposure," 15 (ch. 8, n. 18). See also "Interesting Odds 'n' Ends about Lead," *Lead* 28 (1964): 11.

10. William C. Wilentz, "'Plumbophobia': Occupational Lead Poisoning," *Industrial Medicine* 15 (Apr. 1946): 253-56.

11. A number of studies discuss the transforming effect of germ theory; see, e.g., Meckel, *Save the Babies*, 63 (see ch. 2, n. 39); Brandt, *No Magic Bullet*, 4 (intro., n. 13).

12. Schrader, "Tetra-ethyl Lead and the Public Health," 213-16 (see ch. 7, n. 12).

13. British anti-Lewisite, as its name suggests, was developed by British physicians, and was intended to combat poisoning by Lewisite, an arsenic derivative used in World War I and also known as phenyldichlorarsine; Harry Eagle, "The Effect of BAL on Experimental Lead Poisoning," *Proceedings of Lead Hygiene Conference* (New York: LIA, 1948): 81-92. As Eagle recalled, "Those of us working with the compound . . . were always a little puzzled at the secrecy which surrounded the material. It required little more than a sub-average sense of smell and a laboratory able to do some elementary analyses to arrive at the complete structure of BAL" (83-84).

14. Ibid., 85. According to the "Summary of Reports Received by the Committee on Medical Research of the Office of Scientific Research and Development" (28 Jan. 1946, 1-14-46-1-26-46, bulletin 73, 785-89, typescript labeled *Restricted*, folder "Office of Research and Development #2," Archive B: St65p, Joseph Stokes, Jr., Papers, American Philosophical Society Library, Philadelphia; henceforth "Stokes papers"), four research teams experimented with BAL: Cornell researchers studied the drug's general effects, two teams at Johns Hopkins studied its effects on arsenic and mercury, and Yale physiologists tested its effectiveness against Lewisite.

15. Henry Ryder, Jacob Cholak, and Robert Kehoe, "Influence of Dithiopropanol (BAL) on Human Lead Metabolism," *Science* 106 (18 July 1947): 63–64. The LIA had funded BAL research since 1944. Preliminary testing of a modified version of the compound appeared promising against lead. In 1944, Alsoph H. Corwin, who had been studying mercury and cadmium poisoning at Johns Hopkins, obtained funding from the LIA-affiliated Research Corporation and eventually developed a form of dithizone for which he received a patent in 1948; see Alsoph H. Corwin, "Development of an Antidote for Lead Poisoning," 5 May 1949 (typescript), in "Bowditch folder" (see ch. 5, n. 62); Manfred Bowditch, LIA's director of health and safety, to Robert Ziegfeld, LIA's secretary-treasurer, 9 May 1949, ibid.

16. James G. Tefler, "Use of BAL in Lead Poisoning: A Preliminary Report on One Case," *JAMA* 135 (29 Nov. 1947): 835–37.

17. "Lead Poisoning Cured by War Gas Antidote," *Science News Letter* (6 Dec. 1947): 355–56. Presumably, it was this second course of injections that led to the assertion that the sailor was cured in five days.

18. The situation as of 1948 is summarized in R. V. Randall, "Medical Progress: BAL," *NEJM* 239 (1948): 1004.

19. From 1943 to 1947, the Baltimore Health Department reported fifty-one cases of pediatric lead poisoning, with fifteen fatalities (29.4%); "Lead Poisoning in Children Is Preventable and Can Be Fatal," *Baltimore Health News* 26 (Apr. 1949): 122–24; Julius M. Ennis and Harold Harrison, "Treatment of Lead Encephalopathy with BAL (2.3-dimercaptopropanol)," *Pediatrics* 5 (1950): 853–67, quote, 855.

20. Garrett Deane, Frederick Heldrich, Jr., and J. Edmund Bradley, "The Use of BAL in the Treatment of Acute Lead Encephalopathy," *Journal of Pediatrics* 42 (Apr. 1953): 409–13.

21. Elston Belknap, "The Diagnosis and Treatment of Lead Poisoning—Some New Concepts" (speech delivered at the annual meeting of the LIA, 19 Apr. 1952), folder "Lead—litigation—1958," Box 5: Invitations-M, Aub files (see ch. 5, n. 62).

22. The earliest published study I have located on the use of Versene in pediatric cases is F. E. Karpinski, F. Rieders, and L. S. Girsch, "Calcium Disodium Versenate in the Therapy of Lead Encephalopathy," *Journal of Pediatrics* 42 (1953): 687–99; see also Byers and Maloof, "Edathamil Calcium-Disodium (Versenate)," 559–69 (see ch. 8, n. 43); Arnold Tanis, "Lead Poisoning in Children, Including Nine Cases Treated with Edathamil Calcium-Disodium," *AJDC* 89 (Mar. 1955): 325–31. Riker Laboratories provided the Versene for the latter two studies. The LIA funded Byers's research and was involved in selecting clinicians to test the compound, as well as in setting and distributing CaEDTA protocols for early clinical studies; see Bowditch to Aub, 6 Aug. 1952, "Bowditch folder."

23. "The Lead Hygiene Conference" (transcript of discussion session in an LIA-sponsored conference, 6–7 Nov. 1958), *Industrial Medicine and Surgery* 28 (Mar. 1959): 133; ibid., 154, for a discussion of the case by Byers.

24. See the series of letters between Kehoe, Bowditch, and Rutherford Johnson concerning one Houston entrepreneur, William U. Giessel, whose company,
Occupational Health Laboratories, offered lead manufacturers free advice on lead
poisoning and sold CaEDTA by mail; "LIA folder" (see ch. 8, n. 39).

25. Paul Woolley, "Lead Poisoning during Infancy and Early Childhood,"
American Journal of Roentgenology 78 (Sept. 1957): 547–49.

26. Ibid., 549; Susan Charest et al., "Community Aspects of Lead Intoxication in Children," *Clinical Proceedings of Children's Hospital of Washington, D.C.*
25 (Nov. 1969): 308–19, quotes, 313.

27. Wormser to Aub, 1 Jan. 1945, folder "Lead—1927–1965" (henceforth
"Lead 1927–65 folder"), Box 5: Invitations-M, Aub files; Wormser to Aub, 31 July
1945, ibid.

28. J. H. Schaefer to Wormser, 27 June 1944, in "LIA folder."

29. To get a sense of how much control over distribution Ethyl enjoyed, see
the records of the 1939 anti-trust suit against them, *United States v Ethyl Gasoline
Corp.*, District Court, S. D. New York 27 F. Supp. 959; 41 U.S.P.Q. (BNA) 772, 19
May 1939.

30. The participants at this planning meeting agreed that standards for diagnosis needed to be refined, but it is notable that in this pre-BAL era, the agenda
did not include the treatment of lead poisoning; typescript memorandum, 7 Feb.
1946, in "Lead 1927–65 folder."

31. Aub to Wormser, 11 Oct. 1946, in "Lead 1927–65 folder." Ethyl physician
Willard Machle argued that Wormser's confrontational approach of debunking
"fallacies" rather than constructing scientific foundations for the "facts" would
perpetuate the ignorance of which he complained. In a remarkable four-page diatribe to Schaefer (14 Jan. 1946, in "LIA folder"), who held executive positions at
both Ethyl and the LIA, Kehoe poured out his frustration at the LIA's long history
of "wildcat" funding at Harvard and Johns Hopkins: "Perhaps he [Wormser]
thinks he is purchasing prestige from Harvard University for his Association."
Worse was the LIA's seeming indifference to the Kettering Laboratory's solid record of basic theoretical research: "I have never been able to understand why
Lead Industries Association through Wormser had not interested themselves directly and financially in the work of this Laboratory."

32. Internal memorandum, 15 Jan. 1948, in "LIA folder."

33. Bowditch was born in Dresden, Germany, in 1890. His was an old Boston
family, deeply involved in the city's medical community. For biographical details,
see *National Cyclopædia of American Biography*, 44: 532–33 (see ch. 5, n. 18).

34. Andrew Fletcher to Chairman Cornelius Haley, Senate Committee on
State Administration, 19 Feb. 1945, in "Bowditch folder."

35. Kehoe to Bowditch, 11 Feb. 1948, in "LIA folder."

36. Jam Handy Organization, "The Safe Use of Lead in Modern Industry"
(script prepared for the LIA, May 1949, labeled "Copyright and Confidential"),
in "Bowditch folder." Bowditch sheepishly distributed copies to Aub, Kehoe, and

others, asking for their advice, "since your opinions will, of course, carry great weight with Felix Wormser and the others who must decide if we are justified in spending the large sum involved" in proceeding with the project. Aub (6 June 1949, in "Bowditch folder") gave the script a thumbs up, although he recommended shortening; Kehoe turned up his nose (memo from Bowditch, 4 Nov. 1949, in "LIA folder"). Although I have found no evidence, the lack of further comment on the project would suggest that the film was never made. Two months after Bowditch submitted the script to Aub, he mentioned other industrial films that could serve LIA's purposes, concluding that "if you size it up as I do, I would think that this about ends the discussion, both from your viewpoint and mine"; Bowditch to Aub, 4 Aug. 1949, in "Bowditch folder." Apparently it did not, because three months later, Bowditch, in the November memo, said he was planning to go ahead with the film.

37. LIA, *Lead in Modern Industry: Manufacture, Applications and Properties of Lead, Lead Alloys, and Lead Compounds* (New York: LIA, 1952), 174–76.

38. Bowditch to Aub, 13 Aug. 1952, in "Bowditch folder."

39. Bowditch to Kehoe, 13 July 1951, folder 2, Box 38, Kehoe Archives; Kehoe to Bowditch, 9 Aug. 1951, ibid. In addition to a number of published reports, Randolph Byers, J. Julian Chisolm, and the LIA collaborated on an exhibit entitled "Childhood Plumbism" for the 1953 AMA convention in New York; Bowditch to members of the LIA, 21 May 1953, Box 47, Kehoe Archives. Among the projects the Kettering Laboratory undertook for the LIA were testing the toxicity of new Dutch Boy Paints, assessing whether new lead-bearing plastics caused dermatitis, and studying the toxicity of paints on toys; see "The Occurrence and Significance of Lead in the Paint Used on Toys" (typescript, 18 Nov. 1957), Box 32, Kehoe Archives.

40. Bowditch to Ziegfeld, 16 Dec. 1952, loose papers, Box 47, Kehoe Archives.

41. Fee, "Public Health in Practice," 570–606 (see intro., n. 10); Blackfan, "Lead Poisoning in Children with Especial Reference to Lead" (ch. 2, n. 9).

42. Between 1931 and 1940 Baltimore reported forty-nine lead-poisoning deaths of children under age 15, or 24.3% of the 202 youngsters killed in all of the United States; John M. McDonald and Emanuel Kaplan, "Incidence of Lead Poisoning in the City of Baltimore," *JAMA* 119 (11 July 1942): 870–72. But Baltimore's population, estimated at 838,700 in 1936, was only 0.65% of the U.S. population of 128,052,000, for a rate that was 48.6 times as high. Mistaking a town's recognition of a lead-poisoning problem for a regional monopoly on the disease is not unique to the twentieth century: outbreaks of lead colic from adulterated wines and ciders often bore the name of the town where the "epidemic" was first discovered. For the examples of the colic of Poitou and the Devonshire colic, see Wedeen, *Poison in the Pot* (see intro., n. 6); Eisinger, "Lead and Wine," 296 (ch. 1, n. 41).

43. Fee, "Public Health in Practice," 586–87.

44. *Baltimore City Annual Report of the Department of Health* 1940–50 (see

ch. 2, n. 5). In 1940, the department had 280 specimens tested, including those from sixty-one children; in 1950, it tested 493 samples from 253 children.

45. *Baltimore City Annual Report of the Department of Health* 1940–50.

46. Anna M. Baetjer to Bowditch, 19 Oct. 1949, in "Bowditch folder."

47. Ibid.

48. Bowditch to Aub, 21 Dec. 1949, in "Bowditch folder."

49. Lawrence Fairhall to Bowditch, 6 Jan. 1950, in "Bowditch folder."

50. Aub seconded Fairhall's expectation that lead poisoning should rise in the winter when children are more confined: "One wonders whether this wasn't a summer complaint ascribed to lead"; Aub to Bowditch, 16 Jan. 1950, in "Bowditch folder."

51. Fairhall to Bowditch, 6 Jan. 1950, in "Bowditch folder"; May Mayers to Bowditch, 12 Jan. 1950, ibid. Both Mayers and Fairhall might have looked up another study from Baltimore, published eight years earlier in *JAMA*—i.e., John M. McDonald and Emanuel Kaplan, "Incidence of Lead Poisoning in the City of Baltimore," *JAMA* 119 (11 July 1942): 870–72—which described forty-nine fatal pediatric lead poisonings between 1931 and 1940, approximately 25% of those reported in the nation. Why the disproportionate amount? "Largely as a result of studies associated with a follow-up of a blood lead laboratory service" (in Baltimore). In other words, they were looking for it.

52. Bradley et al., "The Incidence of Abnormal Blood Levels of Lead," 1–6, data from table, 5 (see intro., n. 3). It is significant that this study was supported by a grant from the U.S. Department of Health, Education and Welfare.

53. Records of many of these analyses appear in folder "Distribution—Children," Box 19, Kehoe Archives. Kettering billed the hospitals for these analyses, but since most lead-poisoning patients were indigent, he expected payment only if the hospital billed the family. Kehoe's policy on billing in indigent cases appears in Kehoe to Dr. H. G. Thornton, 6 Feb. 1934, ibid.

54. Hugo Dunlap Smith, "Lead Poisoning in Children, and Its Therapy with EDTA," *Industrial Medicine and Surgery* 28 (Mar. 1959): 148–51.

55. Kehoe to Bowditch, 9 Aug. 1951, and reply, 27 Aug. 1951, in "LIA folder." Kehoe and Smith were instrumental in establishing Cincinnati's Lead Control Center. Kehoe published little on pediatric lead exposure—only Kehoe, Thamann, and Cholak, "On the Normal Absorption and Excretion of Lead," 301–305 (see intro., n. 25)—but he was active in Cincinnati's lead-control program. However, he recalled in 1971 that his ties to the program had been largely through Smith. He expressed regret that after Smith's death he was "pretty much out of contact with the problem of lead poisoning among children." He expected that, with few of the old guard left, younger physicians would have to "repeat some of our experience of the past before they go much further"; Kehoe to Chisolm, 18 Oct. 1971, "Correspondence," Box 98, Kehoe Archives.

56. "His Little Girl Screamed in Pain for 6 Weeks but 4 MDs Said, 'Don't

Worry'; Then She Died," *New York World Telegram and Sun,* 23 Aug. 1951. The next day, the *New York Daily Mirror, Daily Compass,* and *New York Times* ran the story; see photocopies of clippings in "LIA folder."

57. E.g., Baltimore's Huntington Williams described Celeste's case at a conference in 1959; see transcript of discussion at an LIA-sponsored conference, 6–7 Nov. 1958, following Anna Baetjer, "Effects of Season and Temperature on Childhood Plumbism," *Industrial Medicine and Surgery* 28 (Mar. 1959): 143.

58. See, e.g., "City Unit to Seek Lead Poison Curb," *New York Times,* 29 Mar. 1953, 66; "Health Board Warns on Lead Paint Poison as 2 More Children in City Are Killed by It," *New York Times,* 30 June 1953, 30.

59. Mary Culhane McLaughlin, "Lead Poisoning in Children in New York City, 1950–1954," *New York State Journal of Medicine* 56 (1 Dec. 1956): 711–14. Celeste's death seems more senseless when one considers that in August 1951, three pediatrics residents at Kings County Hospital were treating twelve children for lead poisoning; see Giannattasio, Bedo, and Pirozzi, "Lead Poisoning: Observations in Fourteen Cases" (see ch. 2, n. 26).

60. Harold Jacobziner, "Lead Poisoning in Children: Epidemiology, Manifestations, and Prevention," *Clinical Pediatrics* 5 (May 1966): 277–86.

61. For St. Louis, see J. Earl Smith, B. W. Lewis, and Herbert Wilson, "Lead Poisoning: A Case Finding Program," *AJPH* 42 (Apr. 1952): 417–21; Benjamin Lewis, Richard Collins, and Herbert Wilson, "Seasonal Incidence of Lead Poisoning in Children in St. Louis," *Southern Medical Journal* 48 (Mar. 1955): 298–301. For Chicago, see David Jenkins and Robert Mellins, "Lead Poisoning in Children: A Study of Forty-six Cases," *Archives of Neurology and Psychiatry* 77 (1957): 70–78; Joseph Christian, Bohdan Celewycz, and Samuel Andelman, "A Three-Year Study of Lead Poisoning in Chicago," *AJPH* 54 (Aug. 1964): 1241–51.

62. The Philadelphia Health Department established a lead-poisoning division in 1950; see Theodore Ingalls, Emil A. Tiboni, and Milton Werrin, "Lead Poisoning in Philadelphia, 1955–60," *AEH* 3 (Nov. 1961): 575–79. The department made periodic statements to the press about the city's lead-poisoning problems but formulated little in the way of a program until the 1960s; see, e.g., "Lead Poisoning Worries City," *Philadelphia Bulletin,* 29 Aug. 1954. No major study of lead poisoning was conducted until 1959; "Study Project, the Detection, Treatment, and Prevention of Lead Poisoning in a Children's Hospital Out-patient Department" (typescript), Archive B: St65p, Stokes papers. From 1952 to 1954, Walter Eberlein had been a fellow at the Harriet Lane Home of Johns Hopkins, where he coauthored a 1955 study of lead poisoning with Chisolm; personnel records, folder "Eberlein, Walter R.," Archive B: St65p, Stokes papers.

63. In 1953, Cincinnati reported twenty cases of childhood lead poisoning, with five fatalities; Hugo Dunlap Smith, "Lead Poisoning in Children, and Its Therapy," 148–51; Kehoe to Herbert Hillman, technical director, Eaglo-Paint and Varnish Corporation, 4 Sept. 1953, loose papers, Box 47, Kehoe Archives.

64. Bowditch to Kehoe, 26 Dec. 1957, folder "Lead Industries Association," Box 47, Kehoe Archives; Kehoe to Hillman, 4 Sept. 1953, loose papers, Box 47, Kehoe Archives.

65. All but two of the committee members voted in favor of the standard; the representatives of the AMA and the National Safety Council abstained. The AMA representative explained that this "was not so much a question of objecting to the proposed standard as their desire to be over cautious rather than not being cautious enough." The AMA objected to the subcommittee's reliance on Kehoe's work because it relied too heavily on adult standards; "Minutes: Sectional Committee on Hazards to Children, Z66" (typescript), Children's Bureau, "Disease Due to Poisons" (see ch. 8, n. 3).

66. American Standards Association, "American Standard Specifications to Minimize Hazards from Residual Surface Coating Materials" (Standard Z66.1–1955; unbound booklet). The standard also restricted antimony, arsenic, cadmium, mercury, selenium, and barium.

67. ASA, Standard Z66.1–1955; Foulger, "Precautionary Labeling," 122–25, quote, 123 (see ch. 6, n. 30). Compare this standard with that advocated by the American Academy of Pediatrics in 1972, which argued that interior paint should contain no more than 0.06% lead to protect the child who ate only 1 square centimeter of a ten-layer thick paint chip per day, far less than would be consumed by the typical child with pica; Senate Committee, *Lead-Based Paint Poisoning Amendments of 1972,* 49 (intro., n. 22).

68. "Minutes: Sectional Committee on Hazards to Children, Z66," 5.

69. White lead accounted for 13.3% of all lead consumed in the United States in 1926; by 1955 its share had fallen to 1.5%. Red lead and other oxides fared a bit better, if we ignore the war years, when about 40% of all lead consumed went into red lead (presumably to protect steel battleships and tanks); by 1955 red lead and litharge accounted for 7.25% of total lead consumption. See Bureau of Mines, *Minerals Yearbook* 1926–55.

70. ASA, Standard Z66.1–1955.

71. Albert Solomon to Children's Bureau, 18 May 1952, and reply from Alice Chenoweth, in Children's Bureau, "Disease Due to Poisons." Solomon's letter is typical of dozens in the bureau's files. I made no attempt to quantify these letters from worried parents, but noted that they definitely increased in number through the late 1940s and early 1950s. Marian Crane, assistant director of the bureau's Division of Research in Child Development, explained to Martin Varnish Co. (18 Nov. 1952, ibid.) that "the Bureau has discontinued distribution of any lists of paints 'that are poisonous' or 'usually considered harmless.' Many questions were raised as to the accuracy of the list we published in 1942 and we were forced to the conclusion that toxicity data are lacking for so many paints and pigments that it is not practical to try to classify them as 'poisonous' or 'harmless.'"

72. Crane to Fairhall, 19 July 1949, and reply, 2 Aug. 1949, ibid. This passing

of a very small baton from the Children's Bureau to the PHS is typical of the artificial division of labor that typified relations between federal agencies charged with labor or health issues. For the significance of this division in the Progressive Era, see Sellers, "The Public Health Service's Office of Industrial Hygiene" (see ch. 5, n. 32). David Rosner and Gerald Markowitz trace the impact of this split from the New Deal to OSHA in "Research or Advocacy" (ch. 3, n. 73) and "More Than Economism: The Politics of Workers' Safety and Health, 1932–1947," *Milbank Quarterly* 64 (1986): 331–54.

73. Williams sent the report published in *Baltimore Health News* 28 nos. 8–9 (Aug.–Sept. 1951); J. O. Dean to Katherine Bain, 17 Sept. 1951, in Children's Bureau, "Disease Due to Poisons." "Furthermore," Dean continued, "such a notice from the Service or the Agency would be construed as an endorsement of the action taken by the Baltimore City Health Department, and in the absence of any more widespread evidence than is here available our experts caution against early endorsement of the proposal."

74. "Paint Edict Here to Cut Lead Peril," *New York Times*, 30 Oct. 1954, 19; "Paint Makers Eye City Curbs of Lead," *New York Times*, 24 Dec. 1954, 21; Ziegfeld to LIA members, 7 Feb. 1955, loose papers, Box 47, Kehoe Archives. Almost twenty years later a study by the New York City Health Department found that thirteen of the twenty-four brands of interior paints sold in New York still contained over 1% lead; see Paul Montgomery, "City Finds Illegal Lead Content in 13 Brands of Interior Paints," *New York Times*, 22 Aug. 1971.

75. Fee, "Public Health in Practice," 592.

76. The bill, sponsored by Rep. Paul C. Jones of Missouri, was signed into law as Public Law 87–319 by President Kennedy on 26 Sept. 1961; "National Poison Prevention Week (Mar. 18–24, 1962)" (memorandum), in Children's Bureau, "Disease Due to Poisons."

77. On the importance of the poison-control movement, see John C. Burnham, "How the Discovery of Accidental Childhood Poisoning Contributed to the Development of Environmentalism in the United States," *Environmental History Review* 19 (Fall 1995): 57–81; Katherine Bain, "Death Due to Accidental Poisoning in Young Children," *Journal of Pediatrics* 44 (June 1954): 616–23; Edward Press and Robert Mellins, "A Poisoning Control Program," *AJPH* 44 (Dec. 1954): 1515–25, quote, "poisoning now looms," 1516.

78. Press and Mellins, "A Poisoning Control Program"; Graham DuShane, "Middle Ground," *Science* 132 (Oct. 1960): 1221; Jay Arena, "The Pediatrician's Role in the Poison Control Movement and Poison Prevention," *AJDC* 137 (Sept. 1983): 870–73. The PHS had also volunteered to set up a similar clearinghouse, according to Aims McGuinness, special assistant for health and medical affairs, 15 Feb. 1957 (memorandum), in Children's Bureau, "Disease Due to Poisons."

79. Meyer Berger, "Help for the Poisoned Child," *Saturday Evening Post*, 16 Nov. 1957, 24.

80. For sources on the "Mom-ism" suggested in the *Saturday Evening Post* article, see Elaine Tyler May, *Homeward Bound: American Families in the Cold-War Era* (New York: Basic Books, 1988), esp. notes accompanying 96–97.

81. Bain, "Death Due to Accidental Poisoning in Young Children," 619.

82. Woolley, "Lead Poisoning during Infancy and Early Childhood," 548; Conway, "Lead Poisoning—from Unusual Causes," 471 (see ch. 8, n. 35). In 1924, Ruddock ("Lead Poisoning in Children with Special Reference to Pica," 1682 [ch. 2, n. 12]) had written, "Some children affected with pica have a morbid craving to gnaw painted objects, such as window sills, white furniture, crib railings, porch railings and other articles around the home within their reach." It seems that the lead world that concerned Ruddock had changed very little by Woolley's time.

CHAPTER 10: THE SCREAMING EPIDEMIC

1. During the mayoral campaign in October, the Health Department had announced a crash program to distribute 40,000 test kits, donated by Bio-Rad Laboratories, but never followed through. Prior to the Tuesday visits, the Lords distributed flyers along 112th Street: "We are operating our own lead poisoning detection program with students from New York Medical College. . . . The Young Lords and medical personnel will knock on your door. . . . Do not turn them away, Help save your children"; Jack Newfield, "Young Lords Do City's Work in the Barrio," *Village Voice,* 4 Dec. 1969, 1.

2. Joseph Greengard, "Lead Poisoning in Childhood: Signs, Symptoms, Current Therapy, Clinical Expressions," *Clinical Pediatrics* 5 (May 1966): 269–76; Hyman Merenstein, quoted in Mark Oberle, "Lead Poisoning: A Preventable Childhood Disease of the Slums," *Science* 165 (5 Sept. 1969): 992.

3. Berney, "Round and Round It Goes," 13–14 (see intro., n. 10); for the Scientists' Institute for Public Information, see *Scientist and Citizen* 10 (Apr. 1968); Richard W. Clapp, "The Massachusetts Childhood Lead-Poisoning Prevention Program," in *Low Level Lead Exposure,* 285–92 (intro., n. 9).

4. Paul DuBrul, quoted in Diana R. Gordon, *City Limits: Barriers to Change in Urban Government* (New York: Charterhouse, 1973), 28–29; Jonathan M. Stein, "An Overview of the Lead Abatement Program Response to the Silent Epidemic," in *Low Level Lead Exposure,* 279–84.

5. Walter Eberlein, curriculum vitae, folder "Eberlein, Walter R." (henceforth "Eberlein folder"), Stokes papers (see ch. 9, n. 14).

6. [Walter Eberlein], "Study Project: The Detection, Treatment, and Prevention of Lead Poisoning in a Children's Hospital Out-patient Department" (copy of typed manuscript), [1961–62], in "Eberlein folder."

7. Ibid. On the state of the art in diagnosing childhood lead poisoning in the mid-1960s, see Greengard, "Lead Poisoning in Childhood," 269–76.

8. Randolph Byers ("Introduction," in *Low Level Lead Exposure,* 1) recalled

that when he was studying lead poisoning at Massachusetts General Hospital in the late 1930s, he had to smuggle children's blood samples into a laboratory at the Massachusetts Institute of Technology because the state's equipment was reserved for occupational cases. In 1935, Baltimore's Huntington Williams managed to get equipment to use the dithizone method, enabling the city to offer free blood determinations; see Fee, "Public Health in Practice," 581 (see intro., n. 10). By 1955 New York City had provisions for conducting a limited number of blood-leads but relied on a history of pica for initial screening; see Gary Eidsvold, Anthony Mustalish, and Lloyd Novick, "The New York City Department of Health, Lessons in a Lead Poisoning Control Program," *AJPH* 64 (Oct. 1974): 956–62. Cincinnati's program relied on the Kettering Laboratory.

9. Jacobziner, "Lead Poisoning in Children," 277–86 (see ch. 9, n. 60).

10. Kehoe, "The Harben Lectures, 1960," lecture 3, 185 (see intro., n. 16).

11. Greengard, "Lead Poisoning in Childhood," 269.

12. Nancy Hicks, "Drive to Stop Lead Poisoning Begins," *New York Times,* 10 Oct. 1970, 9. Those familiar with lead's long history as a sweetener in wines and sauces will not find Merenstein's appraisal surprising. My only experiment along these lines did not support Dr. Merenstein's opinion, but according to Vincent Guinee, head of New York's lead-poisoning bureau in the early 1970s, taste was not the issue—habit was. "In fact," he concluded, "I doubt if there would be a market for scotch whiskey if people stopped to taste it first"; David H. Rothman, "Keeping the Lead In," *Progressive,* Nov. 1975, 10.

13. John M. Hunter, "Geophagy in Africa and in the United States," *Geographical Review* 63 (Apr. 1973): 170–95. Hunter studied earth eating throughout Africa and found that in certain regions it provided a critical portion of particular nutrients. He concludes that dirt eating among African Americans resulted from "cultural translocation during the slave trade" and similar translocation in the migration from country to city, which involved substituting "nutritionally viable" forms of geophagy for worthless or poisonous ones (laundry starch vs. paint and plaster). He sees the persistence of clay eating as an element of the "Black Diaspora." For a more general study of health consequences of that diaspora, see Kenneth Kiple and Virginia Himmelsteib King, *Another Dimension to the Black Diaspora: Diet, Disease, and Racism* (Cambridge, UK: Cambridge University Press, 1981).

14. Pranab Chatterjee and Judith Gettman, "Lead Poisoning: Subculture as a Facilitating Agent?" *American Journal of Clinical Nutrition* 25 (Mar. 1972): 324–30; see also Frances Millican, Reginald Lourie, and Emma Layman, "Emotional Factors in the Etiology and Treatment of Lead Poisoning," *AJDC* 91 (Feb. 1956): 144–49.

15. "Hand to Mouth," *Time,* 12 Oct. 1962, 75–76; Daniel Haley, quoted in Robert Reilly, "Lead Poisoning of Children Attributed to Parent Neglect," *Philadelphia Inquirer,* 11 Sept. 1972, folder "Lead Poisoning, Misc. 1971–72," Subject Box 72: "Lead-Liberty," Urban Archives (henceforth "Urban Archives Box 72"),

Temple University Libraries Mounted Clips; Margaret English, "There's a Neglected Epidemic in Our Ghettos, and Its Victims Are Lead-Poisoned Kids," *Look*, 21 Oct. 1969, 114. The *Time* article did not mention race or class, even though it was largely based on research conducted at the Children's Hospital of Washington, D.C., by researchers who associated pica with "cultural problems" and stated that "a major portion of this kind of interest [pica] occurs in the Negro population living in the South where it takes the form of clay eating"; Charest et al., "Community Aspects of Lead Intoxication in Children" (see ch. 9, n. 26).

16. Donald Bremner, "New Method Used to Fight Poison by Lead Paint," *Baltimore Evening Sun*, 17 July 1962, quoted in Fee, "Public Health in Practice," 599. The success of Baltimore's new education program was tested by comparing blood-leads of children living in areas where the education program concentrated with those of children in control areas. No difference was noted.

17. Lloyd A. Thomas, Rita G. Harper, and Dorothy L. Trice, "A Community Centered Approach to the Problem of Lead Poisoning," *Journal of the National Medical Association* 62 (Mar. 1970): 106–108.

18. Guinee's earlier "rubella umbrella" campaign had proved that "you can get a serious health message across through television—and through children"; Jane E. Brody, "City's Rubella Drive on TV 'Sells' Children on Need for Shots," *New York Times*, 14 June 1970, 60.

19. Agnes Lattimer, quoted in English, "There's a Neglected Epidemic in Our Ghettos."

20. Gordon, *City Limits*, 26.

21. David Elwyn, "Childhood Lead Poisoning," *Scientist and Citizen* 10 (Apr. 1968): 53–57; Christian, Celewycz, and Andelman, "A Three-Year Study of Lead Poisoning in Chicago" (see ch. 9, n. 61); "Lead Paint in Chicago," *Time*, 9 Aug. 1963, 36–37. The trial screening program compared findings on five hundred children from each of three sections of Chicago. In the oldest and most poverty-stricken area, ninety-one children (18%) showed positive urinary signs; four were hospitalized and thirty-six received chelation therapy on an out-patient basis.

22. Ann Koppelman Simon, "Citizens vs. Lead in Three Communities: 1. Chicago," *Scientist and Citizen* 10 (Apr. 1968): 58–59.

23. On AFSC involvement, "dedicated group of teenagers," and "reminder to the city," see Simon, "Citizens vs. Lead: Chicago," 59; on Chicago's response to peer pressure, see Berney, "Round and Round It Goes," 12.

24. Elwyn, "Childhood Lead Poisoning," 56.

25. Hicks, "Drive to Stop Lead Poisoning Begins," 9; Henrietta K. Sachs, "Effect of a Screening Program on Changing Patterns of Lead Poisoning," *Environmental Health Perspectives* 7 (May 1974): 41–45.

26. Phillip R. Fine et al., "Pediatric Blood Lead Levels: A Study in 14 Illinois Cities of Intermediate Population," *JAMA* 221 (25 Sept. 1972): 1475–79; Den Elger, "7 Suburbs Listed as Having Many with Lead Poisoning," *Chicago Tribune*, 25 May 1972.

27. "Lead Paint Law Delay Denied," *Chicago Tribune,* 28 June 1972, sect. 1, 3; "Two City Aides Defend Lead Detection System," *Chicago Tribune,* 31 Aug. 1972, sect. 4a (N), 10. Flaschner supplied both quotes; see Marcia Opp, "Increase of Lead Poison Laid to Collusion of Slum Landlords," *Chicago Tribune,* 4 May 1972, sect. 2, 7.

28. David J. Wilson, "Citizens vs. Lead in Three Communities: 2. Rochester," *Scientist and Citizen* 10 (Apr. 1968): 60–63.

29. David A. Anderson, "Public Institutions: Their War against the Development of Black Youth," *American Journal of Orthopsychiatry* 41 (Jan. 1971): 65–73. Anderson was clinical advocate in the Department of Preventive Medicine and Community Health at the University of Rochester's School of Medicine and Dentistry and deputy executive director of Rochester's Urban League.

30. McLaughlin, "Lead Poisoning in Children in New York" (see ch. 9, n. 59); Mary Culhane McLaughlin, "Two Health Problems, One Solution," *Bulletin of the New York Academy of Medicine* 46 (June 1970): 454–56; Jacobziner, "Lead Poisoning in Children"; Eidsvold, Mustalish, and Novick, "The New York City Department of Health." For the numbers of children tested and specific case rates for 1954–64, see Figure 9–2 and Table A–3.

31. For some of the constraints working on the department, including financial limits and bureacratic resistance to additional and untried programs top-loaded from the commissioner, see Gordon, *City Limits,* 32–34.

32. Jacobziner, "Lead Poisoning in Children," 285. For the paint ban, see Guinee's statement in Senate Committee, *Lead-Based Paint Poisoning,* 277 (see ch. 3, n. 61). This ban was never enforced with any vigilance. In 1971, a survey of New York City's hardware stores found eight of seventy-six cans of interior paint produced by such manufacturers as Benjamin Moore and Glidden contained over 1% lead, with some exceeding 10%; David Bird, "Test of Paint Finds 10% Has Illegal Lead Content," *New York Times,* 24 July 1971, 1; David Bird, "High-Level Lead Paints Listed by City," *New York Times,* 4 Aug. 1971, 18.

33. Newfield, "Silent Epidemic in the Slums," 3 (see ch. 2, n. 2). For a detailed narrative of New York activists' campaign to roust the Health Department, drawn from interviews with many of the participants, see Gordon, *City Limits,* 17–62. Most of this chapter's discussion of New York was drawn from Gordon's book and Jack Newfield's *Village Voice* articles.

34. DuBrul, quoted in Gordon, *City Limits,* 29

35. Results of the Einstein Medical School project were never made public. Some participants suspected that a hospital official kept the records in hopes of using them as the basis of a grant proposal; see Gordon, *City Limits,* 40. For comments of Joseph Cimino, see Joseph Fried, "Children Periled by Lead Poisoning," *New York Times,* 15 Dec. 1968, 71. This is the earliest instance I have found of childhood lead poisoning being refered to as *a* silent epidemic. In 1969, it would be promoted to *the* silent epidemic.

36. Three of the sponsors were city-wide organizations: the New York City

Health Department, the Public Health Association of New York City, and the Health Research Council. The Scientists' Committee for Public Information was a state association, and SIPI was national. Dubos won a Pulitzer Prize in 1968 for *So Human an Animal.* His statement has been quoted a variety of ways; this version is the one most frequently printed and appeared the day after his speech in Sandra Blakeslee, "Experts Recommend Measures to Cut Lead Poisoning in Young," *New York Times,* 27 Mar. 1969, 25. In addition to this pithy quote, the conference seems to have made "silent epidemic" a staple moniker for childhood plumbism.

37. "Infant Dies after Eating Paint from Tenement Wall," *New York Times,* 28 May 1969, 18; "Brownsville Plagued by Paint Poisonings," *New York Times,* 1 Sept. 1969, 19. The *Times* was slow to cover lead-poisoning stories. It did not write up the Scurry death, despite the press conference called by Robert Abrams, candidate for Bronx borough president; see Gordon, *City Limits,* 41.

38. Newfield, "Silent Epidemic in the Slums," 3; although the article began on page 3, the headline ran on page 1, in large type above the masthead.

39. Carter Burden, quoted in ibid., 39–40. Burden was a wealthy Democratic candidate for City Council who had tried to make lead poisoning the focus of his campaign but could not elicit the community's or the media's interest.

40. John Sibley, "City Starts Drive on Lead Poisoning," *New York Times,* 19 Oct. 1969, 96; McLaughlin, "Two Health Problems, One Solution," 454–55.

41. Jack Newfield, "My Back Pages," *Village Voice,* 9 Oct. 1969, 9; Jack Newfield, "Fighting an Epidemic of the Environment," *Village Voice,* 18 Dec. 1969, 12; John Sibley, "Criticism Rising over Lead Poison," *New York Times,* 26 Dec. 1969, 19; "New City Unit Moving against Landlords Who Allow Poisonous Paint in Their Buildings," *New York Times,* 18 Mar. 1970, 31. The proximity of Newfield's discussion of institutional racism and Cornely's press conference following a speech before the National Conference on Black Students in Medicine and the Sciences suggests one possible source for Cornely's "set of facts."

42. Once associated with immigrants and slums and "treated" by quarantines and garbage removal, by the 1930s polio had become a scourge of middle-class children; see Naomi Rogers, *Dirt and Disease: Polio before FDR* (New Brunswick, N.J.: Rutgers University Press, 1992). For the later years, see Tony Gould, *A Summer Plague: Polio and Its Survivors* (New Haven, Conn.: Yale University Press, 1995), 120; Jane S. Smith, *Patenting the Sun: Polio and the Salk Vaccine* (New York: Anchor/Doubleday, 1990).

43. Gordon, *City Limits* 38; Newfield, "Young Lords Do City's Work," 1.

44. Newfield, "Young Lords Do City's Work," 1.

45. Henry Intill, letter to the editor, *Village Voice,* 11 Dec. 1969, 3; Sibley, "Criticism Rising over Lead Poisoning"; John Sibley, "City Detects Rise in Lead Poisoning," *New York Times,* 23 Sept. 1970, 53; Nicholas Freudenberg and Maxine Golub, "Health Education, Public Policy and Disease Prevention: A Case History

of the New York City Coalition to End Lead Poisoning," *Health Education Quarterly* 14 (Winter 1987): 387–401.

46. "New City Unit Moving against Landlords," 31

47. Ed Rothschild, quoted in Gordon, *City Limits,* 59.

48. The *Guide* reported five articles in 1968 and six in 1969. Admittedly, this was many more than appeared in most previous years, when poison-control articles always outnumbered lead-poisoning stories. Interest in the general topic of lead poisoning increased during these years, in large part because of the growing controversies over environmental lead pollution.

49. See, e.g., "Welfare Unit Fights Use of Lead Paint," *Philadelphia Inquirer,* 13 Nov. 1969, folder "Lead Poisoning, Phila Prior to 1970," in Urban Archives, Box 72; Gary Brooten, "Lead Paint Poison Cases Quadruple over Last Year," *Philadelphia Evening Bulletin,* 13 Aug. 1970, ibid.

50. Newfield, "Silent Epidemic in the Slums," 39.

51. Public Law 91–695, 91st Cong., 2d sess., 1970, 590. For hearings on the bill, see Senate Committee, *Lead-Based Paint Poisoning.* On worries that Nixon might pocket-veto the law, see "Barrett Asks Nixon to Sign Lead Paint Bill," *Philadelphia Bulletin,* 14 Jan. 1971, folder "Lead Poisoning, Misc. 1971–72" in Urban Archives, Box 72; on Nixon signing the bill, see *New York Times,* 15 Jan. 1971, 10.

52. *Congressional Quarterly Almanac* 1970, 590; "Conference Report, Lead-Based Paint Poisoning Prevention Act," rep. no. 91–1802, 91st Cong., 2d sess.; testimony of Jonathan Fine, Senate Committee, *Lead-Based Paint Poisoning,* 236.

53. U.S. Department of Health, Education and Welfare (HEW), Social and Rehabilitation Service, Children's Bureau, [Jane S. Lin-Fu], "Lead Poisoning in Children" (Children's Bureau pub. no. 452), (Washington: GPO, 1967). In 1970 HEW also distributed 10,000 reprints of another Lin-Fu article, "Child Lead Poisoning—An Eradicable Disease" from *Children* 17 (Jan. 1970): 2–9. Figures regarding distribution by HEW and the LIA appear in Senate Committee, *Lead-Based Paint Poisoning,* 188–89.

54. HEW, "Medical Aspects of Lead Poisoning" (statement of the Surgeon General, 8 Nov. 1970); Senate Committee, *Lead-Based Paint Poisoning,* 45–49.

55. For early research into chronic low-level lead exposure, see J. Julian Chisolm, "Chronic Lead Exposure in Children," *Developmental Medicine and Child Neurology* 7 (1965): 529–36; J. Julian Chisolm, "Lead Poisoning," *Scientific American* 224 (Feb. 1971): 15–23; Lin-Fu, "Modern History of Lead Poisoning," 34–43 (see intro., n. 9).

56. HEW, "Medical Aspects of Lead Poisoning"; HEW, PHS, Environmental Health Service, Environmental Control Administration, Bureau of Community Environmental Management, "Control of Lead Poisoning in Children" (prepublication draft, July 1970, quote, VI-1), in Senate Committee, *Lead-Based Paint Poisoning,* 51–174, quote, 135.

57. Testimony of John Hanlon, deputy adminstrator of the PHS's Environmental Health Service, Senate Committee, *Lead-Based Paint Poisoning,* 188.

58. Ibid.; Kennedy quote, 175. The information Hanlon provided noted a number of research programs funded by the Bureau of Occupational Safety and Health and animal and clinical studies conducted by the NIH and other research bodies; see 179. See also Fine testimony, 237.

59. Newfield, "Let Them Eat Paint," *New York Times,* 1 June 1971, 45.

60. Heated debate between legislators, public health officials, pediatricians, and consumer groups accompanied each step in this process; for a summary of the hearings and bills see Susan Bailey, "Legislative History of the Lead-Based Paint Poisoning Prevention Program," in House Committee on Energy and Commerce, *Lead Poisoning and Children,* 97th Cong., 2d sess., 2 Dec. 1982, 3–12.

61. Jane S. Lin-Fu, "Undue Lead Absorption and Lead Poisoning in Children—An Overview," Proceedings of the International Conference on Heavy Metals in the Environment (Toronto: Public Institute of Environmental Studies, 1975), 29–52; George Hardy, assistant director, CDC, in House Committee, *Lead Poisoning and Children,* 28.

62. The 1973 amendment to the LBPPPA (Public Law 93–151) lowered the permissible content of lead in dried films to 0.5%. The CPSC was assigned to determine whether there was a safe level somewhere below 0.5% but above 0.06%, the lowest level test equipment could detect. The commissioner announced that 0.5% was safe, but the Consumer's Union and the Philadelphia Citywide Coalition against Childhood Lead Poisoning sued the CPSC to force a reevaluation. Before the case was settled, another amendment to the LBPPPA (via Public Law 94–317) mandated a full review by the CPSC. This time the commission had to prove that any amount of lead constituted no "unreasonable risk." Put this way, the National Academy of Science study commissioned by the CPSC was "unable to determine that 0.5 percent lead in paint is safe." The ban was promulgated in late 1977 and went into effect six months later; see *Federal Register* 42 (1977): 9405–6; "Lead Poisoning from Paint: The End of the Road?" *Consumer Reports* 42 (Mar. 1977): 124.

63. Based on HEW reports of grants (department data do not cover abatement programs that HUD might have funded); Bailey, "Legislative History of the Lead-Based Paint Poisoning Prevention Program," 12; Douglass W. Green, "The Saturnine Curse: A History of Lead Poisoning," *Southern Medical Journal* 78 (Jan. 1985): 48–51.

CHAPTER 11: FACING THE CONSEQUENCES OF LEADED GASOLINE

1. Sherwin-Williams has used the cover-the-earth trademark since at least 1910. George D. Wetherill and Company, makers of Atlas paint, employed a similar logo at the beginning of the century; see company letterhead used in corre-

spondence from George Wetherill to Rep. W. H. Graham, 20 Feb. 1908, Records of the U.S. House of Representatives, Committee on Interstate and Foreign Commerce, file HR60A-H16.12: "Naval Stores and Paint," Box 707, RG 233, National Archives. Three brush-wielding cherubs busily painted a globe held aloft by the long-suffering Titan. The company's motto? "We Must Cover the Earth with Atlas Paint."

2. Patterson, "Contaminated and Natural Lead Environments" (see ch. 7, n. 3).

3. My young teenage friends and I used to fill the margins of our history notebooks with drawings of Volkswagen "beetles," perversely modified with enormous rear wheels and elongated hoods that barely contained the 400-horsepower supercharged V-8s powering these definitive "funny cars."

4. The committee formulated a set of regulations (essentially those developed by Kehoe) and presented them to the Conference of State and Territorial Health Officers as guidelines for state legislation. Ethyl issued a statement of intent promising to adhere to the guidelines, and no state or federal regulations ever appeared. Into the early 1970s, the industry average for tetraethyl lead per gallon of gasoline was approximately 2 cc; see Jack Lewis, "Lead Poisoning: A Historic Perspective," *EPA* Journal 11 (May 1985): 16.

5. "Surgeon General's Ad Hoc Committee on Tetraethyl Lead" (typed report), 8–9 Jan. 1959, 9, in PHS, Division of Special Health Services, Occupational Health Program, folder "Air Pollution—Tetraethyl Lead-1959," Air Pollution Engineering Branch Correspondence 1959–60, Records of the PHS, RG 90, National Archives (henceforth "Air Pollution folder").

6. PHS, *Conference to Determine Public Health Question of Tetraethyl Lead Gasoline* (see ch. 7, n. 23); PHS, "The Use of Tetraethyl Lead Gasoline in Its Relations to Public Health," bulletin 163 (Washington: GPO, 1926); PHS, "Public Health Aspects of Increasing Tetraethyl Lead" (press release, HEW-L26), 12 Sept. 1959, "Air Pollution folder," vii.

7. In a vain attempt to establish such a baseline, Joseph Aub tried from late 1959 and into 1961 to track down a box of New York City street dust collected in the 1920s by Dr. Paul Reznikoff of the PHS. Reznikoff and Sidney Kaye had published a paper in 1947 comparing this dust with samples collected in 1934 and finding a 50% increase in lead. Kaye reported that the samples were left in his old laboratory in 1941. Another PHS chemist recalled the box of dust, but reported it had been discarded. See Aub to John Goldsmith, 13 Sept. 1959, in folder "Air Pollution Conference, Dec 8–9 1960," Box1: A–Bauer, Archives GA-4, Aub files (see ch. 5, n. 62); Harold Magnuson to Aub, 16 Sept. 1959, ibid.; Kaye to Aub, 31 Jan. 1961, ibid.

8. PHS, "Public Health Aspects of Increasing Tetraethyl Lead," viii; for the full report, see Working Group on Lead Contamination, *Survey of Lead in the Atmosphere of Three Urban Communities,* pub. no. 999-AP-12 (Washington: GPO, 1965).

9. D. R. Diggs et al., "Program for the Survey of Lead in Three Urban Communities," *Journal of the Pollution Control Association* 13 (May 1963): 228–32.

10. Kehoe would have preferred a much less public affair. In a letter to J. H. Ludwig, the working group's chairman (23 Mar. 1961, folder "Some of the Material from the Three Cities Survey," Box 12, Kehoe Archives [see ch. 7, n. to epigraph]), he expressed his disappointment in the press release announcing the project's initiation: "In my philosophy, the less said about investigations of this type, the better it is for every one [sic]. . . . Publicity is an unmitigated nuisance."

11. Quote, "the new chemical compounds," Press and Mellins, "A Poisoning Control Program," 1516 (see ch. 9, n. 77); Burnham, "Discovery of Accidental Childhood Poisoning Contributed to the Development of Environmentalism" (ch. 9, n. 77); Samuel Hays, "Three Decades of Environmental Politics: The Historical Context," in *Government and Environmental Politics: Essays on Historical Developments since World War Two,* ed. Michael J. Lacey (Baltimore: Johns Hopkins Press, 1989), quote, "chemical world out of control," 37. According to Hays (ibid., 25–26), "the notion of pollution as a problem arose far more from new attitudes that valued both smoothly functioning ecosystems and higher levels of human health." He stresses changing demographics, notably the new "mass-middle class." Larger disposable incomes led to "environmental consumption," such as the increased popularity of outdoor recreation far from the city. And for an increasingly health-conscious generation, "control of air and water pollution came to be thought of as an aspect of advance in human health protection."

12. Nriagu, *Lead Poisoning in Antiquity,* 313–14 (see intro., n. 4). In *The Principles and Practice of Medicine,* his nineteenth-century treatise, William Osler (quoted in Carey P. McCord, "Lead and Lead Poisoning in Early America: Lead Mines and Lead Poisoning," *Industrial Medicine and Surgery* 22 [Nov. 1953]: 536) observed, "Miners usually escape [plumbsim], but those engaged in the smelting of lead ores are often attacked. Animals in the neighborhood of smelting furnaces have suffered with the disease and even the birds that feed on the berries in the neighborhood may be affected." For a late nineteenth-century example, see *W. G. Price et al. v George Grantz,* 118 Pa. 402 11 A. 794 (1888), a case brought against a smelter operating in Pittsburgh.

13. The symposium was held 25–27 Feb. 1963, and selected papers appeared in *AEH* 8 (Feb. 1964).

14. Bureau of Mines, "Lead," in *Minerals Yearbook* (see ch. 3, n. 6). Of the 1,389 tons of lead consumed in the United States in 1969, only seven went to white lead and eighty to other lead pigments.

15. The advocates for lead-poisoning victims have adopted aggressive strategies against manufacturers, bringing class-action and market-share liability suits that are far more threatening to lead-industry coffers than were earlier suits aimed at negligent landlords or dilatory government agencies; see Thomas Grillo, "Lead-Paint Makers Become a Target," *Boston Globe,* 4 Sept. 1993, 35; Diane Cabo Freniere, "Private Causes of Action against Manufacturers of Lead-Based Paint: A Re-

sponse to the Lead Paint Manufacturers' Attempt to Limit Their Liability by Seeking Abrogation of Parental Immunity," *Boston College Environmental Affairs Law Review* 18 (Winter 1991): 381–422.

16. About 160 tons of lead went into gasoline in 1959, as compared with ninety-five tons in 1949; Bureau of Mines, "Lead," in *Minerals Yearbook.*

17. Kehoe, "The Harben Lectures, 1960," Lecture 3, 177 (see intro., n. 16).

18. Kehoe, "The Harben Lectures, 1960," Lecture 1, "The Normal Metabolism of Lead," 96.

19. Testimony of Dr. Richard Prindle of the PHS, in Senate Committee, *Air Pollution—1966,* 130–31 (see ch. 8, n. 12); J. H. Ludwig et al., "Survey of Lead in the Atmosphere of Three Urban Communities: A Summary," *American Industrial Hygiene Association Journal* 26 (May/June 1965): 270–84.

20. Prindle testimony, in Senate Committee, *Air Pollution—1966,* 130–31.

21. Dr. Thomas Haley, a toxicologist at the University of California, Los Angeles, found in 1966 that despite a 250% increase in leaded gasoline consumption in the previous thirty years, Americans' average body burdens had remained the same; see "Chronic Lead Poisoning No Threat to U.S.," *Science News* 90 (24 Sept. 1966): 225. He concluded that most of the earlier lead intake was from food. An international study conducted by Columbia University professor Leonard Goldwater for the World Health Organization reached the same conclusion in 1967; see David Alan Ehrlich, "Lead Hazard Discounted," *Science News* 92 (16 Sept. 1967): 278. Samples of blood and urine were collected from hundreds of subjects; the highest levels were found in Finland, the lowest in Peru.

22. Patterson (1922–95) was raised in rural Iowa, received his M.S. from the University of Iowa in 1944, and his Ph.D. in 1951 from the University of Chicago. He was elected to the National Academy of Sciences in 1987, and both an asteroid and an Antarctic peak bear his name. "Pat" claimed that his environmental research and activism were in part penance for his wartime work on the atomic bomb in Chicago and Oak Ridge, Tennessee: "It was the greatest crime that science has committed yet. . . . [I] helped to burn [100,000 Japanese] alive." He was harsh in his criticism of applied scientists, whom he labeled "degenerative hominids," but his concerns ran deeper than professional chauvinism. He worried about the effects of science and technology upon human evolution: "10,000 years of perverted utilitarian rationalization" had produced "a diseased *Homo sapiens* mind." In words reminiscent of William Blake, he stated that humans had lost their ability to enjoy "ethereal visions"; an overload of facts had "virtually eliminated nonutilitarian vision." For more on Patterson, see the special issues of *Geochimica et Cosmochimica Acta* 58 (1994) (source of the above quotes) and *Environmental Research* 78 (1998). The Saul Bellow character based on him, Sam Beech in *The Dean's December* (New York: Harper & Row, 1982), is apparently a faithful and sensitive portrait of the man and his ideas.

23. Patterson, "An Alternative Perspective," 285 (see ch. 1, n. 14).

24. T. J. Chow and Patterson ("The Occurrence and Significance of Lead Iso-

topes in Pelagic Sediments," *Geochimica et Cosmochimica Acta* 26 [1962]: 262) found that lead concentrations in ocean waters decreased with depth, the reverse of the case for barium and other nonindustrial metals. Since lead presumably would settle over time, this inversion indicated that lead pollution was increasing, but moving ocean waters prevented establishing an accurate chronology.

25. Patterson, "An Alternative Perspective," 289.

26. Clair Patterson and Joseph Salvia, "Lead in the Environment: How Much Is Natural?" *Scientist and Citizen* 10 (Apr. 1968): 68.

27. Patterson, "Contaminated and Natural Lead Environments," 350. Many later studies have confirmed Patterson's hypothesis. Ancient bones from Peru were found to contain average lead concentrations of five parts per million, while modern-day skeletons average about fifty parts per million; see "Lead in Ancient and Modern Bones," *Scientist and Citizen* 10 (Apr. 1968): 89 (report on Robert O. Becker and Joseph A. Spadaro, "The Trace Elements of Human Bone," *Journal of Bone and Joint Surgery* 50 [Mar. 1968]: 326–34). A similar study found that bone lead in Danes remained at about three parts per million until around A.D. 1000, when lead glazing was introduced to the region, and from then the level rose until the mid-1940s; P. Grandjean and B. Holma, "A History of Lead Retention in the Danish Population," *Environmental Physiological Biochemistry* 3 (1973): 268–73, cited in Purves, *Trace Element Contamination*, 12 (see ch. 1, n. 11).

28. Patterson, "Contaminated and Natural Lead Environments," 358.

29. The first article, *AEH* 8 (Feb. 1964): 202–12, was LIA's vice president Robert Ziegfeld's "Importance and Uses of Lead."

30. In her introduction, editor Katharine Boucot wrote, "We would have enjoyed publishing the colorful first draft of his manuscript but some of our reviewers felt that a more sober approach was fitting for a scientific journal. So be it!"; see "He Has a 'Concern,'" *AEH* 11 (Sept. 1965): 262. In a detailed reply to a complaint that Patterson's article should never have been published, Boucot described the long process of getting competent evaluations and defended her decision to print such a controversial and unusual piece. Of the six readers for the first draft, only one rejected it outright, but the others recommended sharp reductions in length. Boucot assigned seven readers for the revised draft: four approved, two recommended further revision, and one remained opposed to publication. "It seemed wise," Boucot concluded, "to publish this paper in a medical journal rather than to allow Dr. Patterson and the public to believe that we, as physicians, were not interested in the public health problem"; see *AEH* 12 (Jan. 1966): 138–39.

31. Aub to Patterson, 25 Sept. 1964, folder "Lead—1927–1965," Box 5: "Invitations-M," Archives GA-4, Aub files; Kehoe to Herbert Stokinger (chief, Toxicology Section, Occupational Health Research and Training Facility, Cincinnati), 14 Dec. 1964, folder "Loose Papers," Box 70, Kehoe Archives.

32. Kehoe to James Boudreau, 18 Mar. 1965, folder "LIA," Box 56, Kehoe Archives. Patterson did not make this last comment directly to Kehoe. Word of

the rebuff, Kehoe wrote, "came to me reliably, but at second-hand"; Kehoe to Stokinger, 14 Dec. 1964, ibid.

33. Walter Sullivan, "Lead Pollution of Air 'Alarming,'" *New York Times,* 8 Sept. 1965, 1.

34. Walter Sullivan, "Warning Is Issued on Lead Poisoning," *New York Times,* 12 Sept. 1965, 71.

35. Dr. Cooper, "Some Comments on 'Contaminated and Natural Lead Environments of Man' by Dr. C. C. Patterson" (typed manuscript), 15 May 1965, folder "Loose Papers," Box 70, Kehoe Archives. Patterson was concerned that over the centuries, lead from surrounding soils might leach into bones in uncharacteristic concentrations.

36. A. S. Hawkes to R. K. Scales (internal memorandum, Ethyl Corp.), 26 May 1965, folder "Loose Papers," Box 70, Kehoe Archives; Dr. Cooper, 26 May 1965, ibid.; Cooper, "Some Comments."

37. Kehoe to Boudreau, 18 Mar. 1965.

38. Kehoe, letter to the editor, *AEH* 11 (Nov. 1965): 736–39.

39. Ibid.; Robert Ziegfeld, "Importance of Lead" [letter to the editor], *AEH* 12 (Jan. 1966): 134; Henry A. Schroeder, letter to the editor, *AEH* 12 (Jan. 1966): 270–71. See also Schroeder's polemic book, *The Poisons around Us: Toxic Metals in Food, Air, and Water* (Bloomington: Indiana University Press, 1974).

40. Kehoe, letter to the *AEH* editor; Boucot, *AEH* 12.

41. Clair Patterson, reply to letters, *AEH* 12 (Jan. 1966): 138.

42. Byers and Lord, "Late Effects of Lead Poisoning" (see ch. 8, n. 36). For more on Needleman, see his recollections on his encounters with the lead industries, "Salem Comes to the National Institutes of Health: Notes from inside the Crucible of Scientific Integrity," *Pediatrics* 90 (Dec. 1992): 977–81; for Wedeen, see *Poison in the Pot* (intro., n. 6). Of course, the thousands of lawyers, jurists, lawmakers, and environmental activists who now compose lead's opposition must be added to these individual interlopers.

43. Patterson, "Contaminated and Natural Lead Environments," 357; Aub to Patterson, 25 Sept. 1964; Rutherford T. Johnstone, letter to the editor, *AEH* 12 (Jan. 1966): 135.

44. Thomas J. Haley, "Chronic Lead Intoxication from Environmental Contamination: Myth or Fact?," *AEH* 12 (June 1966): 781–85. Haley's research was supported by a grant from the Atomic Energy Commission (which had also supported Patterson).

45. Kehoe, letter to the *AEH* editor; Patterson, "Contaminated and Natural Lead Environments," 356.

46. Harriet Hardy, letter to editor, *AEH* 11 (Dec. 1965): 878; Arie J. Haagen-Smit quote, unlabeled newspaper clipping, folder 196: "Lead, 1952–78" (henceforth "Hardy lead folder"), Box 10, MC 387, Harriet Louise Hardy Papers (henceforth Hardy papers), Schlesinger Library, Radcliffe College.

47. Patterson's work encouraged a number of other scientists. Dozens of studies of polar ice, land-locked lakes, ancient bones, etc., have confirmed all of his preliminary conclusions. For a recent example, see Ingemar Renberg, Maria Wik Persson, and Ove Emteryd, "Pre-industrial Atmospheric Lead Contamination Detected in Swedish Lake Sediments," *Nature* 368 (24 Mar. 1994): 323–26.

48. PHS, *Symposium on Lead Contamination* (see ch. 7, n. 53); Kehoe's talk appears on 51–58.

49. Charles Schaeffer, "Body's Slow Accumulation of Lead under Scrutiny as Disease Factor," *Staten Island Advance,* n.d., clipping in "Hardy lead folder"; Harriet Hardy, "Lead," in *Symposium on Lead Contamination,* 73–83; Alice Hamilton and Harriet Hardy, *Industrial Toxicology* (New York: P. B. Hoeber, 1949). Patterson wrote to Hardy (23 Mar. 1966, folder 229: "P, 1962–78," Box 11, Hardy papers), "It is strange and somewhat humorous, I think, after all this talk about how people should only speak in those areas for which they are qualified, that I should indignantly reject for the most part such an attitude, except in this particular instance where I feel somewhat presumptuous in wanting to commend you for the difficult and challenging role you have chosen." Hardy's talk, Patterson insisted, was "principally responsible for the fact that the meeting was productive."

50. Discussion, *Symposium on Lead Contamination;* Hermann's comments appear on 147; the Kehoe quote, 149.

51. In a letter to Hardy, Kehoe (13 Apr. 1965, folder "HA–HZ," Box 100, Kehoe Archives) wondered "what manner of man [Patterson] may be," and suggested that "with blood in his eye toward lead, and, it seems, toward me and my associates and all of our works," Patterson (with well-intentioned help from Hardy, who had shared her extensive literature on lead hygiene) had brought Kehoe's downfall.

52. Bellow, *The Dean's December,* 222–23; Patterson, "Contaminated and Natural Lead Environments," 358.

53. S. Colum Gilfillan, "Lead Poisoning and the Fall of Rome," *Journal of Occupational Medicine* 7 (Feb. 1965): 53–60. Gilfillan, who died at age 97 in 1987, was somewhat eccentric. For years he advocated reforms in standard spelling, and he practiced what he preached (writing *thru, altho, thot,* etc.) For more on Gilfillan, see the introduction to his *Rome's Ruin by Lead Poison* (Long Beach, Cal.: Wenzel Press, 1990). It is unclear whether Patterson was aware of Gilfillan's research when he was writing his article, but he often cited it later. Kehoe was certainly aware of Gilfillan; the Kettering Laboratory analyzed (at $12.50 per sample) the lead content of many of the Roman bones Gilfillan located. Correspondence between Kehoe and Gilfillan after his retirement includes the latter's plea for some University of Cincinnati letterhead, so that he could have "connexion" to a research institution when writing to museums for bones. Kehoe declined, but his letters contain good-humored advice to someone for whom Kehoe clearly had no professional respect. See Gilfillan to Kehoe, 2 Sept. 1964, replies, sample analyses, and correspondence in folder "S. Colum Gilfillan," Box 9, Kehoe Archives. To

one of his analysts, Kehoe wrote, "You should have a look at this entire correspondence. Our correspondent is something of a screwball, but perhaps his idea has some merit. At any rate I thought it worth following"; [Robert Kehoe] to Jacob Cholak (unsigned memorandum), n.d., ibid.

54. Bellow, *The Dean's December*, 222. Gilfillan's thesis is at least as well grounded in empirical evidence as many of the dozens scholars have successfully put forth to explain the decline of Rome; Nriagu (*Lead Poisoning in Antiquity*, 415) claims that Gibbon alone cited over two dozen reasons for Rome's fall.

55. Kehoe to Boudreau, 18 Mar. 1965, 2.

56. Kehoe, "The Harben Lectures, 1960," passim.

57. A good summary of the course of deleading America's gasoline stores from the perspective of institutional bias is Hays, "The Role of Values," 267–83 (see intro., n. 17); Melnick (*Regulation and the Courts*, 261–81 [intro., n. 11]) covers much of the same territory but focuses on the courts' influence on federal environmental policy.

58. Senate Committee, *Air Pollution—1966*. Kehoe's testimony (203–28) frequently devolved into arguments with the feisty but frustrated senator from Maine. At one point Muskie conducted a prolonged campaign to nail down anything more than evasions from Kehoe in answer to the question whether "from the point of view of health" it would be desirable "if a substitute for lead could be found for gasolines." Kehoe finally budged—a bit: "I would be glad to dispense with a risk, the question of risk—not the question of hazard because I think it does not exist—but the question of potential hazard, which is risk. I would be glad to trade any risk for the certainty of safety." At which Muskie retreats, saying "I guess that is as close as we can come together." Patterson also testified before Muskie (311–44), as did Felix Wormser (233–43).

59. David Bird, "Industry Defends Lead in Gasoline," *New York Times*, 25 Sept. 1970, 34.

60. J. L. Kimberly, "Controversy: Lead in Gasoline," *Lead* 33 (1970): 5–7.

61. Lewis, "Lead Poisoning: A Historic Perspective," 15–18; John Quarles, *Cleaning up America: An Insider's View of the Environmental Protection Agency* (Boston: Houghton Mifflin, 1976), 119.

62. Quarles, *Cleaning up America*, 122–23; Lewis, "Lead Poisoning: A Historic Perspective," 18; National Research Council, Committee on Biologic Effects of Atmospheric Pollutants, *Lead: Airborne Lead in Perspective* (Washington: National Academy of Sciences, 1972). On the question of industry's influence in setting lead standards, see Robert Gillette, "Lead in the Air: Industry Weight on Academy Panel Challenged," *Science* 147 (19 Nov. 1971): 801–802; Phillip M. Boffey, *The Brain Bank of America: An Inquiry into the Politics of Science* (New York: McGraw-Hill, 1975), 227–44.

63. Lewis, "Lead Poisoning: A Historic Perspective"; on the Natural Resources Defense Council's many lawsuits against the EPA in this period, see Melnick, *Regulation and the Courts*.

64. For a sense of these conflicts, see Quarles, *Cleaning up America.*

65. See Melnick, *Regulation and the Courts,* 269–71; Eric A. Goldstein, "Toxic Lead Aftermath," *Environment* 25 (Mar. 1983): 2–4. Goldstein was at the time an NRDC attorney.

66. In 1971, 264 tons of lead went into tetraethyl lead, but by 1980, the number was 141 tons, and in 1985, 50.4 tons; Bureau of Mines, "Lead," in *Minerals Yearbook.* In the 1980s, the EPA shifted to a regulation regime that concentrated on leaded gasoline only, requiring a maximum of 1.1 grams per gallon of leaded gas; Lewis, "Lead Poisoning: A Historic Perspective," 18.

67. I. Romieu et al., "Vehicular Traffic as a Determinant of Blood-Lead Levels" (see ch. 2, n. 15).

68. The survey was conducted by the National Center for Health Statistics; see Kathryn R. Mahaffey et al., "National Estimates of Blood Lead Levels: United States, 1976–1980," *NEJM* 307 (2 Sept. 1982): 573–79. Detailed medical information was collected from a sample population of about 28,000. Not all the news was as good as the trend in average blood-lead levels. Twelve percent of all black children studied (as compared with 2% of white children) had blood-leads over 30 g/dL, the threshold level for treatment in 1980. Among black children whose family incomes fell below $6,000, the average rose to 18%. According to a PHS report, "the correlation of blood lead levels with national estimates of the amount of lead used in gasoline production was highly significant ($p < 0.001$)" (see Figure 11–2); see J. L. Annest and Kathryn R. Mahaffey, "Blood Lead Levels for Persons Ages 6 Months–74 Years, United States, 1976–1980" in PHS, *Vital and Health Statistics,* series 11, no. 233, pub. no. 84–1683 (Washington: GPO, 1984).

CHAPTER 12: THE RISE AND FALL OF UNIVERSAL CHILDHOOD LEAD SCREENING

1. Quote, "the most prevalent," Roper, cover letter (see intro., n. 8); Ruddock, "Lead Poisoning in Children," 1682–84 (ch. 2, n. 12).

2. Steven Waldman, "Lead and Your Kids," *Newsweek,* 15 July 1991, 42–48.

3. U.S. Senate, Labor and Public Welfare Committee Senate, Senate Report No. 91-1432, 14 Dec. 1970, in *United States Code: Congressional and Administrative News,* 91st Cong., 2d sess., 1970 (St. Paul, Minn.: West Publishing, 1971), 3: 6130–39.

4. Stein, "An Overview of the Lead Abatement Program Response" (see ch. 10, n. 4).

5. National Committee for the Prevention of Childhood Lead Poisoning, "Lead: Still No Action," *Environment* 17 (Sept. 1975): 36–38.

6. Donna Barnako, "Childhood Lead Poisoning," *JAMA* 220 (26 June 1972): 1737–38; Fine et al., "Pediatric Blood Lead Levels" (see ch. 10, n. 26).

7. Since the early 1950s, Chisolm (1924–) has been a pediatrician at the Johns Hopkins Hospital and professor of pediatrics at the Johns Hopkins University School of Medicine, where he received his M.D. in 1946. He has served as a consultant to the EPA since 1976 and on many government-sponsored review panels on lead poisoning. Since 1975 he has directed the lead program at the Kennedy-Krieger Institute.

8. J. Julian Chisolm, "Is Lead Poisoning Still a Problem?," *Clinical Chemistry* 23 (1977): 252–55.

9. Florini, Krumbaar, and Silbergeld, *Legacy of Lead,* 12 (see intro., n. 9). These levels defined only "undue absorption"; in 1975 the threshold for "lead poisoning" remained 80 μg/dL.

10. Annest and Mahaffey, "Blood Lead Levels, 1976–1980" (see ch. 11, n. 68).

11. Among black children from families with an annual household income less than $6,000, 18.5% had blood-leads over 30 μg/dL, compared with 0.7% of white children with family incomes above $15,000; the rate for inner-city blacks was 18.6% as compared with 1.2% for rural whites; ibid., Table B, 10.

12. Ibid., 3.

13. Joseph Stokes, noted on "Qualifications Inquiry on Applicant for a Public Health Service Fellowship" (typewritten form), folder "Needleman, Herbert L. #1," Stokes papers (see ch. 9, n. 14); Herbert L. Needleman et al., "Subclinical Lead Exposure in Philadelphia Schoolchildren: Identification by Dentine Lead Analysis," *NEJM* 290 (1 Jan. 1974): 245–48; see also the first report on the study in Herbert Needleman, Orphan Tuncay, and Irving Shapiro, "Lead Levels in Deciduous Teeth of Urban and Suburban American Children," *Nature* 235 (14 Jan. 1972): 111–12.

14. Herbert L. Needleman et al., "Deficits in Psychologic and Classroom Performance of Children with Elevated Dentine Lead Levels," *NEJM* 300 (29 Mar. 1979): 689–95; Herbert L. Needleman et al., "The Long-Term Effects of Exposure to Low Doses of Lead in Childhood: An 11-year Follow-up Report," *NEJM* 322 (11 Jan. 1990): 83–88; on the unfriendly reception to this article, see Needleman, "Salem Comes to the NIH" (see ch. 11, n. 42).

15. For an assessment of the relationship between lead and IQ, see Herbert L. Needleman, "Low-Level Lead Exposure and the IQ of Children: A Meta-Analysis of Modern Studies," *JAMA* 263 (2 Feb. 1990): 673–78; for his earliest findings on lead and IQ, see Herbert L. Needleman, Alan Leviton, and David Bellinger, "Lead Associated Intellectual Deficit," *NEJM* 306 (11 Feb. 1982): 367.

16. Muriel D. Wolf, "Lead Poisoning from Restoration of Old Homes," *JAMA* 225 (9 July 1973): 175–76.

17. Robert Feldman, "Urban Lead Mining: Lead Intoxication among Deleaders," *NEJM* 298 (18 May 1978): 1143–45; J. W. Sayre et al., "House and Hand Dust as a Potential Source of Childhood Lead Exposure," *AJDC* 127 (1974): 167–70.

18. The toxicity of paint chips versus lead dust is a clear case where size matters: lead in finely ground dust is far more bioavailable than that in intact chips; see Sayre et al., "House and Hand Dust." The significance of low-level exposure for abatement can be seen in the ongoing studies by Mark Farfel, an environmental scientist at the Johns Hopkins University, many of which are coauthored by Julian Chisolm. Farfel has sought to find safe, effective, and inexpensive abatement methods; see Farfel and Chisolm, "Health and Environmental Outcomes of Traditional and Modified Practices for Abatement of Residential Lead-Based Paint," *AJPH* 80 (1990): 1240–45. See also Evan Charney et al., "Childhood Lead Poisoning: A Controlled Trial of the Effect of Dust-Control Measures on Blood Lead Levels," *NEJM* 309 (3 Nov. 1983): 1089–93.

19. On cleanliness in the postwar era, see Hoy, *Chasing Dirt* (see intro., n. 5); Ruth Schwartz Cowan, *More Work for Mother: The Ironies of Household Technology from the Open Hearth to the Microwave* (New York: Basic Books, 1983); Stephanie Coontz, *The Way We Never Were: America's Families and the Nostalgia Trap* (New York: Basic Books, 1992). These historians make it clear, as Coontz's title states explicitly, that the baby boomers were seeking to re-create a world that never was, but historical realities seldom slowed the guilt trip.

20. "Millie Sick with Lead Poisoning," *Washington Post,* 3 Aug. 1990, B1. An important case report of "yuppie plumbism" noted that canine lead poisoning (puppy plumbism?) could be a potential "sentinel event" for detecting pediatric and occupational cases arising from renovations, and urged greater "communication between primary care physicians and veterinary medicine practitioners"; see Phylis Marino et al., "A Case Report of Lead Paint Poisoning during Renovation of a Victorian Farmhouse," *AJPH* 80 (1990): 1183–85.

21. Stapleton went on to be a public affairs chief for the EPA and published a self-help book for parents on prevention of lead poisoning; see Stapleton, *Lead Is a Silent Hazard* (see intro., n. 9).

22. Jane Lin-Fu, "Children and Lead: New Findings and Concerns," *NEJM* 307 (2 Sept. 1982): 615–17.

23. CDC, *Preventing Lead Poisoning in Young Children* (1985), 8, and (1991), 39 (see ch. 1, n. 3).

24. Richard Sorian, *The Bitter Pill: Tough Choices in America's Health Policy* (New York: McGraw-Hill, 1988), 40–51.

25. National Coalition for Lead Control, "Children, Lead Poisoning, and Block Grants: A Year-End Review of How Block Grants Have Affected the Nation's Ten Most Crucial Lead Screening Programs" (Washington: Center for Science in the Public Interest, 1982); repr. in House Committee on Energy and Commerce, *Lead Poisoning and Children: Hearing before the Subcommitee on Health and the Environment,* 97th Cong., 2d sess., 2 Dec. 1982.

26. Jerome F. Cole to Senator Harry Reid, 20 Mar. 1990; reprint, Senate Committee on Environment and Public Works, Health Effects of Lead Exposure:

Hearning before the Subcommittee on Toxic Subtances, Environmental Oversight, Research and Development, 101st Cong., 2d sess., 8 Mar. 1990.

27. For a concise review of the charges against Needleman, see Paul Mushak, "The Lead Debate" [letter to the editor], *Pediatrics* 91 (Apr. 1993): 856–57. Although the University of Pittsburgh panel cleared Needleman of all "misconduct" charges—and in fact lauded the criteria and execution of the study described in Needleman et al., "Deficits in Psychologic and Classroom Performance"—it concluded that he had committed one act of misrepresentation, related to a single discrepancy in published versus actual selection criteria. According to Mushak, who participated in an earlier investigation of the study and testified on behalf of Needleman in the 1992 hearing, the discrepancy was minor— routine, given the "pioneering" nature of the study—and, in any case, worked to weaken the association Needleman hoped to show, undercutting any argument that the researcher had nefarious motives. Needleman recounts his experiences with these investigations in "Salem Comes to the NIH," 171–74, and in several letters to the editors of *Pediatrics;* see 91 (Jan. 1993): 171–74; (Feb. 1993): 519–20; (Apr. 1993): 856–57.

28. Herbert Needleman and Constantine A. Gatsonis, "Low-Level Lead Exposure and the IQ of Children," 673–78.

29. Needleman was an activist even before he took on his lead crusade. In 1966 he served on the steering committee of the Philadelphia chapter of Physicians for Social Responsibility; Needleman to Stokes, 5 Oct. 1966, folder "Needleman, Herbert L. #1," Stokes papers. See also Eliot Marshall, "EPA Faults Classic Lead Poisoning Study," *Science* 222 (25 Nov. 1983): 906–7.

30. From the early 1970s, when Silbergeld was a graduate student at the Johns Hopkins University, she studied lead's effects on neurological development. In the 1980s, she served as adjunct senior scientist at the Environmental Defense Fund and was a member of the EPA's science board and the National Academy of Science's toxicology board. Since 1991 she has been professor of epidemiology and preventive medicine in the Department of Pathology and Toxicology at the University of Maryland. In 1993 Silbergeld was awarded a MacArthur fellowship. See Ellen K. Silbergeld, prepared statement, House Committee on Energy and Commerce, Lead Poisoning: Hearings before the Subcommittee on Health and the Environment, 102d Cong., 2d sess., 25 April and 26 July 1991, 448; *American Men and Women of Science and Medicine* 1998–99 (New Providence, N.J.: R. R. Bowker, 1999), s.v. "Silbergeld, Ellen K."

31. For a contemporary assessment of these reasons for optimism, see Berney, "Round and Round It Goes" (see intro., n. 10).

32. In 1986, the EPA reduced permissible lead in drinking water from fifty parts per billion to ten; Ellen K. Silbergeld, "Preventing Lead Poisoning in Children," *Annual Review of Public Health* 18 (1997): 187–210. Consequently, Congress amended the Clean Water Act to ban lead fixtures and lead solder in new

water-supply systems. In 1988, Congress passed the Lead Contamination Control Act (P.L. 100–572) to ban the use of lead in the manufacture of drinking fountains and to help local schools eliminate lead in existing fountains. Unfortunately, participation in the programs enacted under this act has been spotty; see Stapleton, *Lead Is a Silent Hazard,* 129–30.

33. In 1997, HUD was still funding eighty-five state and local abatement programs in low-income housing, but new standards from the EPA are expected to shift responsibility for directing these programs to state and local governments; Dorothy Morrison, "Lead-Based Paint Hazards: Funding Opportunities Provide Control," *Public Management* 80 (Jan. 1998): 16–19; Senate Committee on Banking, Housing, and Urban Affairs, Subcommittee on Housing and Urban Affairs, *The Residential Lead-Based Paint Hazard Reduction Act of 1992: Hearing before the Subcommittee on Housing and Urban Affairs,* 102d Cong., 2d sess., 19 Mar. 1992.

34. Eric Manheimer and Ellen K. Silbergeld, "Critique of CDC's Retreat from Recommending Universal Lead Screening for Children," *Public Health Reports* 113 (Jan./Feb. 1998): 38–46; Miriam Bar-on and Russel Boyle, "Are Pediatricians Ready for the New Guidelines on Lead Poisoning?" *Pediatrics* 93 (Feb. 1994): 176–82; Susan Ferguson and Tracy Lieu, "Blood Lead Testing by Pediatricians: Practice, Attitudes and Demographics," *AJPH* 87 (Aug. 1997): 1349–51.

35. CDC, *Preventing Lead Poisoning in Young Children* (1991), 39; Howard Pearson, "Stepped-up Lead Screenings Urged," *AAP News* 9 (Apr. 1993): 1.

36. Edgar J. Schoen, "Lead Toxicity in the 21st Century: Will We Still Be Treating It?" *Pediatrics* 90 (Sept. 1992): 481–82; Edgar J. Schoen, "Childhood Lead Poisoning: Definitions and Priorities," *Pediatrics* 91 (Feb. 1993): 504–505.

37. George Gellert et al., "Lead Poisoning among Low-Income Children in Orange County, California," *JAMA* 270 (7 July 1993): 69–71. Researchers at the University of Utah conducted a similar but smaller study of poor children in Salt Lake City; see William Banner, Barbara Vugnier, and Jannette Pappas, "Mythology of Lead Poisoning" [letter to the editor], *Pediatrics* 91 (Jan. 1993): 161. None of the 261 children screened had blood-leads above 15 μg/dL, and only 4.2% measured above 10 μg/dL. The authors called for "public policy to reflect regional priorities to avoid divisive arguments within the community of pediatricians."

38. Birt Harvey, "Should Lead Screening Recommendations Be Revised?" *Pediatrics* 93 (Feb. 1994): 201–204.

39. Debra Brody et al., "Blood Lead Levels in the US Population: Phase 1 of the Third National Health and Nutrition Examination Survey (NHANES III, 1988 to 1991)," *JAMA* 272 (27 July 1994): 277–83; James Pirkle et al., "The Decline in Blood Lead Levels in the United States: The National Health and Nutrition Examination Surveys (NHANES)," *JAMA* 272 (27 July 1994): 284–92.

40. Ellen Ruppel Shell, "An Element of Doubt: Disinterested Research Casts Doubt on Claims That Lead Poisoning from Paint Is Widespread among American Children," *Atlantic Monthly* 276 (Dec. 1995), 24–39; see also Peter Samuel, "Lead Hype," *National Review* 47 (1 May 1995), 69–71.

41. National Center for Environmental Health, *Screening Young Children for Lead Poisoning: Guidance for State and Local Public Health Officials* (Atlanta: CDC, 1997); Nancy M. Tips, Henry Falk, and Richard Jackson, "CDC's Lead Screening Guidance: A Systematic Approach to More Effective Screening," *Public Health Reports* 113 (Jan./Feb. 1998): 47–51.

42. Tips, Falk, and Jackson, "CDC's Lead Screening Guidance," 49.

43. Lynn Goldman and Joseph Carra, "Childhood Lead Poisoning in 1994," *JAMA* 272 (27 July 1994): 315–16; Shell, "An Element of Doubt," 39.

44. While conducting investigations in Salt Lake City for her federal lead survey, Alice Hamilton was astonished when a druggist said he had never seen a case of lead poisoning from the local factories. Then the druggist qualified his statement: "Oh, maybe you are thinking of the Wops and Hunkies. I guess there's plenty among them. I thought you meant white men"; see Hamilton, *Exploring the Dangerous Trades*, 151–52 (see intro., n. 17).

45. Alicia Caldwell, "Hidden Hazard," *St. Petersburg Times*, 27–30 May 1998, section "Neighborhood Times," 1.

46. Caldwell reports that the local health department identified ten zip codes in St. Petersburg with large numbers of pre-1950 homes. I was raised in St. Petersburg, and although most of the homes in my neighborhood were built after World War II, every block had one or two grand old bungalows or "Mediterranean revival" villas built before the Great Depression ended the city's real estate boom; see Raymond Arsenault, *St. Petersburg and the Florida Dream, 1888–1950* (Norfolk, Va.: Donning, 1988). Indeed, our family home was a white lead–painted wood and stucco house built in the 1910s. In the 1960s, the painter my family usually hired, like most in Florida, loved white lead—nothing stood up better to mildew and Florida's heat and humidity, he said. Nonetheless, my neighborhood was not in one of the at-risk zip codes.

47. Caldwell, "Hidden Hazard."

CHAPTER 13: REGULATING "LOW-LEVEL" LEAD POISONING

The epigraph to this chapter is taken from Philip J. Landrigan, "Lead in the Modern Workplace," *AJPH* 80 (Aug. 1990): 907–908.

1. Abdel R. Omran, "Epidemiologic Transition," *International Encyclopedia of Population*, ed. Johan A. Ross (New York: Free Press, 1982), 172–83; Tesh, *Hidden Arguments* (see ch. 2, n. 40).

2. The most recent pediatric lead-poisoning fatality I have identified occurred in 1990; see Schirmer, Anderson, and Saryan, "Fatal Pediatric Poisoning" (see ch. 2, n. 19).

3. Hu et al., "The Relationship of Bone and Blood Lead to Hypertension" (see intro., n. 1); Kim et al., "A Longitudinal Study of Low-Level Lead Exposure" (intro., n. 1); Herbert L. Needleman et al., "Bone-Lead Levels and Delinquent

Behavior," *JAMA* 275 (7 Feb. 1996): 363–69; R. O. Pihl and F. Ervin, "Lead and Cadmium Levels in Violent Criminals," *Psychological Reports* 66 (June 1990): 839–44; Herbert L. Needleman, "The Future Challenge of Lead Toxicity," *Environmental Health Perspectives* 89 (Nov. 1990): 85–89.

4. The story is apocryphal, but variations on it occur daily. For a discussion of disposing of lead paint, see Hazel Bradford, "EPA's RCRA Program Gets Tough," *ENR* 227 (22 July 1991): 8. The description of lead abaters is from Stephanie Pollack, "Solving the Lead Dilemma," *Technology Review* 92 (Oct. 1989): 22–31.

5. George J. Cohen, "Lead Poisoning, 20 Years Later," *Clinical Pediatrics* 19 (Apr. 1980): 250.

6. Herbert L. Needleman, "Low Level Lead Exposure: A Continuing Problem," *Pediatric Annals* 19 (Mar. 1990): 213; Herbert L. Needleman, "The Persistent Threat of Lead: A Singular Opportunity," *AJPH* 79 (May 1989): 643–45.

7. Yona Amitai et al., "Hazards of 'Deleading' Homes of Children with Lead Poisoning," *AJDC* 141 (July 1987): 758–60; Farfel and Chisolm, "Health and Environmental Outcomes of Practices for Abatement of Residential Paint" (see ch. 12, n. 18).

8. A 1994 assessment of HUD projections puts the cost of eliminating lead hazards from the 9.6 million lead-painted units where small children live at $68.8 billion; Meg Koppel and Ross Koppel, "Lead-Based Paint Abatement in Private Homes: A Study of Policies and Costs," briefing paper of the Economic Policy Institute, Aug. 1994. Interim controls and hazard reduction would cost $21.4 billion and $45.0 billion, respectively. Lead Tech '96, a three-day convention sponsored by eight abatement-industry associations, featured 125 speakers and 150 exhibitions from all branches of this burgeoning subfield of the construction trades. The industry also has its own monthly publication, *Lead Detection & Abatement Contractor.*

9. Jeanne Ponessa, "Government's Role in Cleaning up Lead," *Governing* 5 (Aug. 1992): 20–21; Senate Committee, *The Residential Lead-Based Paint Hazard Reduction Act* (see ch. 12, n. 33).

10. Rekus, "Lead Poisoning," 14–16 (see intro., n. 20).

11. On the origins of the OSHA legislation, see Gersuny, *Work Hazards,* 111–15 (see ch. 3, n. 73).

12. For a critique of the OSHA law, see David Rosner and Gerald Markowitz, "Research or Advocacy: Federal Occupational Safety and Health Policies during the New Deal," in *Dying for Work,* 83–102 (see intro., n. 12).

13. Irving R. Tabershaw, preface to papers from the Conference on Standards for Occupational Lead Exposure, *Journal of Occupational Medicine* 17 (Feb. 1975): 75; Jerome F. Cole, transcript of discussion, in U.S. Department of Health, Education, and Welfare, *Health Effects of Lead and Arsenic Exposure: A Symposium,* ed. Bertram W. Carnow (Washington: GPO, 1976), 180. For consistency, I

have translated the values given in these sources from micrograms per 100 grams of blood to the more frequently employed micrograms per deciliter.

14. Lynam, "Lead Industries Association Position," 88 (see ch. 7, n. 53); Cole, transcript of discussion, *Lead and Arsenic Exposure,* 179–82.

15. Sheldon Samuels, "Organized Labor Recommendations," *Journal of Occupational Medicine* 17 (Feb. 1975): 81.

16. Details of the OSHA lead standard are given in CDC, "Surveillance for Occupational Lead Exposure—United States, 1987," *MMWR* 38 (22 Sept. 1989): 642–46. When OSHA officials lowered the TLV in 1978, they cited kidney damage resulting from low-level lead exposure as justification. The work of Richard Wedeen deserves a large measure of credit in prompting OSHA's action; see, e.g., Richard P. Wedeen, letter to editor, *New York Times,* 5 June 1980, section C, 5; see also Wedeen, *Poison in the Pot* (see intro., n. 6).

17. "Justices Let Stand Strict Curbs on Lead in Workplace," *New York Times,* 30 June 1981, section B, 9. Wedeen briefly discusses this episode in *Poison in the Pot,* 3–7. For the persistence of inadequate resources at OSHA, see Don J. Lofgren, *Dangerous Premises: An Insider's View of OSHA Enforcement* (Ithaca, N.Y.: ILR Press, 1989).

18. The UAW brought the suit on behalf of one woman who was demoted, two women who had themselves sterilized, and a man "who had been denied a request for a leave of absence for the purpose of lowering his lead level because he intended to become a father"; U.S. Supreme Court, *Supreme Court Bulletin* 89–1215 (Washington: GPO, 1991), B1159.

19. Linda Greenhouse, "Court Backs Right of Women to Jobs with Health Risks," *New York Times,* 21 Mar. 1991, 1.

20. The information Johnson provided said there was evidence "that women exposed to lead have a higher rate of abortion" but that the evidence was "not as clear . . . as the relationship between cigarette smoking and cancer." All the same, it would be, "medically speaking, just good sense not to run that risk if you want children and do not want to expose the unborn child to risk, however small"; see U.S. Supreme Court, *Supreme Court Bulletin* 89–1215 (Washington: GPO, 1991), B1163.

21. American Cyanamid's policy in 1978 required female employees to prove they were infertile to get or keep jobs in the lead-using areas of the plant—jobs that paid $225 per week plus overtime. Fertile women were offered safer employment at a weekly wage of $175 with no overtime and fewer benefits. Five women in the plant underwent "voluntary" sterilization in order to keep their jobs. The workers' union sued American Cyanamid, but the suit was unsuccessful. One union official complained that the policy showed the old concept, "Alter the worker, don't alter the workplace"; see Greenhouse, "Court Backs Right of Women"; "Pigment Plant Wins Fertility Risk Case," *New York Times,* 8 Sept. 1980, 14.

22. *Supreme Court Bulletin* 89–1215, B1164–6.

23. Ibid., B1159–83, passim.

24. Hamilton, "Women in the Lead Industries," 5 (see ch. 4, n. 51).

25. Greenhouse, "Court Backs Right of Women"; Peter Kilborn, "Employers Left with Many Decisions," *New York Times,* 21 Mar. 1991, A12.

26. *Supreme Court Bulletin* 89–1215, B1180.

27. Thayer, "The Lead Menace," 326 (see ch. 3, n. 18).

28. Hamilton, "What One Stockholder Did" (see ch. 3, n. 21).

29. Senate Committee, *Lead-Based Paint Poisoning Amendments of 1972,* 102 (see intro., n. 22).

30. See, e.g., J. Julian Chisolm and Harold E. Harrison, "The Exposure of Children to Lead," *Pediatrics* 18 (1956): 943–57, where grants from the LIA and NIH are acknowledged.

31. This argument draws upon my interpretation of Tesh's discussion of "science versus values," which she defines as a deadly form of dualism in *Hidden Arguments,* 170; and Sellers, "Manufacturing Disease," 520–25 (see intro., n. 7), where it is posited that scientists working in the public policy arena can demonstrate both bias and objectivity.

32. Testimony of Clair C. Patterson, in Senate Committee, *Air Pollution—1966,* 317 (see ch. 8, n. 12).

33. Oliver, "Industrial Lead Poisoning," 107 (see ch. 3, n. 17).

34. Bellow, *The Dean's December,* 223 (see ch. 11, n. 22).

Index

Library of Congress Cataloging-in-Publication Data
Warren, Christian.
    Brush with death : a social history of lead
    poisoning / Christian Warren.
        p. cm.
    ISBN 0-8018-6289-2 (alk. paper)
      1. Lead poisoning—United States—
    History.   I. Title.

  RA1231.L4 W37 2000
  615.9'25688'0973   99-046329